Dynamics of Combustion Systems

A. K. Oppenheim †

Dynamics of Combustion Systems

2nd ed.

With 230 Figures and 38 Tables

Springer

Antoni K. Oppenheim †
Professor of Engineering
Department of Mechanical Engineering
University of California
5112 Etcheverry Hall
Berkely, CA 94720-1740
USA
e-mail: ako@me.berkely.edu

On the cover
Schlieren photograph of a turbulent combustion field in a cylindrical enclosure (vid. Fig.3.3c)

ISBN 978-3-642-09601-3 e-ISBN 978-3-540-77364-1

DOI 10.1007/978-3-540-77364-1

Cover design: deblik, Berlin

Printed on acid-free paper

9 8 7 6 5 4 3 2 1

springer.com

In Memoriam

A.K. Oppenheim (1915–2008)

On January 12, 2008, our colleague, and friend to many in the combustion community, Tony Oppenheim passed away at his home in Kensington, California. As ever, the "Captain of his Ship," he had opted for hospice care in the comfort of familiar surroundings rather than spend his remaining time in a hospital. In his final days, he continued to work in bed with his laptop on the second edition of his monograph Dynamics of Combustion Systems.

Tony was born in Warsaw, Poland, on August 11, 1915. After graduating as valedictorian from his high school in 1933, he entered the highly competitive Warsaw Institute of Technology, where he studied aeronautical engineering.

His studies were interrupted in 1939 by Nazi Germany's invasion of Poland. Tony fled his home country, making his way through Romania, Greece, France, Spain, and Portugal before arriving in England in June 1940.

Soon after, he enlisted in the Polish Army in Scotland and, like many of his Polish compatriots, taught himself English. In 1942, he was granted leave from the Polish Army to conduct, under the supervision of Sir Owen Saunders, research on improving the performance of piston engines that powered Spitfire and Hurricane RAF fighter planes. Engine modifications flowing from his research resulted in increased aircraft speed and gave RAF fighters a performance edge over German fighters.

Following this work, Tony studied secondary air mixing in the gas turbine combustion chamber and analyzed the combustion dynamics of the pulsed jet engine that powered the German V-1 flying bomb. During the WW II years, he completed his requirements for a Diploma of Engineering degree from the Warsaw Institute of Technology in 1943 by taking courses at the City and Guilds College in London. He also earned a PhD in mechanical engineering at the University of London, and a Diploma of Imperial College degree in 1945.

After his studies, he spent three years as a lecturer in mechanical engineering at City and Guilds College, where he and his postgraduate students built the institution's first supersonic wind tunnel. As a recognized authority on the V1, he was sent to Germany at the conclusion of hostilities as a British intelligence officer to interview the scientists and engineers involved in developing the engine. Oppenheim's interest in detonation phenomenon and combustion grew from this early research in pulsed jet engines.

In 1948, Tony moved to the United States, joining Stanford University as an assistant professor in mechanical engineering. Two years later, he made his way to UC Berkeley as an assistant professor in mechanical engineering. He was promoted to associate professor in 1954 and full professor in 1958.

At Berkeley, Tony and his students worked on, and made seminal contributions to the understanding of, a diversity of problems: development of a network analysis for radiative heat transfer, vector polar methods for the analysis of interactions and intersections of gas dynamic fronts, development of and structure of detonations, blast wave theory, turbulent combustion, plasma jet ignition, turbulent jet plumes, and controlled combustion in engines.

In the 1960s, Tony, along with Rem Soloukhin and Numa Manson, recognized the need to create a forum where recent research advances on the dynamics of explosions and reactive systems could be shared and discussed freely with our colleagues from Russia and Eastern Europe. The first meeting, held in Brussels in 1967, was a great success; the fact that the 22nd ICDERS will be held in Minskis 2009 is a testament to Tony's vision.

Tony's active study of combustion technology spanned more than six decades. His retirement from the University of California-Berkeley in 1986 enabled him to focus his prodigious energy on the exploration and development of his ideas on improving the performance of internal combustion engines. His monograph Combustion in Piston Engines (published by Springer Verlag in 2004) details his concept of how future engines may employ distributed ignition sources to both increase fuel efficiency and reduce the formation of pollutants. Another noteworthy monograph (published by Springer Verlag in 2006), Dynamics of Combustion Systems, details the analysis of confined compressible fields in which exothermic reactions occur. A second edition of this latter monograph is scheduled to appear in 2008.

Among the numerous awards Oppenheim received throughout his career were the Dionizy Smolenski Medal of the Polish Academy of Sciences for outstanding contributions toward advances in the knowledge of combustion and especially to the dynamics of explosions and reactive systems; the

Alfred C. Egerton Medal of The Combustion Institute for distinguished, continuing and encouraging contributions to the field of combustion; and the Berkeley Citation, one of the highest honors bestowed by the university to those who have exceeded the standards of excellence in their fields.

Tony was widely recognized, and earned much formal recognition, for his fundamental contributions to unsteady gas dynamics and combustion. Tony received honorary doctorate degrees from the University of London, the University of Poitiers, and Warsaw's University of Technology. He was also a member of the International Academy of Astronautics, a fellow and honorary member of the American Society for Mechanical Engineers, a member of the U.S. National Academy of Engineering; and a foreign member of the Polish Academy of Sciences.

He was known for his extraordinary energy and involvement in many issues, his capacity to mentor and to inspire his colleagues in their research, his optimism and enthusiasm for many aspects of life, and his ability to present technical information in an entertaining and humorous manner. The combustion community will greatly miss Tony, but we are much richer for his having been among us.

He is survived by his wife, Lavinia (Min), of Kensington, California; their daughter, Terry Ann Cort, of El Cerrito, California; and two grandchildren.

J. Ray Bowen

Originally published as: J.R. Bowen, A.K. Oppenheim (1915–2008), Combust. Flame (2008), doi:10.1016/j.combustflame.2008.04.003

To Min, the great companion of my life for over sixty years, to Terry, our magnificent daughter of over fifty years, and to Jessica and Zachary, our marvelous grandchildren of over twenty years, and wonderful JoAnn of over forty years.

Preface

Of prime concern in this book are combustion systems – confined fields of compressible fluids where exothermic processes of combustion take place. Their purpose is to generate motive power. In their course, exothermic energy* is created by chemical reaction and deposited in the field, both actions carried out concomitantly and referred to popularly as 'heat release.' Particular examples of such systems are cylinders in internal combustion engines, combustors of gas turbines and rockets, as well as explosions engendering blast waves - non-steady flow fields bounded by incident shock fronts that impose on them the constraints of confinement.

The process of combustion is carried out, as a rule, at a high rate, the life time of chemically reacting component being of an order of microseconds, while the exothermic reaction of the whole system is accomplished in few milliseconds. For that reason, its execution has been considered so far to be beyond the intervention of interactive controls – a hindrance that, in our age of microelectronics for which a millisecond is a relatively long time, can be eliminated.

The technological objective of the book is to pave the way towards this end by bringing forth the dynamic features of combustion systems. Their properties are expressed therefore as those of dynamic objects – entities amenable to management by modern tools of control technology.

Sensible properties of combustion systems are displayed in a three-dimensional *physical space*, while their processes are disclosed in a multi-dimensional *thermo-physical phase space*, where the states of components of the working substance are identified and the transformations of its constituents are disclosed. The dimension of the latter is equal to the degrees of freedom - the number of reaction constituents plus two, as specified by the Gibbs phase rule.

---

* potential energy of thermal kind known as 'heat of reaction' and measured in terms of 'heating value' determined by the change in internal energy taking place at NTP (normal temperature and pressure)

Equilibrium states of the working substance and its components are specified in the *thermodynamic phase space* – a three-dimensional subset of the thermo-physical space, provided that the internal energy, e, is one of its coordinates, as pointed out by Gibbs[1] and Poincaré[2]. If e is expressed in units of energy per mole, it is compatible with temperature, T, as its concomitant coordinate. If e is expressed in units of energy per unit mass, as appropriate for mass conserving chemical reactions, its dimensionally compatible coordinate is the *dynamic potential*, $w \equiv pv$ - a parameter providing principal service of liaison between the physical space, where it is established, and the thermodynamic space, where it is employed as a fundamental coordinate of state. The concept of pv is well known in the literature as 'flow work', without realizing its pivotal role in thermodynamics.

The subject matter of the book is exposed in three parts, each consisting of four chapters, Part 1 -. *Exothermicity* – considering the thermodynamic effects due to evolution of exothermic energy in a combustion system; Part 2 – *Field* – exposing the dynamic properties of fields where the exothermic energy is deposited; Part 3 - *Explosion* – revealing the dynamic features of fields and fronts created by deposition of exothermic energy.

In Part 1,

Chapter 1 - *Thermodynamic Aspects* - presents the evolution of the combustion system by a model consisting of two parts: (1) the *dynamic aspects*, dealing with the properties of combustion in the physical space, and (2) the *thermodynamic aspects*, treating the processes of combustion in the phase space.

Chapter 2 - *Evolutionary Aspects* – elucidates the fundamental features of evolution.

Chapter 3 - *Heat Transfer Aspects* – describes experimental and analytical studies of energy loss incurred in a combustion system by heat transfer to its surroundings.

Chapter 4 - *Chemical Kinetic Aspects* – furnishes a résumé of analytical technique for resolution of chemical kinetic processes of combustion.

---

1 Gibbs JW (1875-1878) On the equilibrium of heterogeneous substances. Transactions of the Connecticut Academy, III (1875-76) pp. 108-248; (1877-78) pp 343-524 [(1931) The Collected Works of J.W. Gibbs, Article III, Longmans, Green and Company, New York, 2: 55-353, esp. pp.85-89 and 96-100]

2 Poincaré H (1892) Thermodynamique, Gothiers-Villars, Paris, xix + 432 pp [1908 edition, xix + 458 pp]

In Part 2,

Chapter 5 - *Aerodynamic Aspects* – provides a fundamental background for fluid dynamic analysis of flow fields at the limit of infinite Peclet and Damköhler numbers commensurate with inadequacy of molecular diffusivity and thermal conductivity to affect the rapid process of combustion taking place in *exothermic centers* – sites referred to in the literature as 'hot spots.'

Chapter 6 - *Random Vortex Method* – presents the analytical technique, introduced by Chorin[3, 4], that provides an insight into the mechanism of turbulent flow fields in terms of random vortex motion, mimicking the physical nature of turbulence as a phenomenon due to random walk of vortex elements called blobs.

Chapter 7 - *Gasdynamic Aspects* - describes classical analysis of compressible flow fields, featuring the method of characteristics for solution of hyperbolic equations in terms of which their gasdynamic properties are expressed.

Chapter 8 – *Fronts and Interfaces* - furnishes analytical treatment of gasdynamic effects produced by shock and detonation fronts, as well as by interfaces (impermeable fronts) and simple waves that act as border between different state regimes.

In Part 3

Chapter 9 - *Blast Waves* – provides a fundamental background for the analysis of far fields created by an explosions, with respect to which sizes of their kernels, where the exothermic energy was deposited, is negligibly small.

Chapter 10 - *Self-Similar Blast Wave* - presents salient features of linearly, cylindrically and spherically symmetric fields created by point explosions whose fronts propagate into a vacuum.

Chapter 11 - *Phase Space Method* - ushers in an analytical technique for treating blast waves propagating into atmospheres of finite pressure and density.

Chapter12 - *Detonations* - displays the dynamic properties of fronts associated with exothermic processes.

---

[3] Chorin AJ (1973) Numerical studies of slightly viscous flow. J. Fluid Mech. 57: 785-796

[4] Chorin AJ (1978) Vortex sheet approximation of boundary layers. J. Comp. Phys. 27: 428-442.

Acknowledgement

I am grateful to my collaborators, Professors Harold Schock, Cornel Stan and Andrew Packard, for helpful comments, to my old friends, Professors George Leitmann, Alexandre Chorin and John Lee for valuable advice and inspiration, to my student and associate, Eilyan Bitar, for congenial companionship and valuable assistance in computations and graphics, and to my former students and associates from whom I learned so much.

Contents

PART 1

EXOTHERMICITY

1 Thermodynamic Aspects

1.1 Combustion System

Combustion system, S, is a confined field of a compressible fluid where the exothermic process of combustion takes place, subject to conditions imposed at its boundary. Presented here are global properties of such a system.

The composition of a combustion system is specified by mass fractions of its initial components, Y_K, where K = F, A and B, for, respectively, fuel, air and the non-reacting portion of the fluid, like the recirculated exhaust, or residual gas. The mass fraction of reactants, $Y_R = Y_F + Y_A$, a mixture of fuel and air, is equal to that of the reaction products, Y_P.

The composition of a combustible mixture is expressed by the air/fuel ratio, $\sigma_K \equiv (Y_A/ Y_F)_K$, (K = R, S). For the reactants, σ_R is identified usually with the measured flow rates of air and fuel, whereas for the system, σ_S is deduced from the exhaust gas analysis. The values of these ratios are expressed conventionally in terms of the air-equivalence ratio with respect to the stoichiometric proportion, referred to by subscript "st", $\lambda_K \equiv \sigma_K / \sigma_{st}$, reciprocal of the fuel-equivalence ratio,, ϕ_K.

Thus,

$$Y_K \equiv (1 + \sigma_K) Y_F \equiv (1 + \sigma_{st} \lambda_K) Y_F \tag{1.1}$$

whence, for the same mass fraction of fuel, Y_F,

$$\frac{Y_F}{Y_K} = \frac{1}{1+\sigma_K} = \frac{1}{1+\sigma_{st}\lambda_K} \tag{1.2}$$

A phase diagram of the mass fraction of system components is displayed in Fig.1.1 with respect to the mass fraction of products, y_P, i.e. the substance produced by oxidation of fuel and their mixing with the non-reacted portion of the cylinder charge. The variable mass fractions are denoted in it by small letters and the constants defining the system by capital letters.

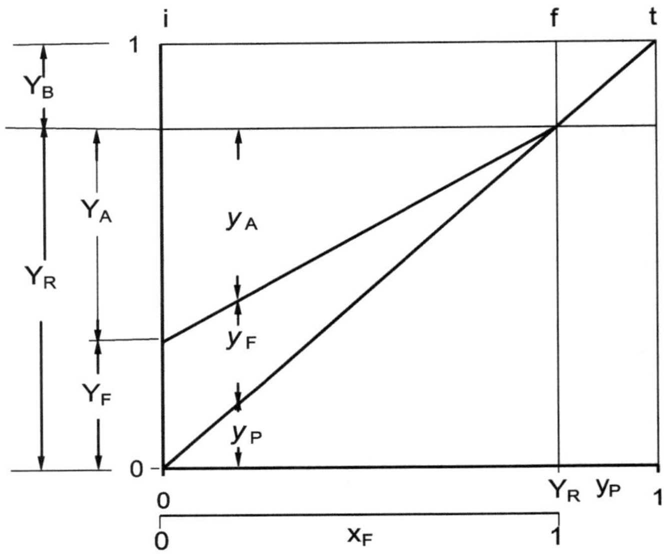

Fig. 1.1. Phase diagram of component mass fractions with respect to mass fraction of products

As displayed by this figure, in the course of combustion the mass fractions of components, y_K (K = F, A), are linear functions of the mass fraction of end products, y_P, namely

$$y_K = \frac{Y_K}{Y_R}(Y_R - y_P) \tag{1.3}$$

while the latter is expressed in terms of the mass fraction of fuel, x_F, as

$$y_P = Y_R x_F \tag{1.4}$$

For a system constrained by the walls of its enclosure, like a cylinder in a piston engine, the sole purpose of combustion is to generate pressure. With a measured pressure profile, the objective of its thermodynamic analysis is the solution of an inverse problem: the deduction on this basis of data on the evolution of the thermodynamic state parameters of the system components, as well as of the mass fraction of combustion products – a quantity proportional, according to (1.4), to that of fuel.

Mass fractions of fuel, x_F, and of products, y_P, consist of the effective parts, x_E and y_E, formed to produce the measured pressure profile, and of the ineffective parts, x_I and y_I, expended by energy loss due primarily to heat transfer to the walls of the enclosure, i.e. $x_F = x_E + x_I$ and $y_P = y_E + y_I$.

Their profiles are displayed by Fig. 1.2 in terms of the normalized time, $\tau \equiv (t - t_i)/T$, where, with $t_f = t_t$, $T = t_f - t_i$ is the lifetime of the exothermic process that shifts the process of expansion away from that of compression to form the work cycle. As demonstrated in Chapter 3, the initial state, **i**, at τ=0, is a fundamental, sharp, saddle type, singularity of combustion, while the final state, **f**, occurring at the same time as the terminal state, **t**, is at τ=1 – the end of the exothermic process identified above – where the effective parts, x_E and y_E, are at their maxima forming a smooth, nodal type, singularity.

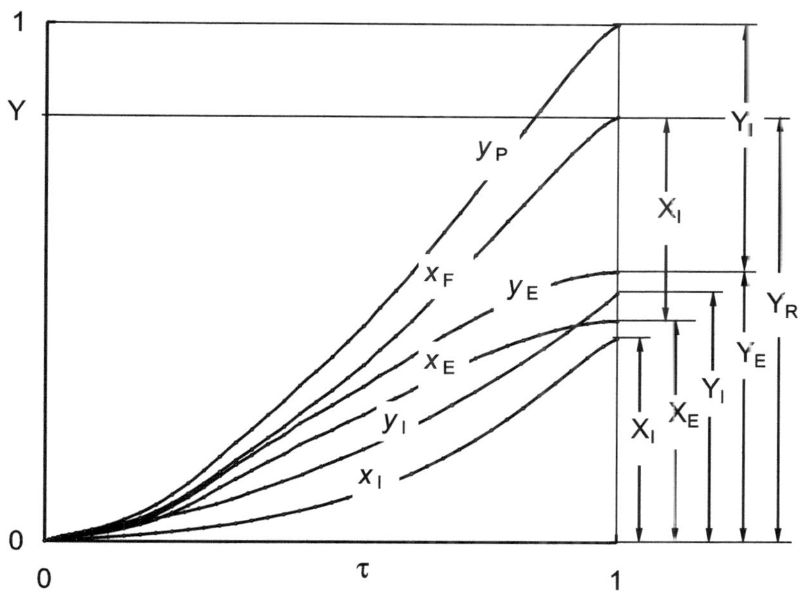

Fig. 1.2. Profiles of effective and ineffective mass fractions of fuel and products

Global properties of a combustion system are established by *pressure diagnostics* – a procedure for solution of an inverse problem based on measured pressure, p, and given mass averaged specific volume, v. The procedure of pressure diagnostics consists of two perspectives:

1. *Dynamic Properties* manifested by the physical space of the work cycle. Their parameters are derived directly from data of the measured pressure profiles, $p(t)$, and the corresponding volume profiles, $v(t)$, obtained from the crank-angle mechanism of the engine. Besides coordinates of the work cycle, $p(v)$, they provide data for the profile of the *dynamic potential*, $w(t) \equiv p(t)v(t)$, exposed in Section 1.2.1, and their expressions in terms of analytic functions, described in Section 1.2.2.

2. *Thermodynamic Properties* manifested by the thermodynamic phase space delineating the three component processes of the: compression, the exothermic process referred to as the *dynamic stage,* and expansion. The fundamental reference coordinate of this space is the dynamic potential, $w(t)$. The trajectories of the processes carried out by the system in the course of the work cycle are delineated on the state diagram, $e(w)$, where e is the internal energy of the working substance. Obtained on this basis are profiles of the temperature, $T(t)$, as well as of the effective mass fractions of products, $y_P(t)$, and, hence, of fuel expended to create pressure, $x_F(t)$.

1.2 Dynamic Properties

The analysis of the dynamic properties yields analytic expressions for the exothermic process (popularly referred to as 'heat release') of combustion. It takes place in the course of, what is here referred to, the dynamic stage. Its life time is identified by the bounds: the rank angle of the initial state, **i**, and that of the final state, **f**. State **i** is an essential singularity of combustion, a saddle point representing the state where the specific volume of the products, $v_i = V_i / M_i = 0/0$ (!) – the *raison d'être* of what is known in combustion literature as the "cold-boundary difficulty," occurring in laminar flames (Williams 1985). Since nature abhors corners, it is bypassed by experimental data and, hence, state i is not identifiable by a data point.

State **f**, is at a singular point of maximum in the evolution of the combustion system, where its support by the deposition of exothermic energy is at equilibrium with the endothermic loss of energy incurred by heat transfer to the walls.

1.2.1 Dynamic Potential

The dynamic potential

$$w \equiv pv \equiv \frac{p}{\rho} \equiv h - e \tag{1.5}$$

is of particular significance to pressure diagnostics by furnishing an essential link between the dynamic Properties, where it is evaluated as a dependent variable, and the thermodynamic Properties for which it provides the fundamental thermodynamic reference parameter. The temperature is delegated thereby to the position of a dependent variable. Its magnitude can be readily determined from the equation of state for given values of p and w, or v, of the substance.

By expressing the specific volume in a non-dimensional normalized form, w is measured in units of pressure. Its value is popularly called flow work - a term demeaning its cardinal nature, as pointed out by Kestin 1966 (Sections 4.1 and 4.2). The significance of the dynamic potential is emphasized by most equations of state. For example, consider the following equations.

Van der Waals: $$w = \frac{\mathrm{R}}{\mathrm{M}} T \frac{v}{v - \mathrm{b}} - \frac{\mathrm{a}}{v}$$

Dietrich: $$w = \frac{\mathrm{R}}{\mathrm{M}} T \frac{v}{v - \mathrm{b}} \mathrm{e}^{-\mathrm{a}/v\mathrm{R}T}$$

Beattle-Bridgeman: $$w = \frac{\mathrm{R}}{\mathrm{M}} T + \frac{\beta}{v} + \frac{\gamma}{v^2} + \frac{\delta}{v^3}$$

BKW[1]: $$w = \frac{\mathrm{R_u}}{\mathrm{M}} T\left[1 + x\mathrm{e}^{\beta x}\right] \qquad x = \frac{\mathrm{k}}{v(T + \vartheta)^{\kappa}}$$

JWL[2]: $$w = \mathrm{A}\left[1 - \frac{v}{\mathrm{c_{v1}} \mathrm{v_c}}\right] v\mathrm{e}^{-\mathrm{R_1}v/\mathrm{v_c}} + \mathrm{B}\left[1 - \frac{v}{\mathrm{c_{v2}} \mathrm{v_c}}\right] v\mathrm{e}^{-\mathrm{R_2}v/\mathrm{v_c}} + \frac{\mathrm{R}}{\mathrm{c_v}} u$$

Virial Equation of State: $$w = \mathrm{A}(\Theta) + \mathrm{B}(\Theta)p + \mathrm{C}(\Theta)p^2 + \ldots$$

In the latter, $\Theta \equiv \frac{w_{\mathrm{m}}}{\mathcal{R}}$, where subscript m implies that the dynamic potential, w, is expressed per mole of the substance, while $\mathcal{R}$ is the universal gas constant.

---

[1] Becker-Kistiakowski-Wilson

[2] Jones-Wilkins-Lee

It is, in fact, for this reason that an equation of state for a system in equilibrium can be determined experimentally without measuring the temperature – a procedure providing the measure for the absolute temperature scale (vid. e.g. Kestin 1966-68).

1.2.2 Data

The procedure for establishment of the dynamic Properties is illustrated here by the specific example of a HCCI (Homogeneous Charge Compression Ignition) engine – a system that today is at the crest of popular research on novel piston engines. The data of the engine adopted for the present purpose are provided by Table 1.1. The operating conditions of its dynamometer test are listed in Table 1.2.

Table 1.1. HCCI engine data

Bore (cm)	7.95
Stroke (cm)	9.56
Length of piston rod (cm)	14.4
Compression ratio	16.5
Intake valve closing, Θ_a	205
p_a(bar)	1.39
T_a(K)	325
Exhaust valve opening Θ_z	512

Table 1.2. Operating conditions

Speed	rpm	1200
Torque	Nm	75
Fuel (gasoline)	Octane No.	87
Stoichiometric air/fuel ratio	ν_{st}, mol/mol	12.305
	σ_{st}, gm/gm	15.064
Air-equivalence ratio, λ_S,	-	2.2
Cylinder pressure at Θ_a, p_a,	atm	1.39
Gas temperature at Θ_a, T_a,	K	325

The air equivalence ratio cited in Table 1.2 was obtained from exhaust gas analysis and pertains therefore to the system, S, rather than the reactants, R.

Profiles of the dynamic parameters in terms of the crank angle, Θ, provided by the dynamometer test data of the engine operating at constant speed, are presented in Figs. 1.3-1.6.

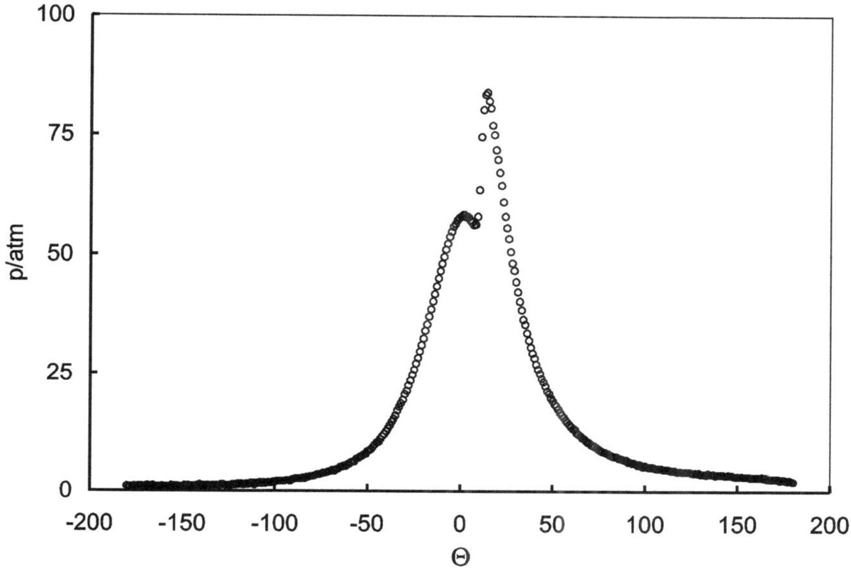

Fig. 1.3. Data of the pressure profile

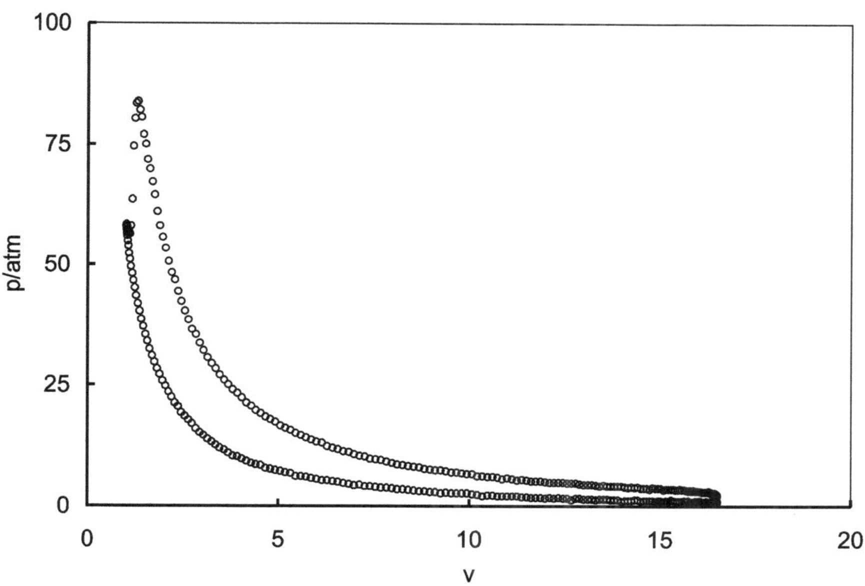

Fig. 1.4. Data of the work cycle in linear scales

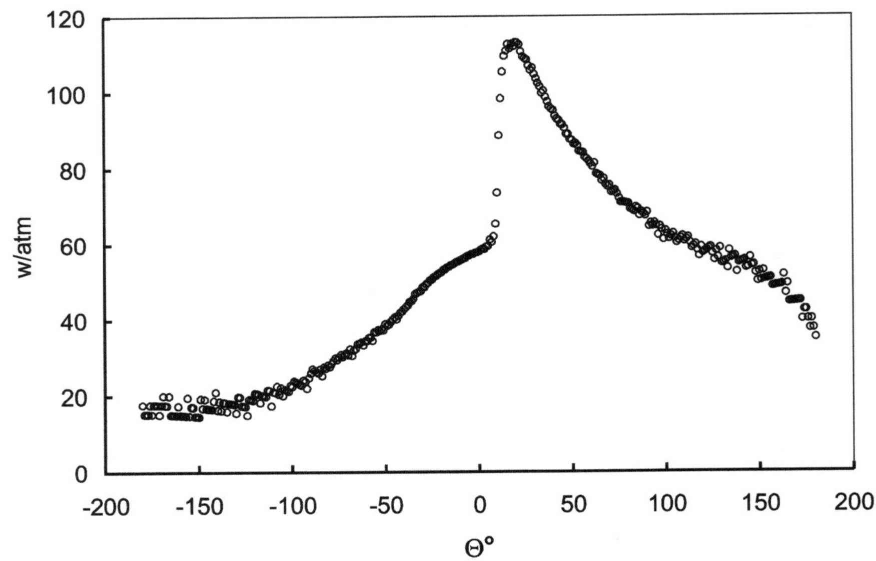

Fig. 1.5. Data of the dynamic potential

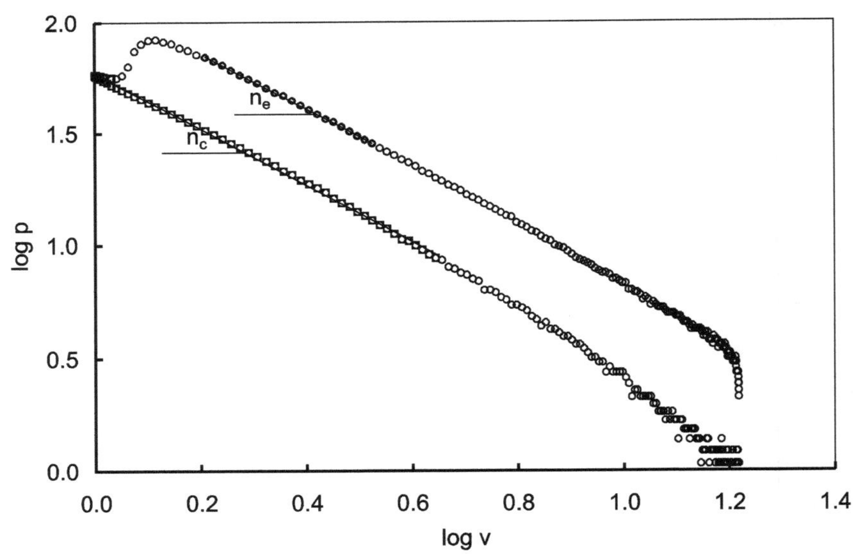

Fig. 1.6. Data of the work cycle in logarithmic scales

One should note that in pressure diagnostics $v \equiv v_S / v_c$, c denoting clearance, and, since in dynamometer tests engines are operated at constant speed, the time, t, is expressed in terms of the crank angle, Θ.

1.2.3 Functions

To provide a rational interpretation of the test data, they are expressed in terms of analytic functions. The key to this task is the polytropic function

$$\pi_k \equiv pv^{n_k} \tag{1.6}$$

where k = c, e, d for, respectively, the processes of compression, expansion, and the dynamic stage between them. In the first edition of this book and in Oppenheim 2004, the polytropic function was referred to by its generic name "polytropic pressure model", because for an enclosure of constant volume $n_k=0$, so that $\pi_k=p$.

Upon selection of the scope of the data for analytic interpretation, bounded by point **b** at the start of the last sector of compression and point **c** at the end of the first sector of expansion, their processes are expressed by polytropes – functions for which n_c and n_e, as well as π_c and π_e, are constant. In Fig, 1.6, these polytropes are identified by straight-linear sectors between the postulated bounds, **b** and **i-** for compression and **f+** and **c** for expansion, where the two exponents are identified by their slopes.

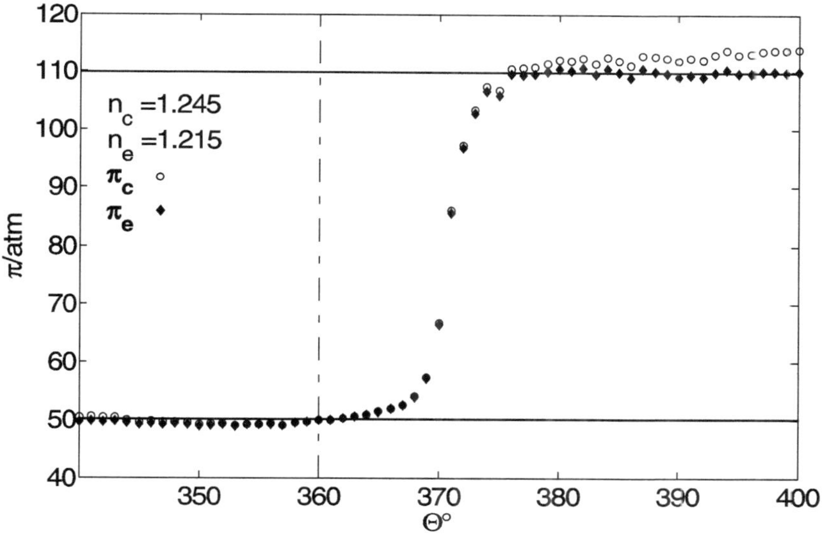

Fig. 1.7. Profiles of data for polytropic functions, evaluated with invariant indexes n_c (open points), and n_e (closed points)

Profiles of data expressed by polytropic functions of the HCCI engine, with n_c for compression and n_e for expansion, are portrayed by Fig. 1.7, where the two polytropes, extended over the width of the diagram, are displayed by horizontal lines.

The a priori unknown bounds of the analytic function for the dynamic stage, x_d, the initial point, **i**, and final point, **f**, must be located on these lines. To satisfy this requirement, the analytic function has to be established by iteration, with an, at least, partially variable exponent n_d.

The initial point is, therefore, first estimated to be ahead of the eventual position of **i**, at point **i-**, where the data of π_c start to rise above the horizontal polytrope for compression, and the final point postulated to be beyond the expected position of **f**, at point **f+**, where the data of π_e start to drop below the horizontal polytrope of expansion. The exponent of expansion, n_e, is retained below point **f** to some intermediate point, **n**, where the data of π_c and π_e are coincident. Up to that point, the polytropic function of the dynamic stage is expressed in terms of $n_d = n_e$. Below this point, it has to culminate at the initial point **i** on the extended polytrope of compression that, as brought out at the outset of Section 1.2, is a sharp singularity obviated by the data, for nature abhors corners.

For this purpose, the data of the dynamic stage are expressed in terms of the progress parameter for the polytropic function

$$x_k(\pi) = \frac{\pi - \pi_o}{\pi_k - \pi_{o-}} \qquad (k = n, f)\ (o = i-, i) \qquad (1.7)$$

in terms of which

$$n = n_c + (n_e - n_c)x_n(\pi) \qquad (1.8)$$

Upon establishing in this manner the progress parameter, $x(\pi)$, made out of two sectors, one from **i-** to n according to (1.8), and the other from **n** to **f+** with $n = n_e$, a unique analytic function for the dynamic stage is, obtained by regression of $x(\pi)$, between **i** and **f** to the life function introduced in Chapter 2 – a normalized exponential

$$x = \frac{e^{\zeta} - 1}{e^{\zeta_f} - 1} \qquad (2.47)$$

whose exponent

$$\zeta = \frac{\alpha}{\chi + 1}[1 - (1 - \tau)^{\chi+1}] \tag{2.48}$$

is a power function of the progress parameter for time that, for an engine operating at constant speed, is measured in terms of the crank angle, Θ, i.e.

$$\tau \equiv \frac{\Theta - \Theta_i}{\Theta_f - \Theta_i} \tag{1.9}$$

For this purpose, the exact position of point **i** is established by iteration to minimize the standard deviation of regression, yielding the two life function parameters, α and χ.

As demonstrated in Chapter 2, the life function thus determined satisfies both conditions imposed by the singular boundaries of the dynamic stage, specified at the outset of Section 1.2.

Analytic interpretation of the profiles of the polytropic functions for the HCCI engine, are presented by Fig. 1.8, in comparison with data of Fig. 1.7.

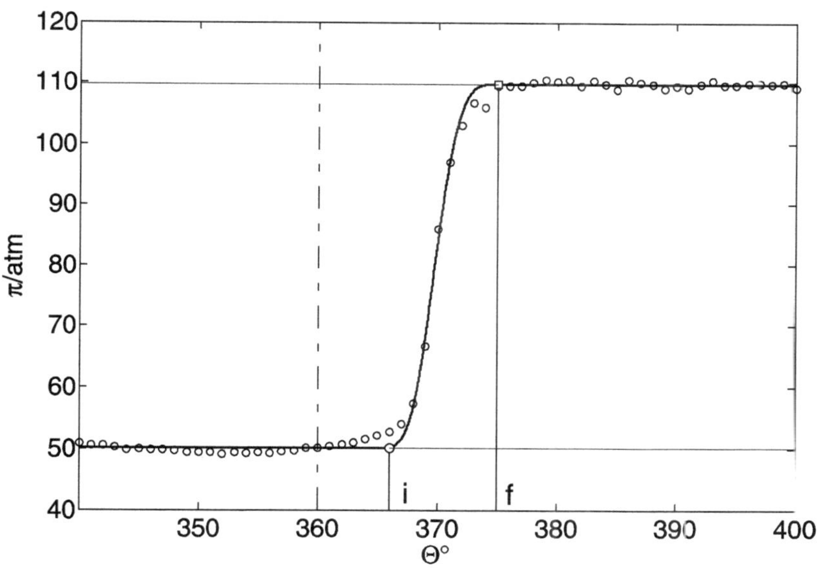

Fig. 1.8. Analytic interpretation of data for the last sector of compression the dynamic stage, and the first sector of expansion of the HCCI engine

Values of all the parameters involved in this procedure for the HCCI engine are listed in Table 1.3, together with the crank angles of all the critical points for pressure diagnostics. The consecutively evaluated analytic interpretation of data for profiles of the dynamic potential, the work cycle, and pressure, are displayed, respectively, by Figs. 1.9, 1.10 and 1.11.

Table 1.3. Parameters of Dynamic Properties

Crank angle of point a	205°
Crank angle of point b	320°
Crank angle of point i	366 °
Crank angle of point f	375°
Crank angle of point c	500°
Crank angle of point z	512°
Life function coefficient, α	13.55
Life function exponent , χ	2.19

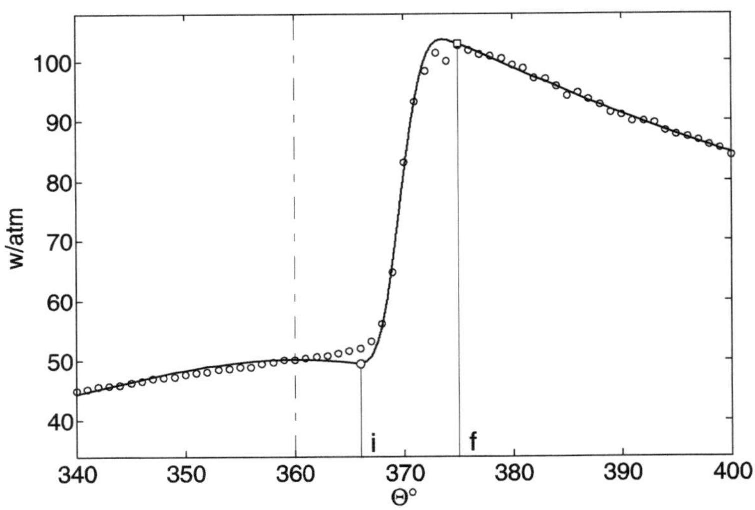

Fig. 1.9. Analytic interpretation of data for the dynamic potential of the HCCI engine

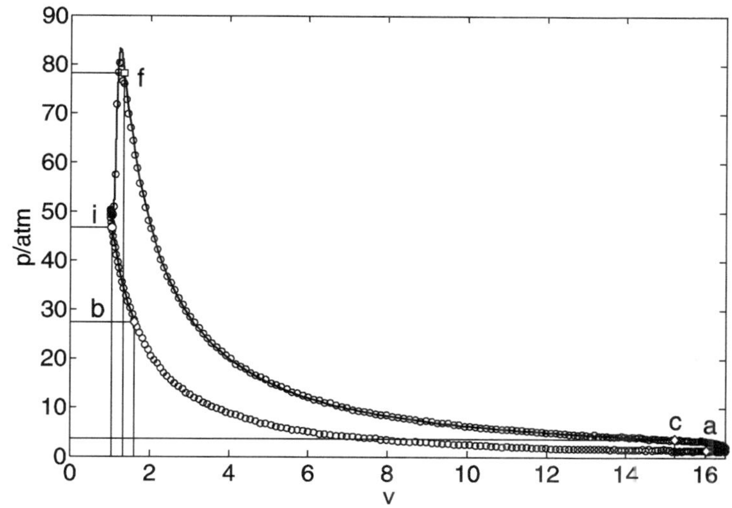

Fig. 1.10. Analytic interpretation of data for the work cycle of the HCCI engine

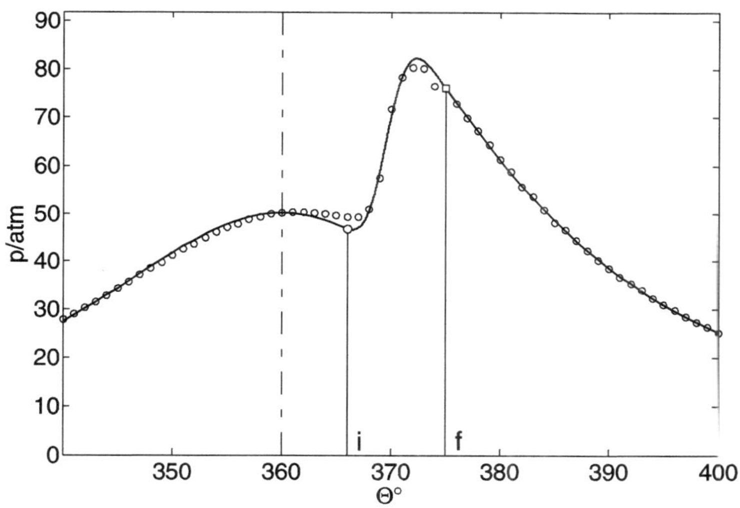

Fig. 1.11. Analytic interpretation of data for the pressure profile of the HCCI engine

1.3 Thermodynamic Properties

In accord with the zero-dimensional nature of dynamic properties, the variables of thermodynamic p are expressed in terms of time-dependent mass averaged thermodynamic parameters of state: the pressure, $p_K(t)$, the temperature, $T_K(t)$, the specific volume, $v_K(t)$, the internal energy, $e_K(t)$, and the dynamic potential, $w_K(t)$, where K = F, A, B, C, R and P, providing a link between the dynamic and thermodynamic properties. Moreover, since all the parameters of the dynamic properties are expressed in terms of analytic functions, the time coordinate can be expressed in terms of any of them, in particular pressure whose time profile provides the basis for their evaluation.

1.3.1 Thermodynamic State

The state of a constituent is identified by three parameters of its own equilibrium. The conventional equation of state, expressing a relationship between pressure, *p*, specific volume, *v*, and temperature, *T*, does not provide a comprehensive specification of state, because, to evaluate internal energy, *e*, specific heats are, moreover, required. If, however, internal energy is included among the parameters of state, then, according to the fundamental principle of the First Law provided by Gibbs (1875-1878) and formulated by Poincaré (1892) and Carathéodory (1909), the state of a constituent is thereby completely specified.

Accordingly, the thermodynamic state of a constituent is expressed by a point in a three-dimensional thermodynamic phase space whose coordinates are specific internal energy, *e*, specific dynamic potential, *w*, and pressure, *p*. If internal energy is expressed in molar (volumetric) basis, the fundamental role of a reference coordinate is played by the temperature, *T*, rather than *w*. However, since in system undergoing a chemical reaction mass is conserved rather than volume, the latter is for this purpose more appropriate.

The reactants, R, are considered as an air/fuel mixture at its local air-equivalence ratio, λ_R. The products, P, are specified by the molecular composition of the system at thermodynamic equilibrium. The states of both of them are established by appropriate algorithms. Their thermodynamic parameters are presented by Table 1.4 upon evaluation by means of STANJAN (Reynolds 1996), and CEA (Gordon and McBride 1994,

McBride and Gordon 1996)[3], on the basis of thermodynamic data provided by the JANAF Tables (Stull and Prophet 1971), as well as the Chemistry WebBook of NIST[4], for a mixture of iso-octane and normal heptane in proportions provided by the octane number and the air-equivalence ratio cited in Table 1.2.

Table 1.4. Thermodynamic parameters of components in the HCCI engine

K	States	p	T	v	e	h	w	M
		atm	K	m^3/kg		kJ/g		g/mol
	a	1.35	325	0.672	-0.12	0.03	0.09	29.50
R	i	49.34	802	0.045	0.27	0.50	0.23	29.50
	f	78.37	1407	0.050	0.86	1.25	0.40	29.50
P	$(hp)_i$	49.34	1832	0.106	-0.03	0.50	0.53	28.73
	$(hp)_f$	78.37	2359	0.086	0.57	1.25	0.68	28.74

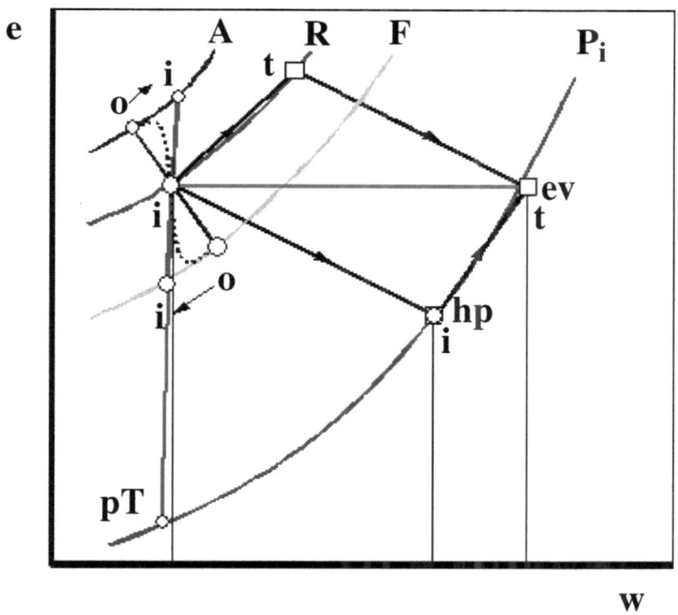

Fig. 1.12. State diagram for adiabatic and isochoric combustion system

[3] <http://www.grc.nasa.gov/WWW/CEAWeb>

[4] <http://webbook.nist.gov>

In a three-dimensional state space, the planar platform of this diagram presented by Fig. 1.12 is at the level of the initial pressure, p_i. For higher pressures, the e-w platform is at higher levels. For the regime of pressures and temperatures existing in internal combustion engines, the reactants are made out of fixed fuel and air fractions, while, according to the available source of thermodynamic data, these components are perfect gases. The locus of states of the reactants, R is, therefore, pressure independent. However, the locus of states of the products, P, is dependent on pressure, because, in order to comply with the condition of thermodynamic equilibrium, their composition is variable. On the e-w plane, for higher pressures, the line of P tends to get straighter, so that, as indicated in Fig. 1.12, it is further away from the internal energy axis.

In the course of an exothermic process, the states of reactants, R, and products, P, move along their loci of states, starting from the initial point, **i,** and ending at the terminal point **t**, where the generation of products, and hence, consumption of fuel, is terminated, that, in principle, is different than the final point, **f.** Nonetheless, as demonstrated in Chapter 3, they are coincident The change of state taking place in the course of exothermic reaction, when the reactants are transformed into products at a fixed pressure, is presented by a straight line between a point on R and a point on P.

The changes of state of the reactants, R, and products, P, taking place in the combustion system are presented for them by vectors **i** – **t** along their loci of states, while the processes of exothermic reactions carrying out the transition between them are depicted by vectors $\mathbf{i}_{\text{on R}} - \mathbf{i}_{\text{on P}}$ and $\mathbf{t}_{\text{on R}} - \mathbf{t}_{\text{on P}}$ across their loci of states.

1.3.2 Processes

1.3.2.1 Mixing

For a chemical reaction to take place, its components must be first mixed to form a molecular aggregate. If, initially, the thermodynamic coordinates of fuel and air are different, they have to be brought to the same state i on R - a task accomplished physically by transport processes of molecular mass diffusion and thermodynamic conduction, assisted by viscosity. In Fig. 1.12, the concomitant changes of state taking place in the course of mixing are expressed by broken curves between points **o** on A and **o** on F to point **i** on R, with their directions indicated by arrows. The effect of mixing is manifested by rotation of the end point of the state vector around point i on R. Irrespectively of the influence of molecular diffusion, which, as a rule, must be involved in forming the reacting mixture, its outcome can be identified right from the outset by the intersection of the straight line between points **o** on A and **o** on F with R.

1.3.2.2 Exothermic Reaction

Chemical reaction of combustion takes place in an exothermic center. The concept of exothermic centers has been known for a long time in detonation literature under the name of "hot spots." Their non-steady behavior under the influence of molecular diffusion has been studied extensively as the process of ignition (e.g. Boddington et al 1971; Gray and Scott 1990; Griffiths 1990). Their non-steady version in a turbulent field is referred to as the 'flamelet model' (Peters 2000). The fluid dynamic features of exothermic centers were investigated experimentally and theoretically in connection with their relevance to detonation and explosion phenomena, leading to the identification of mild and strong ignition centers. In a gasdynamic field where exothermic reaction takes place, exothermic centers occur at discrete sites. Each of them behaves then as a point singularity - a constant pressure deflagration where a finite change of state takes place locally at constant pressure - rather than across a straight line as it does in the classical version of a deflagration front.

1.3.3 Thermodynamic Parameters

The behavior of a combustion system is specified by the balances of mass, volume and internal energy. The mass balance is expressed simply by the fact that $\mathrm{Y_S} = \sum y_\mathrm{K} = 1$, or $\mathrm{X_F} = \sum x_\mathrm{K} = \mathrm{Y_R}$, as depicted in Fig. 1.1. The balance of volumes, $v_\mathrm{S} = \sum y_\mathrm{K} v_\mathrm{K}$ is expressed by the balance of dynamic potential, $w_S = \sum y_\mathrm{K} w_\mathrm{K}$, since by definition $w(t) = p(t)v(t)$, where $p(t)$ is the same for all the components. The balance of energy is similarly given by $e_S = \sum y_\mathrm{K} e_\mathrm{K}$.

The two coordinates of the thermodynamic states are expressed by a single generalized state parameter, $z_K = w_K$, e_K (K = S, F, A, P, B, C, R), in terms of which the balances of volume and energy are cast into a single equation,

$$z_S = y_F z_F + y_A z_A + y_P z_P + y_B z_B \tag{1.10}$$

whence, in view of (1.3) and (1.4),

$$z_S = \mathrm{Y_F} z_F + \mathrm{Y_A} z_A + \mathrm{Y_B} z_B + (\mathrm{Y_P} z_P - \mathrm{Y_F} z_F - \mathrm{Y_A} z_A) x_F \tag{1.11}$$

The state of the cylinder charge, C - a substance whose composition is fixed by initial conditions while its thermodynamic parameters vary,

$$z_C \equiv \mathrm{Y_F} z_F + \mathrm{Y_A} z_A + \mathrm{Y_B} z_B \tag{1.12}$$

The state of reactants, R, considered as a mixture of fuel, F, with air, A, at their fixed ratio of σ_R is determined by the mass average

$$z_R = \frac{\mathrm{Y_F} z_F + \mathrm{Y_A} z_A}{\mathrm{Y_{F}}_F + \mathrm{Y_A}} = \frac{z_F + \sigma_R z_A}{1 + \sigma_R} \tag{1.13}$$

With (1.12) and (1.13), in view of (1.8), (1.13) is reduced to

$$z_S - z_C = (z_P - z_R) y_P \tag{1.14}$$

In a combustion field, an exothermic center is a front across which a jump from **i** to **t** takes place at a constant pressure. Its amplitude is

$$\mathrm{z_{SP}} - \mathrm{z_{SR}} = \mathrm{Y_R}(\mathrm{z_P} - \mathrm{z_R}) \tag{1.15}$$

Presented in Fig. 1.13 are changes of state taking place in the general case of an exothermic system that, unlike that of Fig.1.12, is affected by energy losses occurring in the course of the transition from R to P. Under such circumstances, the end equilibrium states on P are, below those reached at constant enthalpy.

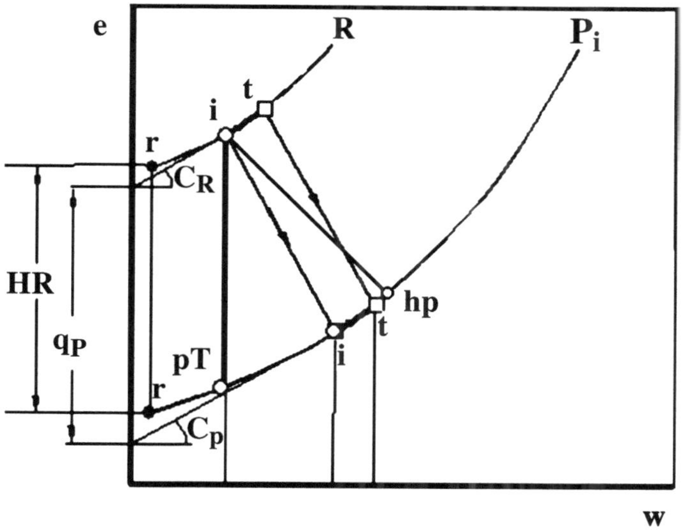

Fig. 1.13. Diagram of states for an exothermic system with energy loss

1.3.4 Coordinate transformation

As portrayed Fig. 1.13, the slope at any point of state trajectory

$$C_K \equiv (e_t - e_i)_K/(w_t - w_i)_K \qquad (K = R, P) \tag{1.16}$$

Consequently

$$e_R = C_R w_R \tag{1.17}$$

while

$$e_P = C_P w_P - q_P \tag{1.18}$$

where $q_P = e_{Ro} - e_{Po}$, subscript "o" referring to the intersections of the lines of K with the axis of ordinates. The quantity q_P, referred to here as the exothermic energy, provides a geometric measure of a reference parameter that replaces the conventional concept of 'heat release' (HR), used as a constant in conjunction with fixed specific heats. Its awkward role when specific heats are variable is evident in Fig. 1.13, where the reference states of HR are denoted by r.

According to (1.17) and (1.18),

$$e_R - e_P = \mathrm{C_R} w_R - \mathrm{C_P} w_P + \mathrm{q_P} \tag{1.19}$$

The coordinates of internal energy, $e_K(w_K)$, can be thereby expressed in terms of the local trajectory slope, C_K, and the exothermic energy, q_P, as illustrated in Fig. 1.14 for the initial state, **i**, and the terminal state, **t**, of transition from a reactant, R, to product, P, taking place in an adiabatic exothermic center.

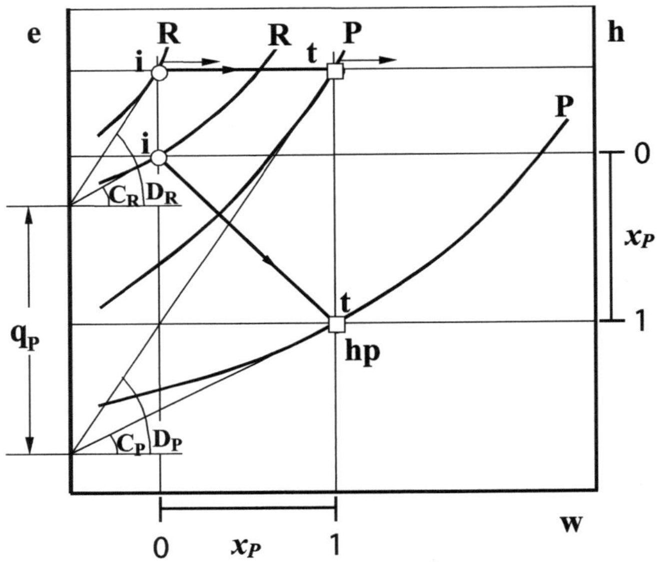

Fig. 1.14. Diagram of states for an adiabatic exothermic center in e(w) and h(w) coordinates

The loci of states for R and P are displayed in this figure in e(w) and h(w) coordinates. A change of state taking place in the course of an adiabatic exothermic reaction proceeds along a constant enthalpy path – a diagonal on the plane of e-w expressed by $e = h_i - w$, and a horizontal line on the h-w plane. As implied by the geometry of this diagram, the slope $D_K = C_K + 1$ (K = R,P). If the reactants are considered as perfect gases with constant specific heats, $(c_v)_K = de_K/dT$ and $(c_p)_K = dh_K/dT$, then, in terms of $R_K \equiv R/M_K$, where R is the gas constant and M_K the molar mass, whence $\gamma_K \equiv c_{Kp}/c_{Kv}$, $C_K = c_{Kv}/R_K = 1/(\gamma - 1)$ and $D_K = c_{Kp}/R_K = \gamma/(\gamma - 1)$.

Since, generally, the curvature of a state trajectory, R or P, is relatively small, the changes of state performed by reactants and products in the course of an exothermic process are expressed with remarkable accuracy by straight lines. The relationships for trajectory slopes specified in the previous section can be applied, therefore, to the sectors pertaining to the process of an exothermic center.

1.3.5 Isentropic Relationships

Of particular significance to the coordinates of the state diagram are the isentropic relationships. Irrespectively whether the substance is a perfect gas or not, nor whether its process is isentropic or not, the parameter

$$\Gamma \equiv (\frac{\partial h}{\partial e})_{\mathrm{p}} = 1 + (\frac{\partial w}{\partial e})_{\mathrm{p}} = \frac{\mathrm{C_k} + 1}{\mathrm{C_k}} = \frac{\mathrm{D_k}}{\mathrm{D_k} - 1} \tag{1.20}$$

while the polytropic index

$$\mathrm{n} \equiv \frac{\mathrm{dln}\, p}{\mathrm{dln}\, \rho} = -\frac{v}{p}\frac{\mathrm{d}p}{\mathrm{d}v} \tag{1.21}$$

Then, by virtue of the Second Law, according to which

$$\begin{aligned} T\mathrm{d}s &= \mathrm{d}e + p\mathrm{d}v \\ &= \mathrm{d}h - v\mathrm{d}p \end{aligned} \tag{1.22}$$

the isentropic index, with subscript "s" denoting an isentropic process,

$$\gamma \equiv \mathrm{n_s} = (\frac{\partial h}{\partial e})_{\mathrm{s}} = -\frac{v}{p}(\frac{\partial p}{\partial v})_s = \Gamma_{\mathrm{s}} \tag{1.23}$$

Thus, noting that $\frac{p}{\rho} \equiv pv \equiv w$, the velocity of sound

$$a \equiv \sqrt{(\frac{\partial p}{\partial \rho})_s} = \sqrt{\gamma \frac{p}{\rho}} = \sqrt{\gamma \mathrm{w}} \tag{1.24}$$

whence

$$(\frac{\partial a}{\partial p})_s = \frac{\gamma}{2a}[\frac{\partial (w)}{\partial p}]_s \tag{1.25)}$$

while, in view of (1.21) and (1.23),

$$[\frac{\partial (w)}{\partial p}]_s = v + p(\frac{\partial v}{\partial p})_s = v(1 - \frac{1}{\gamma}) = \frac{\gamma - 1}{\gamma} v \tag{1.26}$$

so that, by virtue of (1.24),

$$(\frac{\partial p}{\partial a})_s = \frac{2}{\gamma - 1}\rho a = \frac{2\gamma}{\gamma - 1}\frac{p}{a} \tag{1.27}$$

- a relationship of particular significance to gasdynamics presented in Chapter 7.

Since, by definition, $dw = w(\mathrm{dln}p + \mathrm{dln}v)$, it follows from (1.26) that

$$\mathrm{d}w = \frac{\gamma - 1}{\gamma} w\mathrm{dln}p \tag{1.28}$$

and, according to (1.20),

$$\mathrm{d}e \equiv \mathrm{C_k d}w = \frac{1}{\gamma - 1}\mathrm{d}w = \frac{p}{\gamma \rho}\mathrm{dln}p \tag{1.29}$$

- a relationship of particular significance to blast wave theory presented in Chapter 9.

1.3.6 Closed System

A closed system is one whose volume, $v(t)$, is restricted by prescribed boundary conditions. In the cylinder of an internal combustion engine, for instance, the boundary conditions are imposed by piston motion established by the kinematics of crankshaft mechanism. Its volume and energy balances are specified by (1.14), for which $w_S(t)$ is prescribed.

The volume balance is then obtained for $z = w$,, while $p_R = p_P$, so that

$$(w_P - w_R)\, y_P = w_S - w_C \tag{1.30}$$

The energy balance is obtained similarly for $z = e$, whence

$$(e_P - e_R)\, y_P = e_S - e_C \tag{1.31}$$

for which

$$e_S = e_{Si} - e_e \tag{1.32}$$

while

$$e_e \equiv w_w + q_w \tag{1.33}$$

expresses the energy expenditure that, in an internal combustion engine, consists of piston work,

$$w_w \equiv \mathrm{v_c} \int_{\mathrm{i}}^{\mathrm{t}} (p - p_b)\, \mathrm{d}v_{\mathrm{S}} \tag{1.34}$$

and the energy loss, q_w.

In the above, subscript "c" denotes clearance volume per unit mass, while "b" refers to the backpressure, whereas q_w, is incurred primarily by heat transfer to the walls that, besides some usually negligible effects of leakage and mixing, is the principal irreversibility of a combustion system.

With (1.19) and (1.32), the energy balance specified by (1.31) becomes

$$(\mathrm{C_R} w_R - \mathrm{C_P} w_P + \mathrm{q_P})\, y_P = \mathrm{C_C}(w_C - w_{Si}) + e_e \tag{1.35}$$

Then, eliminating w_P from (1.30) and (1.35), the mass fraction of products

$$y_P = \frac{\mathrm{C_P}(w_S - w_C) + \mathrm{C_C}(w_C - w_{Si}) + e_e}{\mathrm{q_P} + \mathrm{C_R} w_R - \mathrm{C_P} w_R} \tag{1.36}$$

In terms of normalized variables, $\mathrm{W_K} \equiv \mathrm{w_K} / \mathrm{w_{Si}}$ (K = S, R, P) and $W_e \equiv e_e / w_{Si}$, while $\mathrm{Q_P} \equiv \mathrm{q_P} / \mathrm{w_{Si}}$, (1.36) yields

$$y_P = \frac{C_P(W_S - 1) - (C_P - C_C)(W_C - 1) + W_e}{Q_P - (C_P - C_R)W_R} \tag{1.37}$$

Since, according to JANAF tables – the conventional sources of thermodynamic data – all the components are treated as perfect gases, while the thermodynamic properties of the reactants and the system charge – their mixtures at fixed composition – are practically the same,

$$W_C = W_R = P^m \tag{1.38}$$

where $\mathrm{m} \equiv 1 - \mathrm{n}_\mathrm{c}^{-1}$.

Thus, according to the volume balance expressed by (1.30),

$$W_P = W_R + \frac{W_S - W_C}{y_P} = P^\mathrm{m} + \frac{PV - P^\mathrm{m}}{y_P} \tag{1.39}$$

- a relationship emphasising the singular nature of state **i,** at which the numerator $W_S - W_C = 0$, while the denominator $y_P = 0$.

According to the results of a comprehensive, semi-empirical study of heat transfer in a closed combustion vessel presented in Chapter 3, the mass fraction of products, $y_\mathrm{P}(\Theta)$ is equivalent to $x(\tau)$, the latter identified by (2.49) with (2.48).

The mass fraction of the generated products, as pointed out in (1.15),

$$y_\mathrm{P} = y_\mathrm{E} + y_\mathrm{I} \tag{1.40}$$

where y_E is its effective part and y_I is the ineffective part. Then, by virtue of (1.37) and (1.38), with $W_w \equiv \mathrm{w}_w / \mathrm{w}_\mathrm{Si}$, while, by definition, $W_S = PV$ where $P \equiv p / \mathrm{p_i}$, $V \equiv v_S / v_{Si}$,

$$y_\mathrm{E} = \frac{\mathrm{C_P}(PV - 1) - (\mathrm{C_P} - \mathrm{C_C})(P^\mathrm{m} - 1) + W_w}{\mathrm{Q}_P - (\mathrm{C_P} - \mathrm{C_R})P^\mathrm{m}} \tag{1.41}$$

while, in turn, the effective part, y_E, consists of the products generated for internal energy, y_ε, and to produce piston work , y_ω, so that

$$y_{\mathrm{E}} = y_{\varepsilon} + y_{\omega} \tag{1.42}$$

where

$$y_{\varepsilon} = \frac{\mathrm{C_P}(PV-1)-(\mathrm{C_P}-\mathrm{C_C})(P^m-1)}{\mathrm{Q_P}-(\mathrm{C_P}-\mathrm{C_R})P^m} \tag{1.43}$$

and

$$y_{\omega} = \frac{W_w}{Q_P-(C_P-C_R)P^m} \tag{1.44}$$

while the ineffective part

$$y_{\mathrm{I}} = \frac{Q_{\mathrm{I}}}{\mathrm{Q_P}-(\mathrm{C_P}-\mathrm{C_R})P^m} \tag{1.45}$$

where $Q_I \equiv q_w / \mathrm{w_{Si}}$.

It is of interest to note that, if $\mathrm{C_P} = \mathrm{C_C} = \mathrm{C_R}$, (1.41) is reduced to

$$y_E = \frac{(PV-1)+W_w}{\mathrm{Y_R Q_P / C_P}}$$

- a relationship of the well established "heat release analysis", based on the classical paper of Rassweiler and Withrow (1938).

For an isochoric and isentropic system of $v = 1$, $W_w = 0$ and $Q_I = 0$, obtained thereby is the popular proportionality between the mass fraction of products, y_P, and the measured overpressure, $(P - 1)$ [vid. e.g. the classical text of Lewis and von Elbe (1987)].

The procedure of pressure diagnostics is completed by taking into account the results obtained from heat transfer study described in Chapter 3. According to them,

(1) the time coordinate of the final state is identified with that of the terminal state, so that $\Theta_{\mathrm{f}} = \Theta_{\mathrm{t}}$

(2) the mass fraction of products is identified with the progress variable for the polytropic function expressed in terms of the life function, while $y_{\mathrm{P}}(\Theta) = x(\Theta)$, determined for the excthermic process, whose profile is displayed by Fig. 1.8.

Obtained therefore are means for evaluating the ineffective mass fraction of products $y_I = y_P - y_E$, identifiable, as demonstrated in Chapter 3, with energy loss incurred by heat transfer to the walls. The relationship between these variables is displayed on the phase diagram illustrated by Fig. 1.15, corresponding to their profiles displayed by Fig. 1.2.

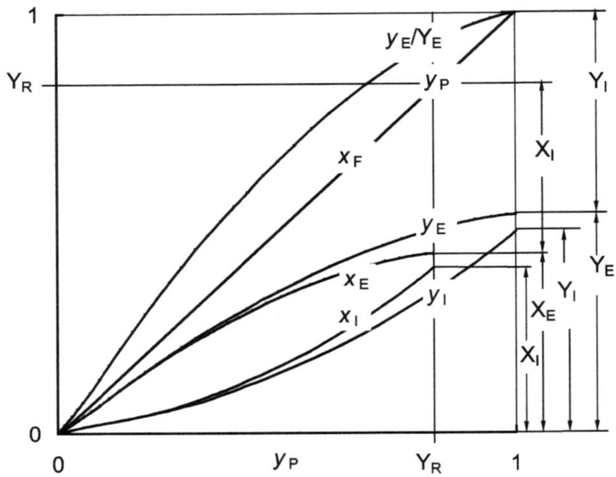

Fig. 1.15. Phase diagram of the effective and ineffective parts of products

Concomitantly, (1.39) yields the profiles of the dynamic potential and, hence, the temperature profiles of the reactants and the products. The results obtained thereby from dynamometer tests of a diesel engine were found to be in satisfactory agreement with the well-known Woschni correlation [Woschni 1966/67, 1967, 1970, Woschni and Anisits 1973, 1974].

1.3.7 Procedure

As for the elucidation of dynamic properties, the procedure for implementation of the exothermic process is presented here for the specific example of a HCCI engine.

The state diagram of its exothermic process is presented by Fig. 1.16. Its parameters are presented by Table 1.5. Profiles of the mass fraction of products, $y_P = x$, its effective part, y_E, consisting of y_ε and y_ω, are determined with the use of these parameters, according to (1.40), by (1.41), that, according to (1.42), by (1.43) and (1.44). The results are displayed by Fig. 1.17. The dynamic potential in its normalized form, W_K, is thereupon obtained by virtue of (1.39), and its profile, $w_K(\Theta) = W_K(\Theta)w_{Si}$, where K = R, P, S, is presented by Fig. 1.18, whence the temperature profiles depicted by Fig. 1.19, are evaluated the use of the equation of state whence $T_K(t) = w_K/R_K$.

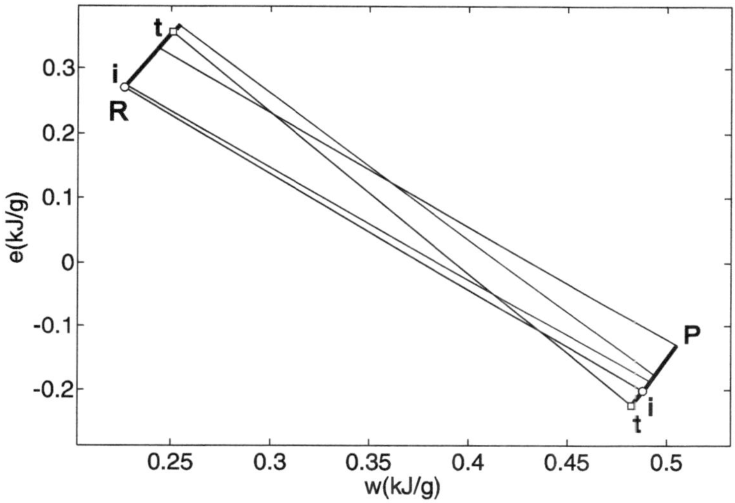

Fig. 1.16. State diagram of the exothermic process in the HCCI engine

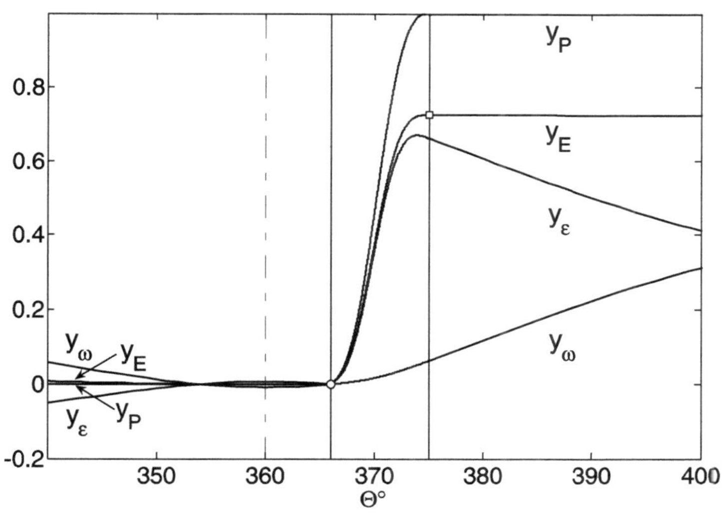

Fig. 1.17. Profiles of the mass fraction of products and its components of the HCCI engine

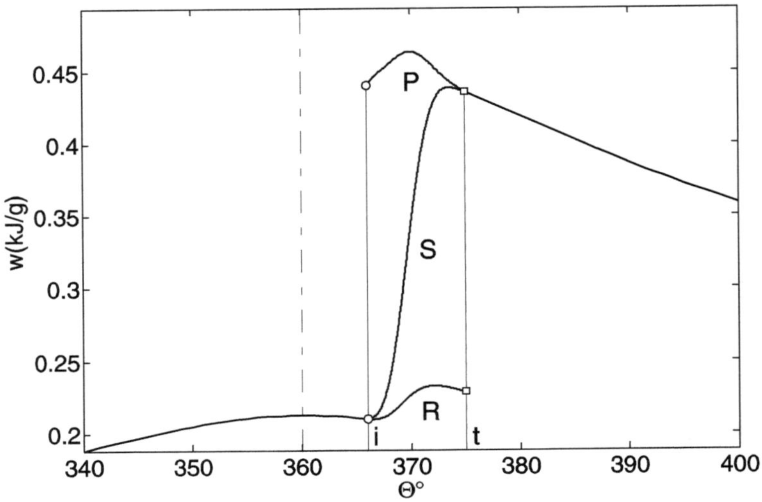

Fig. 18. Profiles of the dynamic potential of the HCCI engine

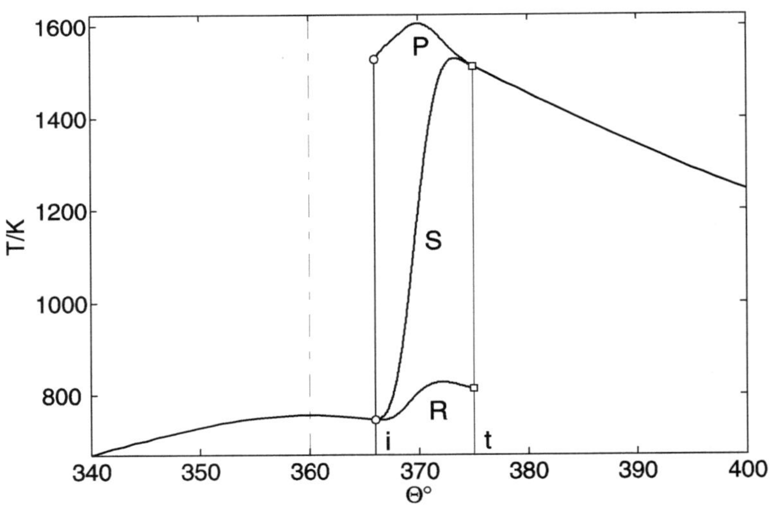

Fig. 1.19. Temperature profiles in the HCCI engine

1.3.8 Control Logistics

Pressure diagnostics provides the analytical method for ushering in micro-electronic control system to optimize the execution of the exothermic process of combustion and improve thus significantly its performance. Put forth thereby is the essential objective of what is referred to as 'internal treatment' – a technology that has been aborted by industry in favor of 'external treatment' as a consequence of the fact that, at the time when clean air laws were introduced, micro-electronics was not yet available.

The benefits of internal treatment are realized by implementing the principles of Aero-thermo-chemistry introduced by von Karman: *fluid dynamics* to create turbulent plumes as far away from the walls as possible, *thermodynamics* to reveal their physical properties and assess their effectiveness, and *chemistry* to modulate the chemical kinetic process of the exothermic reaction of combustion so that it takes place at the lowest temperature allowable at the verge of extinction.

In effect, one is treating, therefore, an inverse problem of the inverse problem of pressure diagnostics: upon assessing the performance of a combustion system on the basis of the measured pressure profile, its execution is optimized by controlling its operating conditions. A closed-loop micro-electronic control system suitable for this purpose is described in Oppenheim 2004. Its primary feedback is a pressure transducer; its principal actuators are provided by a pulsed jet injector and a pulsed flame jet igniter.

Presented here for illustration is the simplest example of optimization attainable by micro-electronic control, one particularly appropriate for HCCI engines: shifting of the exothermic process of combustion so that it takes place as close as possible to the top dead center – a task akin to engine tuning for maximum torque.

The logistics of control are implemented in two steps:

(1) shift of the exothermic process to the top dead center

(2) reduction of the thus augmented IMEP of the work cycle to its standard value

The application of this procedure is illustrated by the subsequent figures where the reference case is depicted by continuous lines, while the two steps are delineated, respectively, by broken and dotted lines, and in the labels they are marked by single and double primes.

As depicted in Fig. 1.20, the profile of the polytropic function is extended, first, by shifting the life function of the dynamic stage along the horizontal polytropes of the processes of compression and expansion, so that its point of inflection is at TDC (top dead center), and, thereupon, by decreasing the level of the expansion polytrope to provide the required work output.

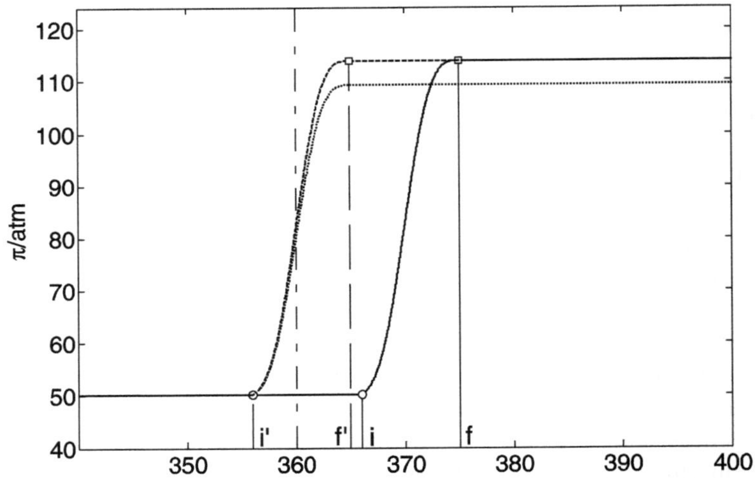

Fig. 1.20. Profiles of the polytropic functions for the HCCI engine in the original and controlled mode of operation

On this basis, all the parameters for the controlled mode of the engine combustion system are evaluated. The results are displayed by Figs. 1.21 – 1.26 in comparison to the original performance.

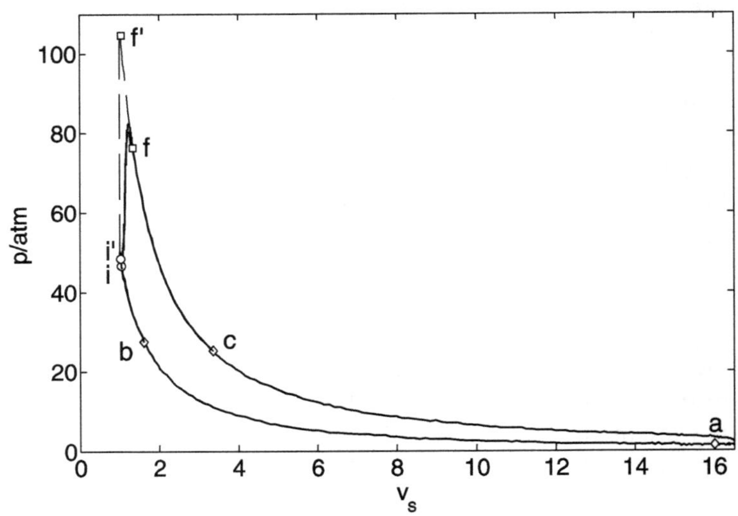

Fig. 1.21. Work cycle of the HCCI engine in the original and controlled mode of operation

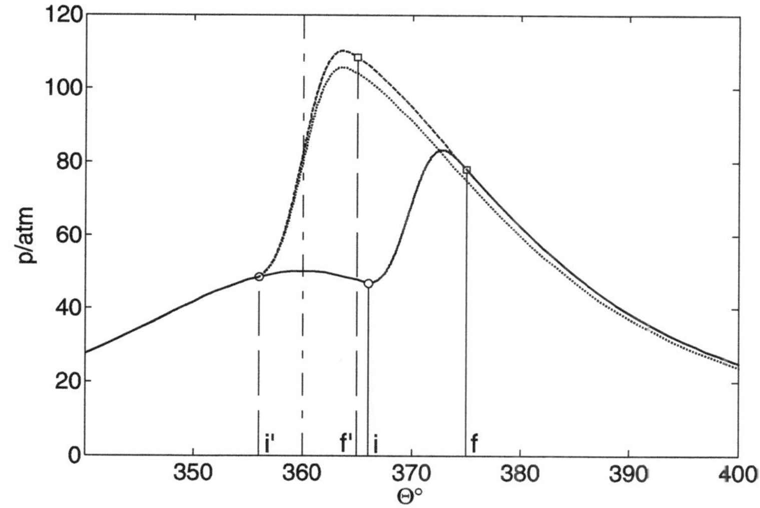

Fig. 1.22. Pressure profiles of the HCCI engine in the original and controlled mode of operation

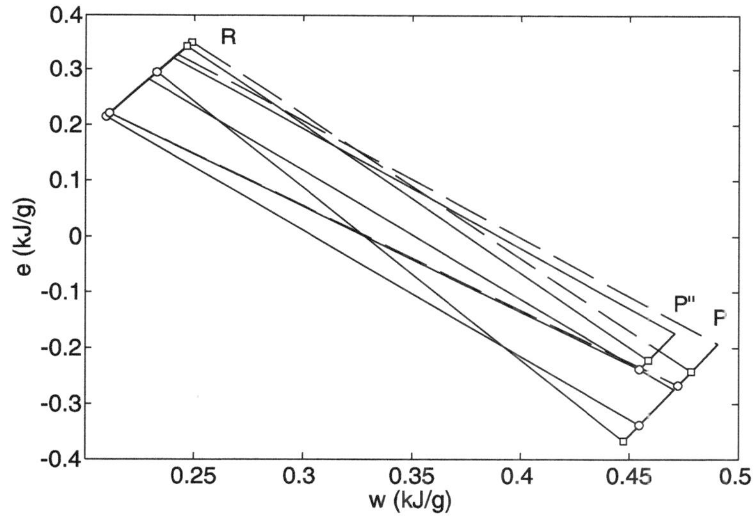

Fig. 1.23. State diagram of the HCCI engine in the original and controlled mode of operation

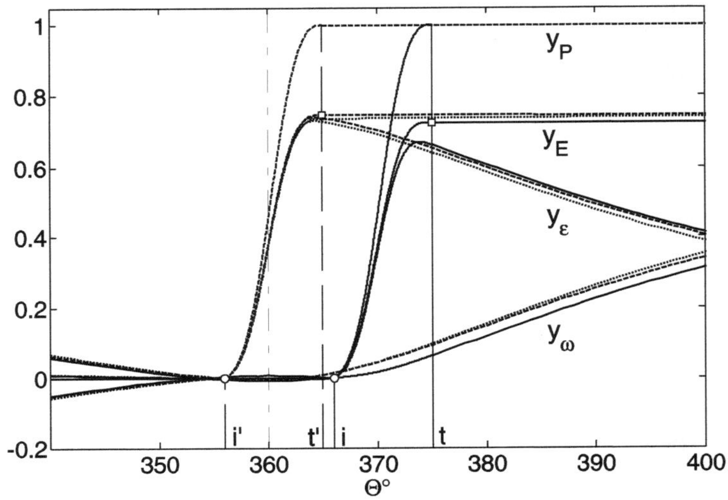

Fig. 1.24. Profiles of the mass fraction of products and its components of the HCCI engine in the original and controlled mode of operation

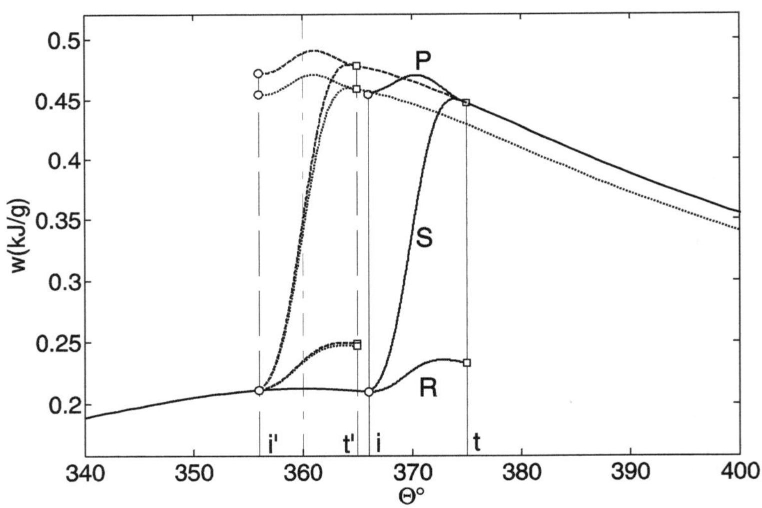

Fig. 1.25. Profiles of dynamic potential for the HCCI engine in the original and controlled mode of operation

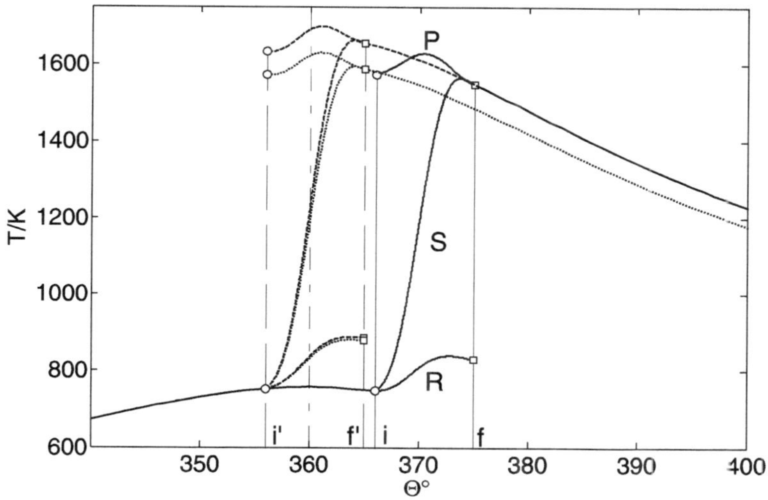

Fig. 1.26. Temperature profiles of the HCCI engine in the original and controlled mode of operation

Since the work cycle of the HCCI engine is kept invariant in the original and controlled mode of operation, it follows from (1.2) for K = R that, with the same exothermic energy per unit mass (referred to popularly as calorific or "heating" value) and the same mass fraction of reactants,

$$\frac{\text{IMEP}''}{\text{IMEP}'} = \frac{Y_F''}{Y_F} = \frac{1+\sigma_{st}\lambda}{1+\sigma_{st}\lambda''} \tag{1.46}$$

whence

$$\lambda'' = [(1+\sigma_{st}\lambda)\frac{\text{IMEP}'}{\text{IMEP}''} - 1]/\sigma_{st} \tag{1.47}$$

yielding

$$\frac{Y_F''}{Y_F} = \frac{1+\sigma_{st}\lambda}{1+\sigma_{st}\lambda''}\frac{Y_E}{Y_E''} \tag{1.48}$$

Provided thus is an expression for the mass fraction of fuel in the controlled and uncontrolled cases. The results based on the data of Figs. (1.21) and (1.24), are presented by Table 1.5.

Table 1.5. Parameters of thermodynamic properties

Step	**0**		**1**		**2**
Θ_i	366			3.56	
Θ_f	375			365	
IMEP/atm	7.77		8.36		7.77
λ		2.2			2.37
C_R		3.4839			3.4839
C_P		4.0145			3.9448
Q_P		7.3649			6.7781
Y_E	0.7247		0.7465		0.7361

According to the above, the ratio of mass fractions of fuel expended in

$$\frac{Y_F''}{Y_F} = \frac{(1+12.305\cdot 2.2)\cdot 0.7247}{(1+12.305\cdot 2.37)\cdot 0.7361} = \frac{20.3431}{22.2029} = 0.916$$

The saving in fuel consumption obtained by controlled operation of the engine is therefore 8.4% – a gain associated with equivalent reduction in the formation of pollutants. This type of improvement can be advanced significantly further by charge stratification, so that, within the relatively short lifetime of the dynamic stage, the combustion zone is kept away from the walls, whereby the energy loss incurred by heat transfer to the surroundings is diminished, as demonstrated in Chapter 4.

1.4 Production Engine

To illustrate the application of pressure diagnostics to an industrial product, consider a Renault spark ignition engine operating at full and part loads, the latter being most often encountered in the European driving cycle [Gavillet et al. 1993; Oppenheim et al 1997, Oppenheim 2004]. Its specification is provided by Table 1.6.

Table 1.6. Engine Data

Model	F7P-700
Bore(mm) x stroke(mm)	82.0 x 83.5
Cylinders	4
Piston rod length (mm)	144
Compression ratio	10

1.4.1 Full Load

The operating conditions of the dynamometer test at full load are specified by Table 1.7.

Table 1.7. Operating conditions of the Renault engine operating at full load

Speed (rpm)	2000
Torque (Nm)	128
BMEP (kPa)	912
Fuel	RON 95
Fuel flow (gm/min)	32
λ	1
σ	15.0
P_i /atm	1
T_i /K	300
n	1.323

1.4.1.1 Dynamic Properties

Parameters of the dynamic properties and the life function are presented by Table 1.8.

Table 1.8. Parameters of dynamic properties and the life function for Renault engine operating at full load

θ_ι	353
θ_f	384
α_π	9.38
χ_π	1.29

The characteristic features of the dynamic properties are displayed by Figs. 1.27-1.30 in terms of, respectively, the measured pressure profile, $p(\Theta)$, the work cycle in linear scales, $p(v)$, the profile of the polytropic function, $\pi(\Theta)$, and the progress parameter of fuel consumption in the course of the dynamic stage, $x_\pi(\Theta)$, together with the analytic expression for the pressure profile based on the life function.

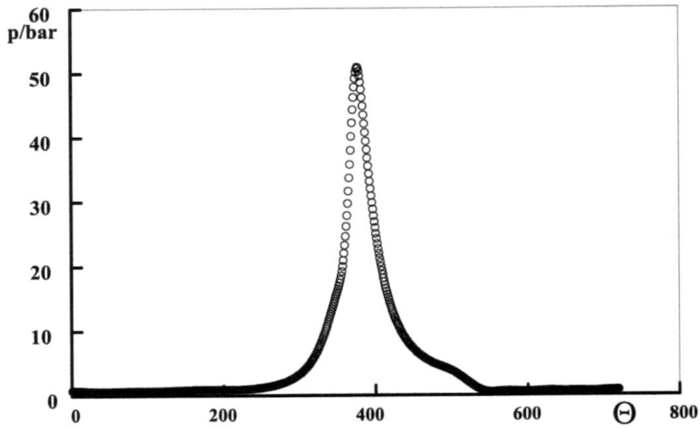

Fig. 1.27. Measured pressure profile of the Renault engine operating at full load

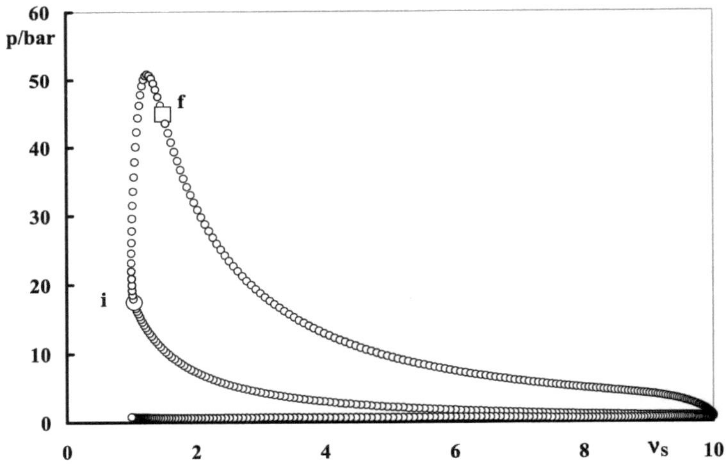

Fig. 1.28. Work cycle in linear scales of the Renault engine operating at full load

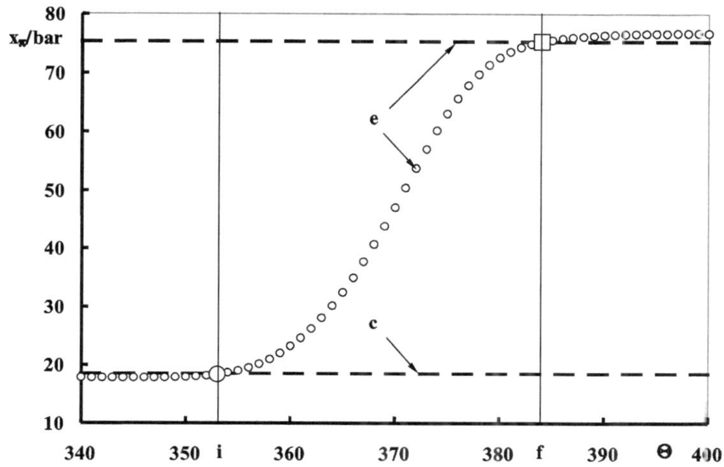

Fig. 1.29. Profile of polytropic functions for the Renault engine operating at full load

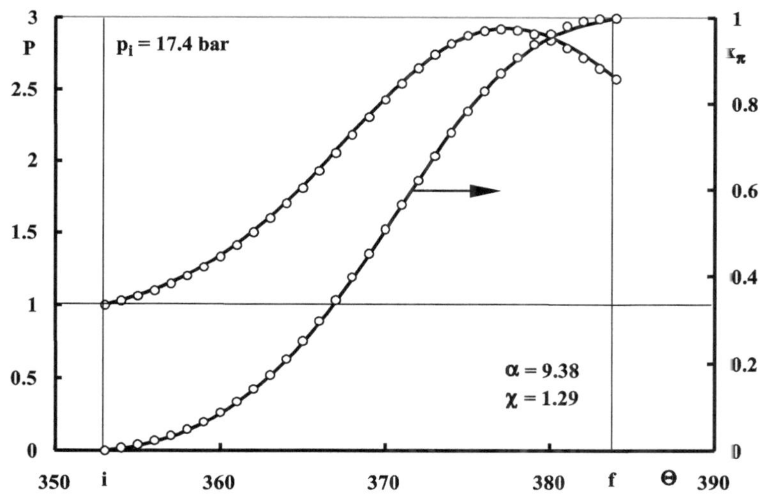

Fig. 1.30. Profiles of the measured pressure data in normalized form, $P(\Theta) \equiv p/p_i$, and the progress parameter of consumed fuel, $x_\pi(\Theta)$, (presented by circles) and the polytropic function expressed in terms of the life function (displayed by continuous lines) for the dynamic stage in the Renault engine operating at full load

1.4.1.2 Thermodynamic Properties

The thermodynamic parameters of the working substance are presented by Table 1.9.

Table 1.9. Thermodynamic parameters of Renault engine operating at full load

		p	T	v	u	h	w	M
	States	atm	K	m^3/g		kJ/g		g/mol
F	i	17.18	623	26.206	1.2294	-1.1838	0.0456	113.53
	f	44.24	637	10.412	1.1868	-1.1401	0.0467	
R	i	17.18	623	98.328	0.0693	0.24046	0.1712	30.258
	f	44.24	772	47.295	0.2092	0.42124	0.2120	
P	uv	141.65	2870	98.328	0.0693	0.9176	0.8483	28.131
	hp	17.18	2515	423.87	0.4975	0.2405	0.7380	28.330
	pT	17.18	623	104.01	2.7255	-2.5444	0.1811	28.606

The coordinates of the thermodynamic states of the components are listed in Table 1.10.

Table 1.10. State parameters of Renault engine operating at full load

K	C_K	u_{K0}	q_K
A	2.7673	-0.3412	0
F	38.7273	-2.995	2.654
R	3.4281	-0.5176	0.1764
P	5.1373	-4.2886	3.9474

The state diagram of the exothermic process in the course of the dynamic stage is displayed by Fig. 1.31.

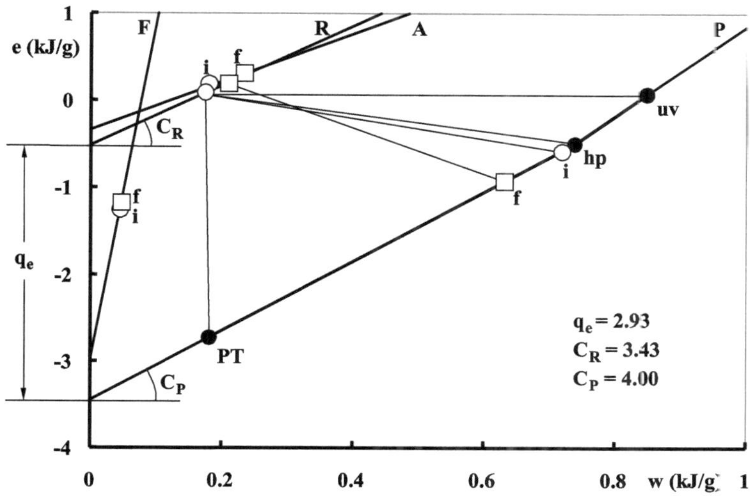

Fig. 1.31. State diagram of the exothermic process in the course of the dynamic stage for the Renault engine operating at full load

Profiles of the normalized temperatures, $\widetilde{T}_K \equiv T_K / T_i$, and specific volumes, $v_K \equiv v_K / v_i$, are depicted by Fig. 1.32.

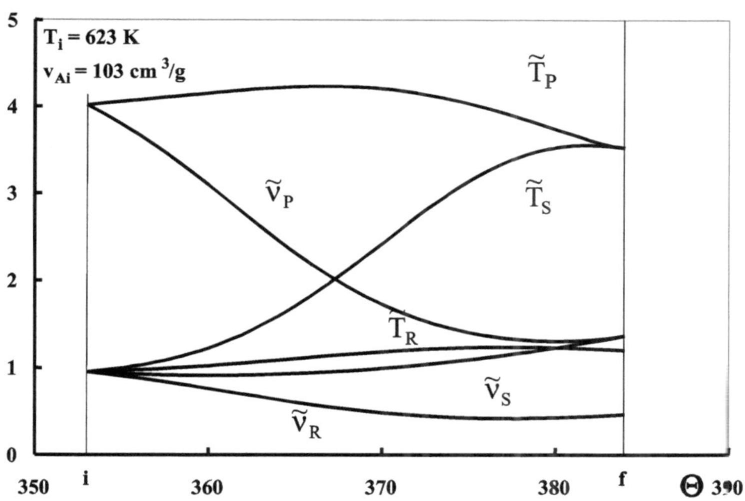

Fig. 1.32. Profiles of normalized temperatures and specific volumes in the course of the dynamic stage in the Renault engine operating at full load

1.4.2 Part load

The operating conditions and life function parameters of the Renault engine run at part load are as follows.

1.4.2.1 Dynamic Properties

Table 1.11. Operating conditions and life function parameters of Renault engine run at part load

Speed (rpm)		2000
Torque (Nm)		9.83
BMEP (kPa)		70
Fuel		RON 95
Fuel flow (gm/min)		32
λ		1
σ		15.0
P_i /atm		0.6
T_i /K		300
n		1.365
	α	14.2
x_π	χ	2.78

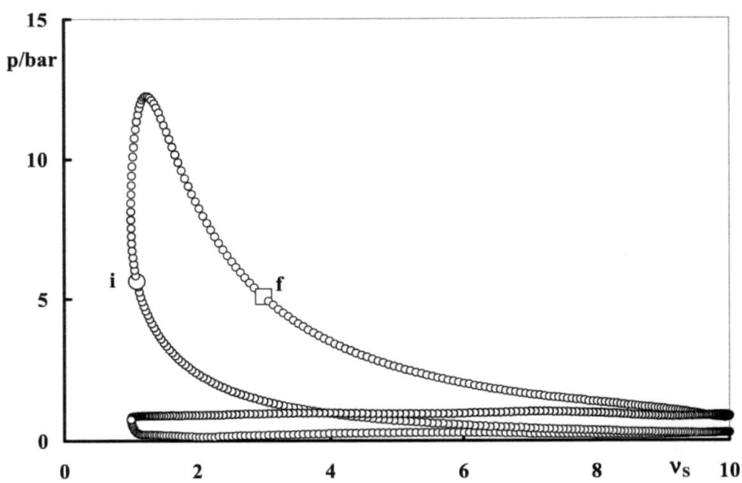

Fig. 1.33. Work cycle of the Renault Engine operating at part load

The work cycle of the Renault engine operating at part load is presented by Fig. 1.33. The profile of the polytropic function for the dynamic stage is displayed in Fig. 1.34, while profiles of the progress parameter $x_\pi(\Theta)$ and of the normalized pressure, $P(\Theta) \equiv p(\Theta)/p_i$, represented by its measured data (marked by circles) and its analytic expressions (delineated by continuous lines), is depicted in Fig. 1.35.

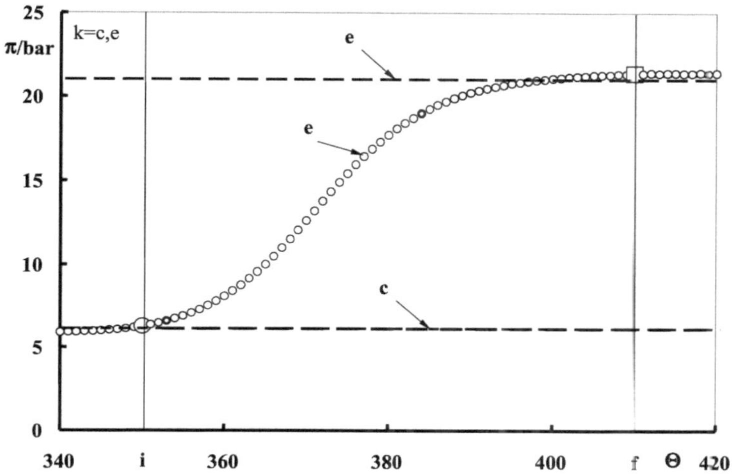

Fig. 1.34. Profile of the polytropic function for the dynamic stage in the Renault engine operating at part load

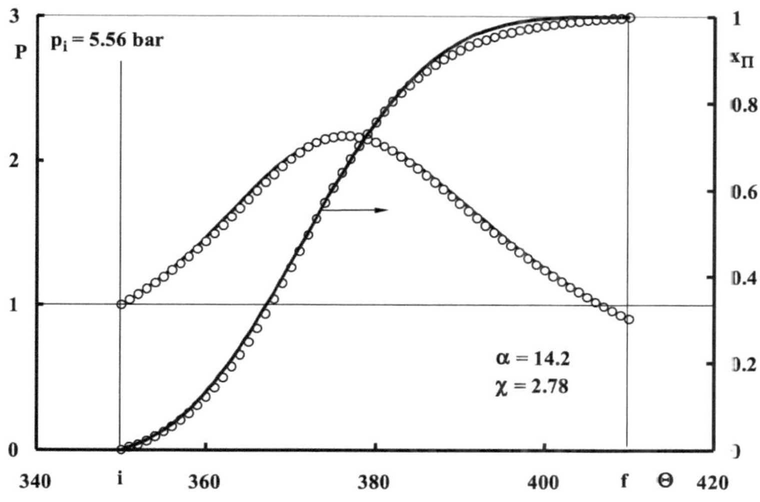

Fig. 1.35. Profiles of the measured pressure data in normalized form, $P(\Theta) \equiv p / p_i$, and the polytropic function of the dynamic stage for the Renault engine operating at part load

1.4.2.2 Thermodynamic Properties

The thermodynamic state parameters of the exothermic process in the Renault engine operating at part load are specified in Table 1.12.

Table 1.12. State parameters of the dynamic properties of the Renault engine operating at part load (Gavillet et al. 1993)

K	States	C	u_o kJ/g
A	i f	2.7857	0.3459
F	i f	54.0000	3.7857
R	i f	3.0192	0.4364
P	uv hp	5.4725	4.4670
	PT	4.2093	3.5230

Profiles of the normalized temperature, $\tilde{T}_K \equiv T_K / T_i$, and specific volume, $v_K \equiv v_K / v_i$, are depicted in Fig. 1.36.

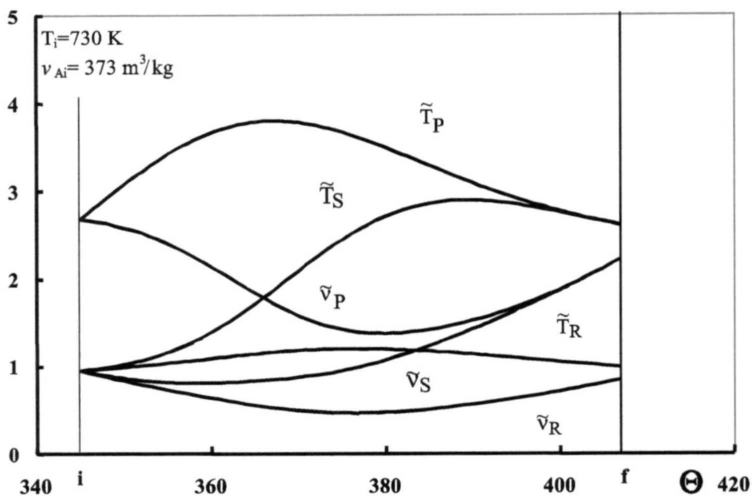

Fig. 1.36. Profiles of normalized temperatures and specific volumes in the course of the dynamic stage for the Renault engine operated at part load

2 Evolutionary Aspects

2.1 Introduction

In essence, the process of combustion is a chemical reaction in the course of which an evolution, associated with the transformation of reactants into products, takes place – a metamorphosis affecting usually a part of the system, while the rest changes its thermodynamic state as a consequence of its passive presence in the system without altering its identity. The system exhibits then all the properties of a dynamic object: its state is displaced from a definite starting point to an end point – a process carried out at a rate, or velocity, whose variation plays the role of acceleration.

Progress of this evolutionary process is recorded, as a rule, by measurements of its symptoms, like concentration of certain species, temperature or pressure, at discrete instances of time, delineating thereby its trajectory determined by interpreting the sampled data points in terms of an integral curve. One is confronted thus with an inverse problem: evaluation of the dynamic properties of the system from the record of the discrete data marking its evolution. Since the dynamic properties are expressed by differentials, finite differences between such data are inappropriate for this purpose. The progress of evolution has to be expressed, therefore, in terms of ordinary differential equations (ODE's). Equations pertaining to evolutionary processes occupy a prominent position in the theory of ODE's. Their formal exposition was provided, among others, by Sell in *Dynamics of evolutionary equations* (1937) and by Hofbauer in *The theory of evolution and dynamical systems* (1956).

Life function and its predecessors presented here are pragmatic examples of solution of evolution equations pertaining to specific problems in biophysics, physical chemistry and combustion.

2.2 Biophysical Background

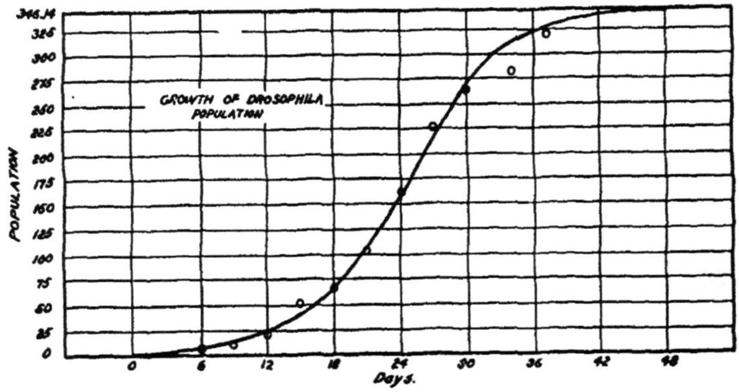

Fig. 2.1. A growth curve of population of fruit flies described by Lotka (1924)

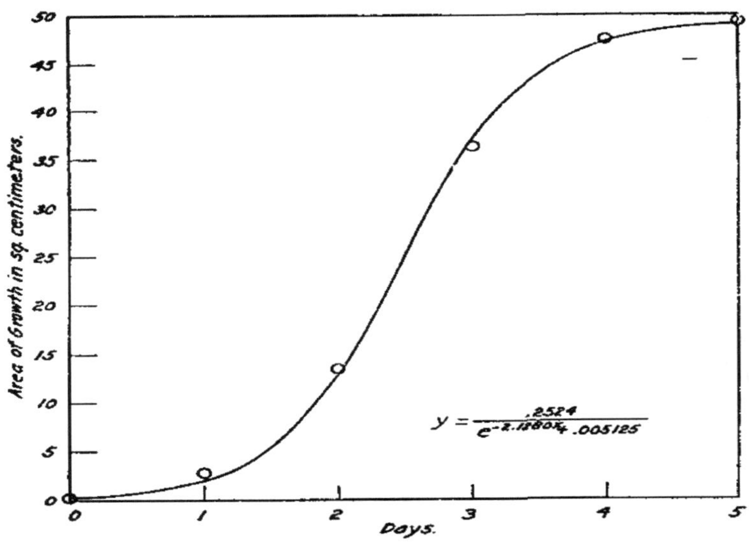

Fig. 2.2. Growth curve of a bacterial colony displayed by Lotka (1924)

The dynamics of evolution is of direct relevance to the mathematical description of life, as manifested prominently in the biophysical literature.

Fundamental principles for analytic description of life were formulated by Lotka in his book on *Elements of Physical Biology*(1924), where he introduced the "Law of Population Growth" in terms of a general equation

$$\frac{dX}{dt} = F(X,t) \tag{2.1}$$

where F represents a prescribed algebraic function. For illustration, he described then a simple, autonomous case of

$$F(X) = aX + bX^2 \tag{2.2}$$

whence, in terms of λ α a/b, and $X_o = X$ at $t = 0$

$$X = \frac{\lambda}{(1+\lambda X_o^{-1})e^{-\alpha t} - 1} \tag{2.3}$$

By adjusting the origin of time at an appropriately selected state, (2.3) can provide an interpretation to a variety of data, as exemplified by Fig. 2.1, where, at t=0, N = Λ/2, so that the population growth

$$N = \frac{\Lambda}{1+e^{-\alpha t}} \tag{2.4}$$

Another example of the many cases considered by Lotka, is presented here by the growth of a bacterial colony, depicted by Fig. 2.2, with its constitutive equation displayed on the diagram. In a treatise on "Fundamentals of the Theory of the Struggle for Life" ("Les fondements de la théorie de la lutte pour la vie"), Volterra (1937) described the population growth rate of biological species as[1]

$$\frac{dN}{dt} = (a - bN)N \tag{2.5}$$

– an expression that, in terms of a normalized variable x ≡ (b/a) N, is equivalent to

$$\frac{dx}{dt} = \alpha x(1 - x) \tag{2.6}$$

[1] vid. Volterra (1937) eq. (1) on p.4 and its simplified version on p.5

where $\alpha = a^2/b$. At the two boundaries of x = 0 and x = 1, $\dot{x} = 0$, while, at x = ½, $\ddot{x} = 0$ - a point of inflection. The time profile of population is expressed then by an S-curve, just like that of (2.4), with this point adopted as the origin of time, described by

$$x = \frac{1}{1 + e^{-\alpha t}} \tag{2.7}$$

so that at x = 0 at t = - ∞, and x = 1 at t = + ∞, – an infinite lifespan of god-like quality.

Volterra introduced then a concept of the quantity if life, Q(t), defined as the time integral of its population. As a consequence of (2.7) with time measured from the point of inflection,

$$Q(t) \equiv \int_o^t x\mathrm{d}t = t + a^{-1} \ln \frac{1 + e^{-at}}{2} \tag{2.8}$$

- a finite quantity of an infinite life.

Analytical concepts of this kind have been developed also by Rashevski, who, in his book on *Mathematical Biophysics* (1948), formulated the evolution of cell division from the observation observing that, in the simplest case, it can be expressed as[2]

$$\frac{\mathrm{d}\varepsilon}{\mathrm{d}t} = \mathrm{A}\varepsilon - \mathrm{B}\varepsilon^2 \tag{2.9}$$

- the same function as (2.5), where

$$\varepsilon = \frac{r_1 - r_2}{\sqrt[3]{r_1 r_2^2}} \tag{2.10}$$

expresses the cell elongation in terms of r_1 - its half-length and r_2 - its half-width.

The process of cell elongation is prescribed then by

2 vid. Rashevski (1948) eq. (45) on p. 159, with its simplified version on p. 162

$$\varepsilon(t) = \frac{A}{B + A\exp[-A(t - t_0)]} \tag{2.11}$$

- vid. (52) in [9] – an expression equivalent to (2.7).

It is of interest to note that, in their classical paper on the essential mechanism of diffusion, Kolmogorov et al (1937) brought up, as an example of a simplified, one-dimensional problem, the case of the "struggle for life" of a bacteria-like colony. Its evolution was described in terms of the following basic equation[3]

$$\lambda \frac{dv}{dx} = k\frac{d^2 v}{dx^2} + F(v) \tag{2.12}$$

Letting $dv/dx = p$, one gets then

$$\frac{d^2 v}{dx^2} = \frac{dv}{dx}\frac{dp}{dv} = p\frac{dp}{dv} \tag{2.13}$$

and (2.12) becomes

$$\frac{dp}{dv} = \frac{\lambda p - F(v)}{kp} \tag{2.14}$$

- a more general form of the first order ordinary differential equation than (2.1) with (2.2), because here F = F(X,t), rather than just F(X).

2.3 Physico-chemical Background

It was the advent of the chain reaction theory that gave the impetus to for the generation of mathematical expressions to describe the evolution of physico-chemical processes. The founder of this theory, Nickolai Nickolaevich Semenoff (or Semenov), who received a Noble Prize for its development, provided its detailed description in his monographs on chemical kinetics and reactivity (1934). In the simplest case of a gaseous

[3] vid. 1 Kolmogorov et al (1937) eq. (7) on p.243 in Selected Works (1991)

substance, undergoing an exothermic process in an enclosure of fixed volume, his method of approach is as follows[4].

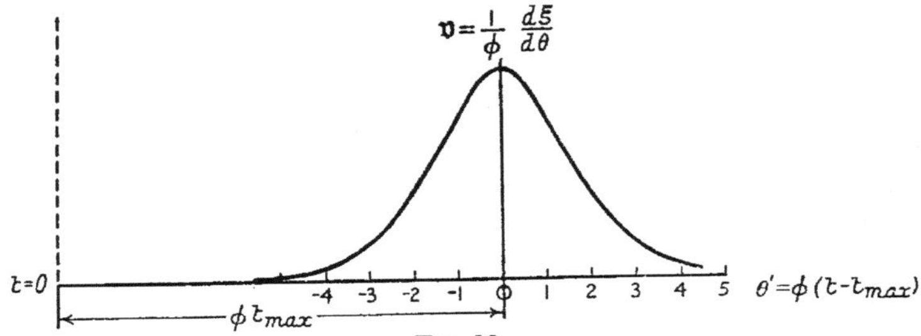

FIG. 13

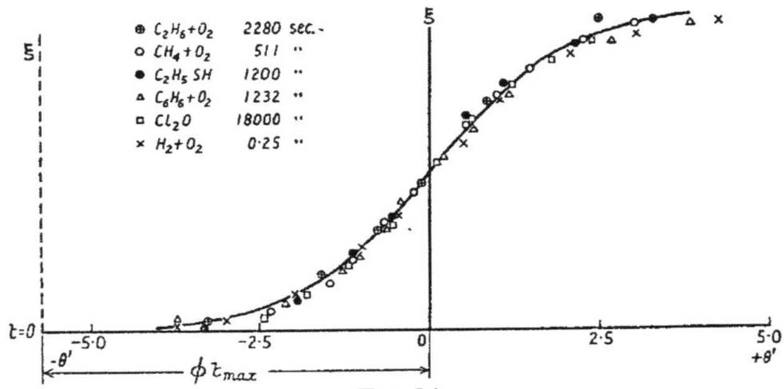

FIG. 14

Fig. 2.3. Profiles of the amount of substance that has reacted and reaction velocity in comparison to experimental data, provided by Semenoff (1934)

As a consequence of chain branching, the rate at which a quantity, x, of the substance reacts in time t can be expressed in terms of

$$\frac{dx}{dt} = Bx(P - x) \tag{2.15}$$

- just like (2.6) and (2.9), whereas B is here a measure of the number of chain branching steps per active center, while P is the normalized pressure change associated with this event, upon the understanding, of course, that the reaction takes place in an enclosure of fixed volume.

Thus, in terms of the time interval to reach maximum reaction rate, $\Theta = t - t_m$, where t_m is t at $(dx/dt)_{max}$, just like (2.4), (2.7) and (2.11), it follows that

[4] vid. Semenoff 1934 eqs. (49)-(50) pp. 57-68 and Figs. 13 & 14

$$x = \frac{P}{1 + e^{-\phi\Theta}} \tag{2.16}$$

where $\phi \equiv BP$. A plot of this function, (in terms of $\xi = 100x$) and its derivative, presented by Semenoff, is reproduced in Fig. 2.3.

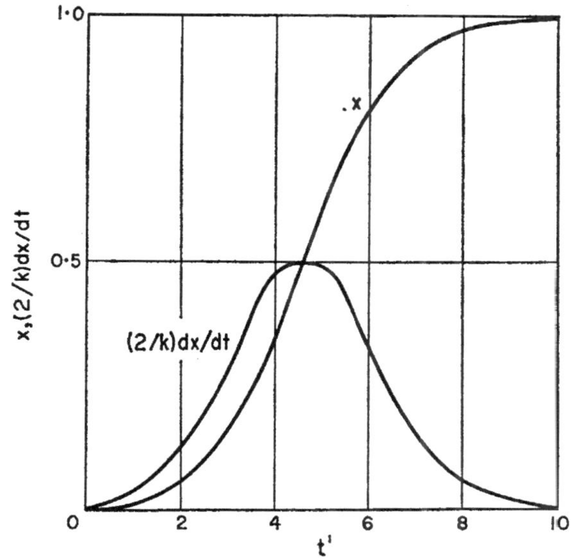

FIG. 9. Kinetic curves of an autocatalytic reaction ($x_0 = 0{\cdot}01$).

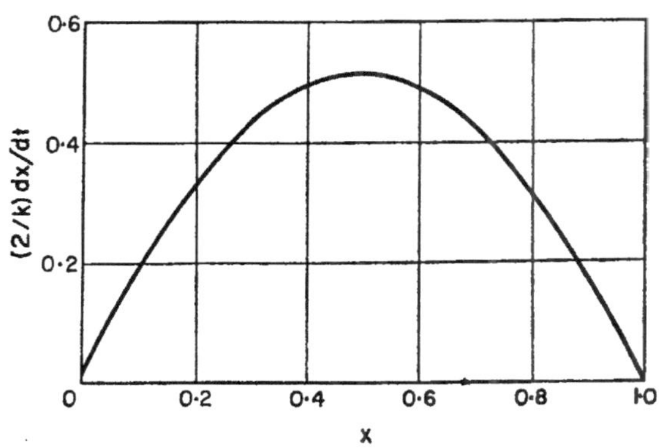

FIG. 10. The rate of an autocatalytic reaction as a function of the relative quantity of reacted substance (x).

Fig. 2.4. Kinetic rules for chemical reactions according to Kondrat'ev (1964)

The dynamic features of the auto-catalytic chain reactions were expressed by Kondrat'ev (1964) in the general form of[5]

$$\frac{dx}{dt} = k(x + x_o)^n (1 - x)^m \tag{2.17}$$

whence, in the particular case of n = m = 1, upon integration with initial condition of x = 0 at t = 0,

$$x = x_o \frac{\exp[(1 + x_o)kt] - 1}{1 + x_o \exp[(1 + x_o)kt]} = x_o \frac{1 - \exp(-t')}{x_o + \exp(-t')} \tag{2.18}$$

where $t' = (1 + x_o)kt$, while x = 1 at t = ∞ - an unlikely feature of real life. The solution is thus expressed by an S curve whose point of inflection is located at $t^* = -\ln x_o / [k(1 + x_o)]$. The curve, its slopes, as well as the phase portrait of the solution, are displayed on Fig. 2.4.

2.4 Combustion Background

A similar situation arose in the theory of flame structure, as exposed first by Zel'dovich and Frank-Kamenetskii (1938). According to the Fourier equation of heat conduction in one-dimensional form, they express the thermal flame propagation as

$$K_1 \frac{d^2T}{dx^2} = -\dot{Q}(T) \tag{2.19}$$

where K_1 denotes the thermal conductivity of reaction products, while $\dot{Q}$ is the volumetric rate of heat release. Then, in terms of $p = dT/dx$, similarly as for (2.13),

---

[5] vid. Kondrat'ev (1964). Chapter 1 General Rules for Chemical Reactions; §3. Catalysis by End Products, pp. 38-43, eqs. (3.22), (3.23); (3.25) and Chapter 9 Chain Reactions; §39. Reaction Kinetics Taking into Account Fuel Consumption. Overall Law of the Reaction; p. 624, eqs. (39.35)

$$\frac{d^2T}{dx^2} = p\frac{dp}{dv} = \frac{1}{2}\frac{dp^2}{dT} \tag{2.20}$$

so that, by quadrature, (2.19) yields

$$\frac{dT}{dx} = \sqrt{\frac{2}{K_1}\int_T^{T_1}\dot{Q}(T)dT} \tag{2.21}$$

Considering T to express the temperature in terms of its progress parameter within the reaction zone, so that at x = 0, T = 0, they derived, by integration of (2.21), the following expression for the mass rate of combustion

$$u = \frac{1}{q}\quad K_1\frac{dT}{dx} = \frac{1}{q}\sqrt{2K_1\int_0^{T_f}\dot{Q}(T)dT} \tag{2.22}$$

where q is the calorific value per unit mass of the mixture.

Later, this argument was cast by Zel'dovich (1941) into a more conventional form of [6]

$$u = \frac{[2\eta\int\dot{Q}(T)dT]^{1/2}}{\rho q} \tag{2.23}$$

where η expresses the thermal conductivity, while ρ is the density of the reactants.

On the basis of this method of approach, Spalding (1957) developed a "temperature explicit" theory of laminar flames. Its principal variable is a generalized progress parameter, known today as the Zeldovich variable,

$$\tau \equiv \frac{T - T_i}{T_f - T_i} = Y \tag{2.24}$$

where subscripts i and f denote, respectively, the initial and final states. As a consequence of it, the diffusion and energy equations collapse into one.

---

[6] vid. Zeldovich (1941) eqs (15) &(16)

Postulating its rate of change across the flame width, ξ, to be described by

$$\frac{d\tau}{d\xi} = \tau(1-\tau^n) \tag{2.25}$$

- an extended version of (2.5) – he obtained by quadrature the following expression for the flame structure

$$\xi - \xi_0 = \ln\frac{\tau}{(1-\tau^n)^{1/n}} \tag{2.26}$$

according to which $\tau = 0$ at $\xi = -\infty$, $\tau = 1$ at $\xi = +\infty$, while, at $\xi = \xi_o$, $\tau = 1/2n^{1/n}$ - again an S-curve of infinite life span, as unrealistic as (2.4), (2.7), (2.11) and (2.16).

Over the years, there were a number of similar functional relationships put forth in the combustion literature. For example, in investigating the influence the burning speed exerts upon the working cycle of a diesel engine, Neumann (1934), proposed two functions, $x(\Theta)$, where Θ denotes the time normalized with respect to life time. The first is

$$x = (2-\Theta)\Theta \tag{2.27}$$

whence

$$\dot{x} = 2(1-\Theta) \tag{2.28}$$

so that, at $\Theta_i = 0$, $x_i = 0$ and $\dot{x}_i = 2$, while, at $\Theta_f = 1$, $x_f = 1$ and $\dot{x}_f = 0$; whereas $\ddot{x} = -2$.

The second is

$$x = (3-2\Theta)\Theta^2 \tag{2.29}$$

whence

$$\dot{x} = 6(1-\Theta)\Theta \tag{2.30}$$

so that, at $\Theta_i = 0$, $x_i = 0$ and $\dot{x}_i = 0$, while, at $\Theta_f = 1$, $x_f = 1$ and $\dot{x}_f = 0$; whereas $\ddot{x} = 6\text{-}7\Theta$, whence at the point of inflection, $\ddot{x} = 0$, $\Theta^* = 6/7$.

Later, in a publication on a "Precise Method for the Calculation and Interpretation of Engine Indicator Diagrams," Gončar (1954) introduced an empirical formula

$$x = 1 - (1 + \Theta)e^{-\Theta} \tag{2.31}$$

where $\Theta \equiv t/t_m$, is the ratio of elapsed time to the time of the maximum rate of combustion (maximum burning speed at the point of inflection).

According to (2.31), the rate of combustion is

$$\dot{x} = \frac{\Theta}{t_m} e^{-\Theta} \tag{2.32}$$

while its rate of change

$$\ddot{x} = \frac{1}{t_m^2}(1 - \Theta)e^{-\Theta} \tag{2.33}$$

whence the maximum burning speed, i.e. the rate of combustion at the point of inflection,

$$\dot{x}^* = 1/et_m \tag{2.34}$$

so that, when $t_m \to 0$, $\dot{x}^* \to \infty$, while, when $t_m \to \infty$, $\dot{x}^* \to 0$.

2.5 Vibe function

Following the elucidation provided by Erofeev (1946) Vibe (1956, 1970), expressed Semenov's method of approach on the basis of the postulate that the rate at which reacting molecules, N, decay, due to consumption by chemical reaction, is directly proportional to the rate at which the effective reaction centers, N_e, are engendered, i.e.

$$\frac{dN}{dt} = -n\frac{dN_e}{dt} \tag{2.35}$$

while the latter is expressed in terms of a relative number density function, ρ, as

$$\frac{dN_e}{dt} = \rho N \tag{2.36}$$

whence, with t expressed by the progress parameter of time and $N = N_0$ at $t = 0$,

$$N = N_0 \exp(-\int_0^t n\rho dt) \tag{2.37}$$

Now, if $\rho = kt^m$, while n = const, the fraction of molecules consumed by chemical reaction[7]

$$x \equiv \frac{N_0 - N}{N_0} = 1 - \exp(-\frac{nk}{m+1} t^{m+1}) \tag{2.38}$$

The function expressed by (2.38) is specified by two positive parameters, $\alpha = nk / (m + 1)$ and $\beta = m + 1$. With their use (2.38) is normalized, so that at $t = 0$, $x_i = 0$, while, at $t = 1$, $x_f = 1 - e^{-\alpha}$, yielding

$$x = \frac{\exp(-\alpha t^{\beta}) - 1}{\exp(-\alpha) - 1} \tag{2.39}$$

- an expression referred to in the English literature on internal combustion engines[8]as the Wiebe function, rather than Vibe - the proper name of its founder, I.I. Vibe (И. И. Вибе), Professor at the Ural Polytechnic Institute in Sverdlosk (now Ekaterinburg).

It was introduced by him in the nineteen fifties (Vibe 1956) and later, upon many publications described its origin, in his book under the title "Новое о Равочем Цикле Двигателей: Скорость Сгорания и Равочий Цикл Двигателя" ("Novel Views on the Engine Working Cycle: Rate of Combustion and Working Cycle in Engines") (Vibe 1970).

---

[7] vid. Vibe (1956) eq. (7), or Vibe (1970) eq. (44)

[8] vid. e.g. Heywood (1988), as well as Horlock and Winterbone (1986), in contrast to the German book of Pischinger et al(1989-2002) where it is spelled correctly.

Early publications of Vibe came to the attention of Professor Jante at the Dresden Technical University, who, upon their translation into German by his associate, Frick, wrote an enthusiastic paper entitled "Das Wiebe-Brenngesetz" (Jante 1960), known in English in translation as "The Wiebe Combustion Law." The name, as well as the initials (J.J. rather than I.I.), of Vibe were then misspelled – a misnomer that became thereby inadvertently introduced into English literature by Heywood et al (1979.

According to (2.39), the rate of progress

$$\dot{x} = \frac{\alpha\beta t^{\beta-1}\exp(-\alpha t^{\beta})}{1-\exp(-\alpha)} \tag{2.40}$$

whence, for $\beta > 1$, it follows that, at the initial state of $t = 0$, $\dot{x}_i = 0$, while, at the final state of $t = 1$, $\dot{x}_f = \alpha\beta e^{-\alpha}/(1-e^{-\alpha}) > 0$.

Its rate of change is

$$\ddot{x} = \frac{\alpha\beta t^{\beta-2}(\beta-1-\alpha\beta t^{\beta})\exp(-t^{\beta})}{1-\exp(-\alpha)} \tag{2.41}$$

so that at $t = 0$, $\ddot{x}_i = 0$, while, at $t = 1$, $\ddot{x}_f = \alpha\beta(\beta-1-\alpha\beta)e^{-\alpha}$, whereas the point of inflection, where $\ddot{x} = 0$, is at

$$t^* = \left(\frac{1-\beta^{-1}}{\alpha}\right)^{\beta^{-1}} \tag{2.42}$$

and

$$x^* = \frac{\exp(\alpha+\beta^{-1}-1)-1}{\exp\alpha-1} \tag{2.43}$$

Examples of this function, with profiles of its slopes, are displayed in Fig. 2.5 for $\alpha = 3$, on terms of β. As brought out by it, for $\beta > 1$ the Vibe function describes an S-curve, starting at $t = 0$ with an exponential growth at, initially, zero slope, traversing through a point of inflection into an exponentially decaying stage, until it reaches $t = 1$ at a finite, positive slope. For $\beta = 1$, it is reduced to an expression for a straightforward exponential growth, whereas, for $\beta \leq 1$ and $t^* \leq 0$, it depicts only a decaying growth.

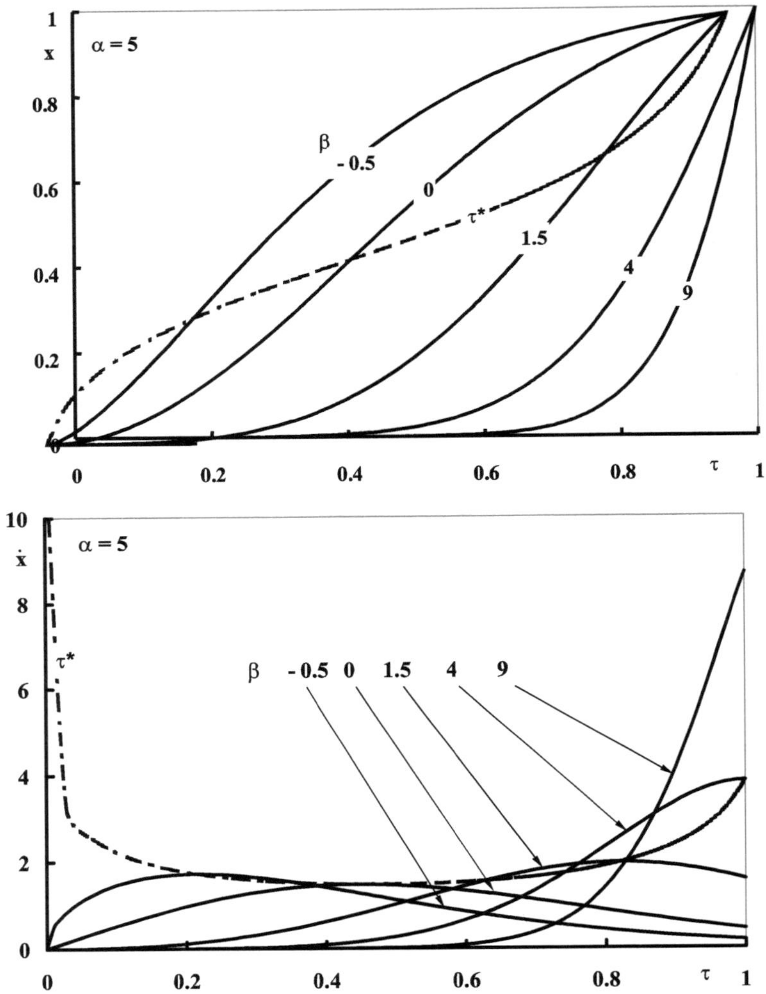

Fig. 2.5. Time-profiles of the mass fraction of fuel consumed by combustion and its rates for $\alpha = 3$ in terms of β, according to the Vibe function

2.6 Life function

The essential purpose of pressure diagnostics is to express the behavior of the system as a dynamic object – a typical kind that a control system can affect. Such an interpretation of the evolution of the exothermic process of combustion is provided by the progress parameter, $x(\tau)$, akin to displacement, or distance of travel, with its slope akin to velocity and the change of slope akin to acceleration - a monotonic, smooth function of time. Between the singularities at its bounds, the initial point, **i**, and the final point, **f,** the trajectory expressed by $x(\tau)$ furnishes an analytic expression for the measured pressure data, as well as of the effective mass fraction of products generated by the exothermic process. It models, in effect, the evolution of life, starting at a finite slope - the condition of birth - and ending at zero slope - the condition of death. For the exothermic process of combustion, it provides an analytic expression for its principal features: history of the evolution of fuel consumption (the rate of burn) and the generation of products. It has to be, therefore, sufficiently smooth.

The rate of change satisfying these conditions is expressed in terms of the progress parameter for time, $\tau \equiv (t - t_i)/T$, where $T \equiv t_f - t_i$ is the lifetime of the exothermic process, by the bi-parametric function

$$\dot{x} \equiv \frac{dx}{d\tau} = \alpha\,(\xi + x)(1 - \tau)^{\chi} \tag{2.44}$$

whence, at $\tau = 0$, $\dot{x}_i = \alpha\xi > 0$, and at $\tau = 1$, $\dot{x}_f = 0$, as postulated at the outset.

For the sake of convenience, (2.44) is split into two parts,

$$\frac{dx}{d\zeta} = \xi + x \tag{2.45}$$

and

$$\dot{\zeta} \equiv \frac{d\zeta}{d\tau} = \alpha(1 - \tau)^{\chi} \tag{2.46}$$

Hence, by quadrature,

$$x = \xi(e^{\zeta} - 1) \tag{2.47}$$

and

$$\zeta = \frac{\alpha}{\chi + 1}[1 - (1 - \tau)^{\chi+1}] \tag{2.48}$$

According to the latter, at $\tau = 1$, $\zeta = \zeta_f = \dfrac{\alpha}{\chi + 1}$, and, to satisfy the boundary conditions of $x = 0$ at $\tau = 0$ and of $x = 1$ at $\tau = 1$, $\xi = \dfrac{1}{e^{\zeta_f} - 1}$.

Obtained thus is the life function

$$x = \frac{e^{\zeta} - 1}{e^{\zeta_f} - 1} \tag{2.49}$$

– a reverse of the Vibe function – an exponential whose exponent is a power function of (2.48).

According to (2.45), the change in the rate of growth

$$\ddot{x} \equiv d^2 x / d\tau^2 = (\xi + x)(\dot{\zeta}^2 + \ddot{\zeta}) \tag{2.50}$$

while

$$\ddot{\zeta} = -\alpha\chi(1-\tau)^{\chi-1} = -\frac{\chi}{1-\tau}\dot{\zeta} \tag{2.51}$$

At the point of inflection (menopause) where $\ddot{x} = 0$, according to (2.50), $\dot{\zeta}^2 + \ddot{\zeta} = 0$, so that, with (2.46) and (2.51),

$$\alpha(1-\tau^*)^{\chi+1} - \chi = 0 \tag{2.52}$$

yielding

$$\tau^* = 1 - \left(\frac{\chi}{\alpha}\right)^{\frac{1}{\chi+1}} \tag{2.53}$$

and

$$\zeta^* = \frac{\alpha - \chi}{\chi + 1} \tag{2.54}$$

whence, according to the life function of (2.49),

$$x^* = \frac{e^{\frac{\alpha-\chi}{\chi+1}} - 1}{e^{\frac{\alpha}{\chi+1}} - 1} \tag{2.55}$$

so that, in view of (2.44),

$$\dot{x}^* = \alpha \frac{e^{\frac{\alpha-\chi}{\chi+1}}}{e^{\frac{\alpha}{\chi+1}} - 1} \left(\frac{\chi}{\alpha}\right)^{\frac{\chi}{\chi+1}} \tag{2.56}$$

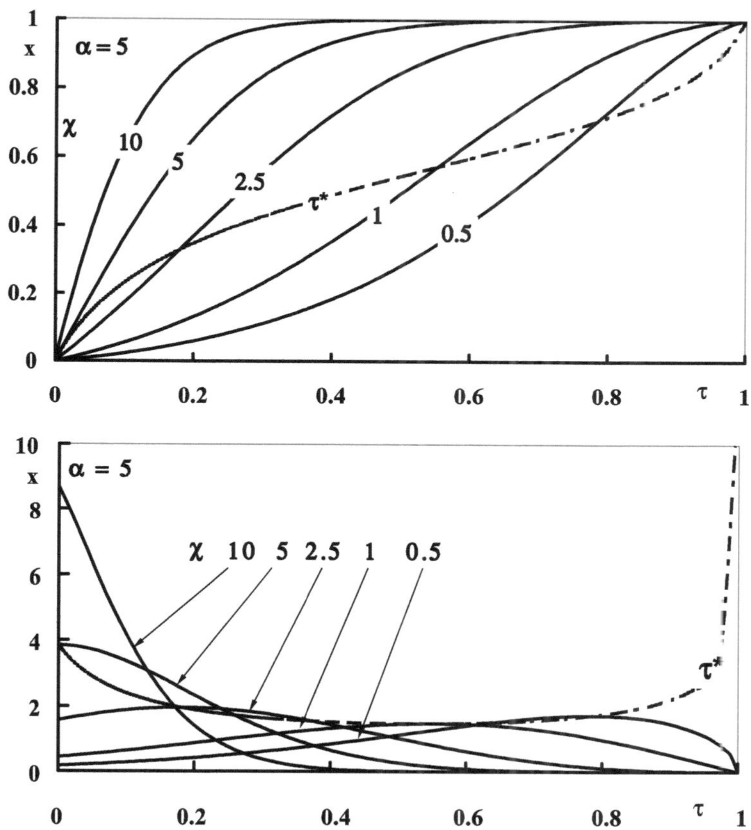

Fig. 2.6. Time-profiles of the mass fraction of fuel consumed by combustion and its rates for $\alpha = 3$, in terms of χ according to the life function

Profiles of a set of life functions and of their derivatives for $\alpha = 3$, in terms of χ, are presented by Figs. 2.6, depicting, in effect, the reverse of the Vibe function displayed by Figs. 2.5, where $\beta = \chi + 1$.

On the basis of (2.56), two types of combustion events can be identified: those for which $\chi > \alpha$, so that $\tau^* > 0$, and those for which of $\chi \leq \alpha$, whence $\tau^* \leq 0$. The former is initiated by mild ignition producing flame kernels, and the latter is created by strong ignition forming blast waves.

3 Heat Transfer Aspects

3.1 Introduction

Pressure diagnostics presented in Chapter 1, provides a rational procedure for determining the mass fraction of the generated combustion products, $y_P,(t)$, with its effective and ineffective parts, y_E and y_I, the latter incurred principally by heat transfer loss to the walls, q_w. Presented here is an experimental and analytical study carried out to evaluate the time profile of this loss, $q_w(t)$, and determine on this basis the profile of the mass fraction of products, y_P(t). The salient features of this study, initially reported by Oppenheim and Kuhl, 2000a, 2000b, are as follows:

The experiments were carried out in a constant volume cylinder where combustion of lean air/propane mixtures was carried out. The cylinder was equipped with thin film surface thermometers for time resolved measurements of local heat fluxes, and high frequency response transducers for recording the concomitant pressure profiles. The measured heat flux profiles, integrated over the wall surface area as a function of time, were found to be self-similar and, on this basis, profiles of the energy loss incurred by heat transfer to the walls of the enclosure, $q_w(t)$, were computed. As expounded in Section 1.3.6, the ineffective mass fraction of products,

$$y_I = \frac{q_w}{q_P - (C_P - C_R)w_R}$$

where q_P denotes the exothermic energy, C_K (K=R,P) delineate the slopes of the trajectories of reactants and products on the state diagram, and w_R is the dynamic potential of reactants.

Time profile of the effective mass fraction of products, $y_E(t)$, was concomitantly deduced from the measured pressure records by the procedure of pressure diagnostics, and the total mass fraction of products, $y_P(t)$, was thereupon determined by the sum of $y_E(t)$ and $y_I(t)$, specifying the mass fraction of reactants consumed in the course of the dynamic stage, $Y_R = y_P(t_f)$, subscript f denoting its final state.

3.2 Experiments

The experimental tests were carried out using a cylinder 3.5" in diameter and 2" deep, amounting to 283 cm^3 in volume, fitted on both sides by optical glass windows for unobstructed insight for schlieren photography. Its size corresponds to that of a CFR engine cylinder at a compression ratio of 8:1, when the piston is at 60 degrees crank angle from the top dead center. For tests, it was filled with a carefully mixed propane-air mixture at equivalence ratio of 0.6, maintained initially at a pressure of 5 bars and a temperature of 65°C.

Four frames of schlieren cinematographic records are presented here for three modes of combustion: F for spark ignited Flame Traversing the Charge; S for combustion initiated by a Single stream Pulsed Flame Jet; and T for a Triple stream Pulsed Flame Jet.

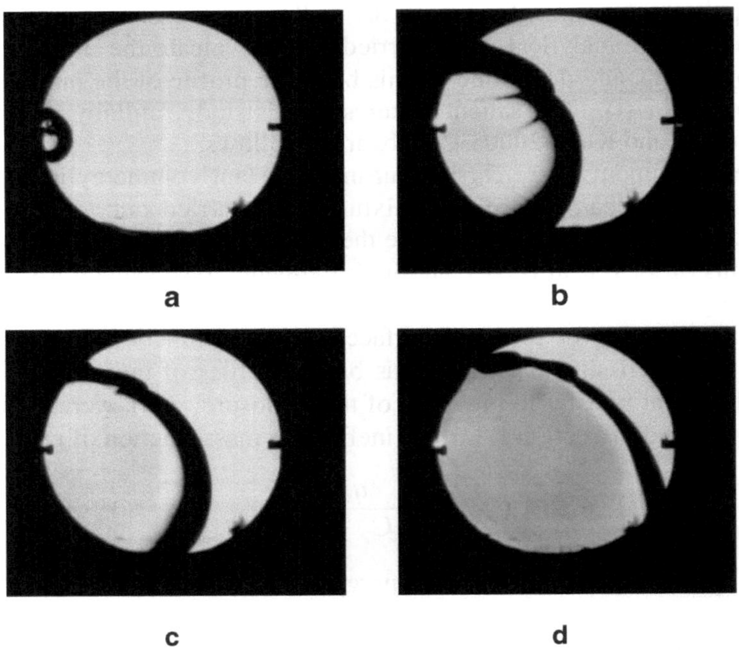

Fig. 3.1. Sequence of schlieren records of mode F (Hensinger et al 1992)

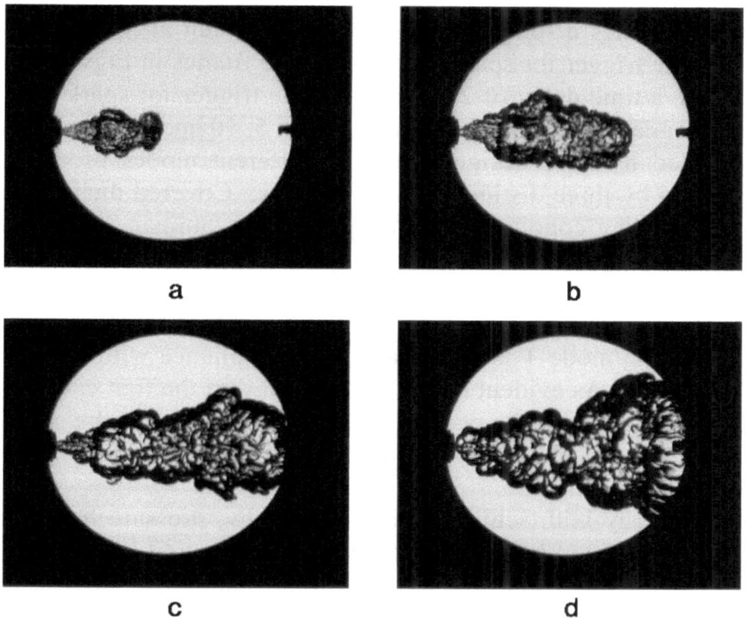

Fig. 3.2. Sequence of schlieren records of mode S (Hensinger et al 1992)

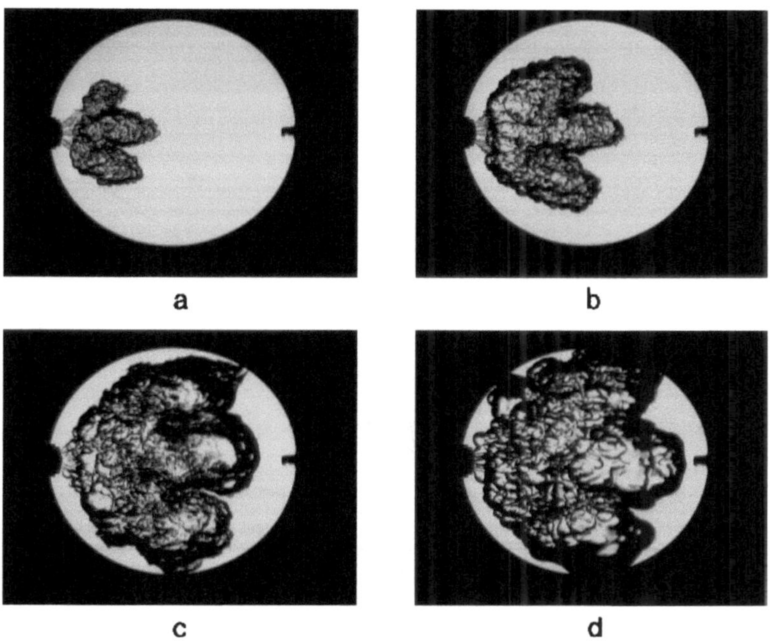

Fig. 3.3. Sequence of schlieren records of mode T (Hensinger et al 1992)

The four frames displayed in Fig. 3.1 were taken at 5, 10, 20 and 50 msec after the trigger for spark discharge. The frames in Figs. 3.2 and 3.3 started upon a time delay of 21 msec after the trigger for spark ignition in the generator cavity and thereupon were taken 5, 10 and 20 msec later.

As implied by their designations, the different modes of combustion were realized by three distinct types of igniters. Covered thus was a wide range of operating conditions, from a practically laminar flame to a fully developed turbulent combustion.

In mode F, the flame front is clearly delineated, acting as the boundary between the reactants and products. Mode S appears as a round turbulent jet plume, while mode T is disk-shaped, in compliance with the geometry of the enclosure. As evident on the photographs, in the first case the products are in contact with the walls right from the outset. In the second, the combustion zone reaches the walls upon a distinct time delay, while in the third this delay is longer.

Simultaneously with schlieren cinematography, pressure transducer records were taken and heat flux profiles were measured by thin film heat transfer gauges, one located at the side wall, marked by letter s, and the other on the back wall, marked by letter b. The temperature sensed by heat transfer gauges, $\Delta T(t)$, and the pressure profiles are presented by Figs. 3.4, 3.5 and 3.6.

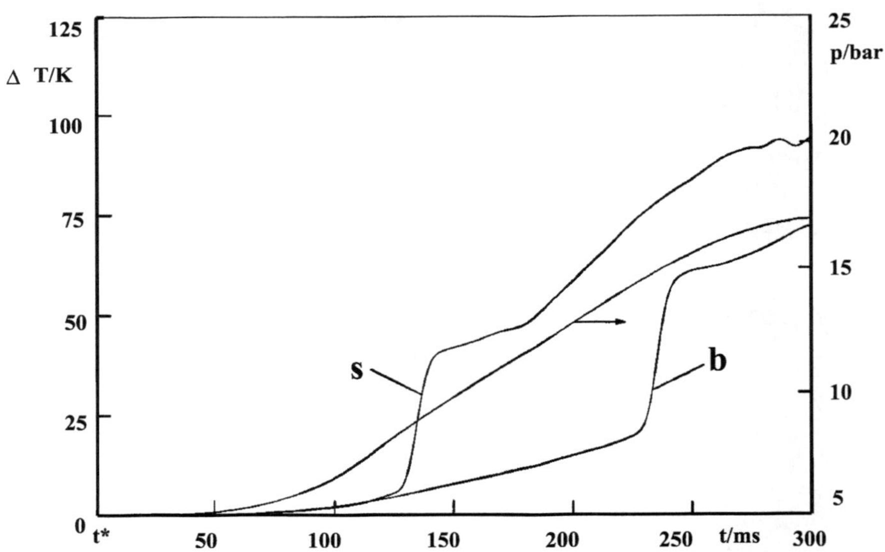

Fig. 3.4. Profiles of data sensed by thin film heat transfer gauges and a piezo-electric pressure transducer in mode F

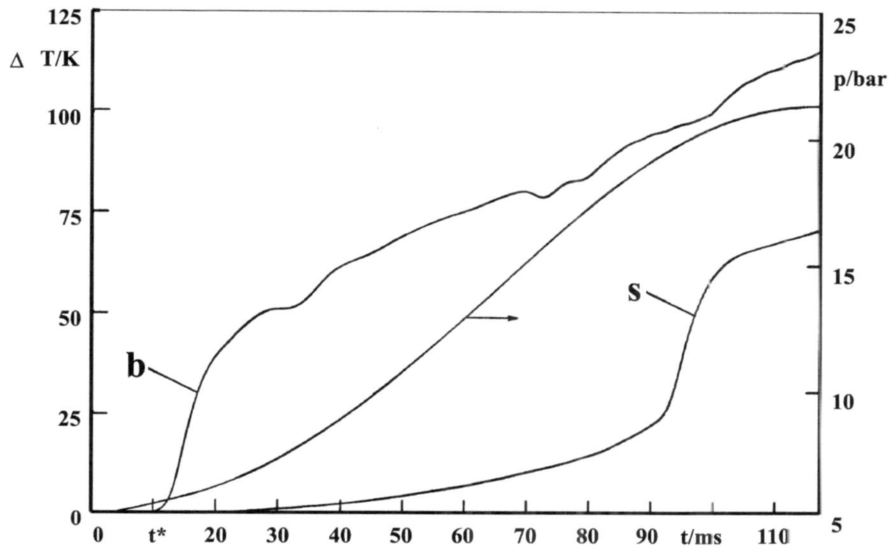

Fig. 3.5. Profiles of data sensed by thin film heat transfer gauges and a piezo-electric pressure transducer in mode S

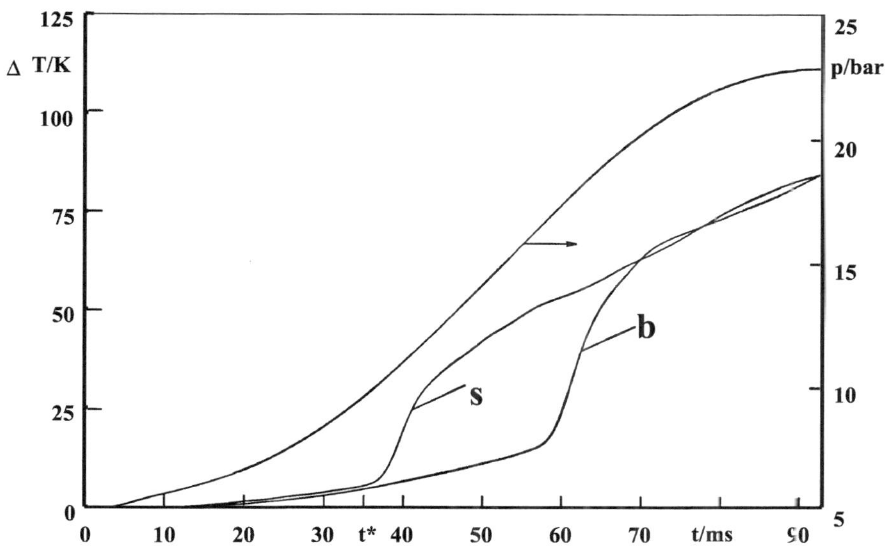

Fig. 3.6. Profiles of data sensed by thin film heat transfer gauges and a piezo-electric pressure transducer in mode T

3.3 Analysis

The heat flux was evaluated from the temperature profile, T(t), by the Duhamel superposition integral (vid. Carslaw and Jaeger 1948)

$$\dot{q}''(t) = \beta \int_0^t \frac{\mathrm{d}T(t)}{\mathrm{d}t_i} \frac{\mathrm{d}t_i}{(t-t_i)^{1/2}} \tag{3.1}$$

where $\beta=(k\rho c/\pi)^{1/2}$, while k is the thermal conductivity of the substrate, ρ its density and c the specific heat. For Macor that was employed for this purpose, at the temperature of 300 K, k = 12.87 MJ/(s-cm-K), ρ = 2.52 g/cm^3 and c = 0.795 J/(g-K), so that β = 2.866 MJ/(s$^{1/-}$-cm^2-K).

The integration was performed using the algorithm of Arpaci (1966). The results are presented by Figs 3.7, 3.8, and 3.9.

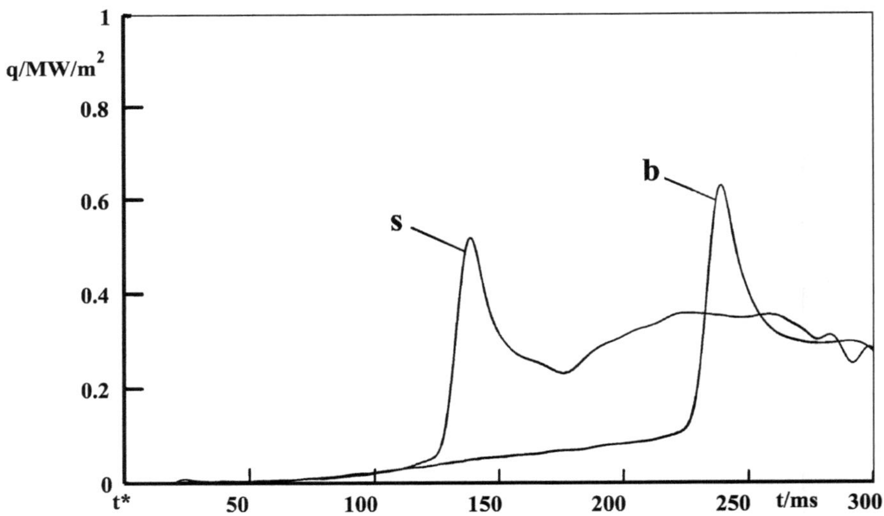

Fig. 3.7. Heat flux profiles in mode F deduced from the data of Fig.3.1

All the heat flux profiles display a remarkably similar pattern: a ramp followed by a sharp pulse with a wavy decay. The ramp is due to radiation from the high temperature combustion zone, replete of such strong radiators as H_2O and CO_2, while the rapid increase of heat flux is engendered by the combustion zone touching the walls - an event starting at time t*.

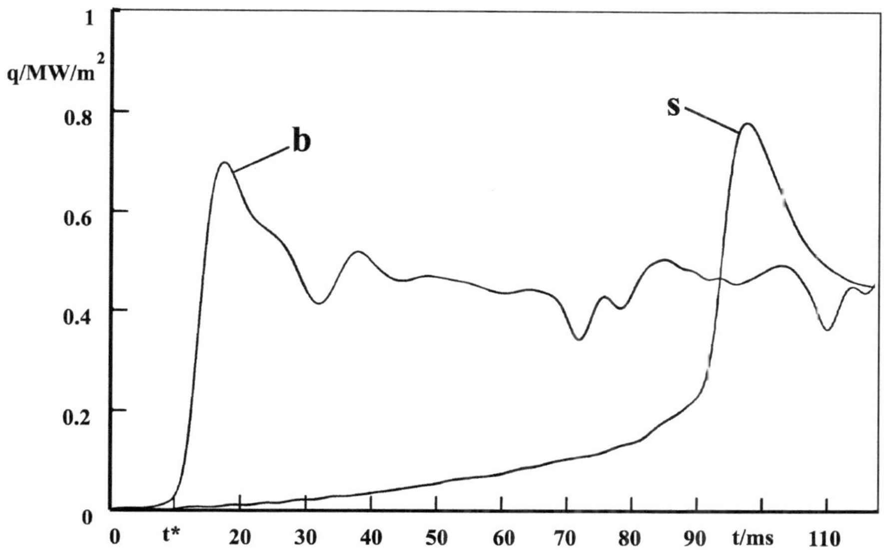

Fig. 3.8. Heat flux profiles in mode S deduced from the data of Fig. 3.2

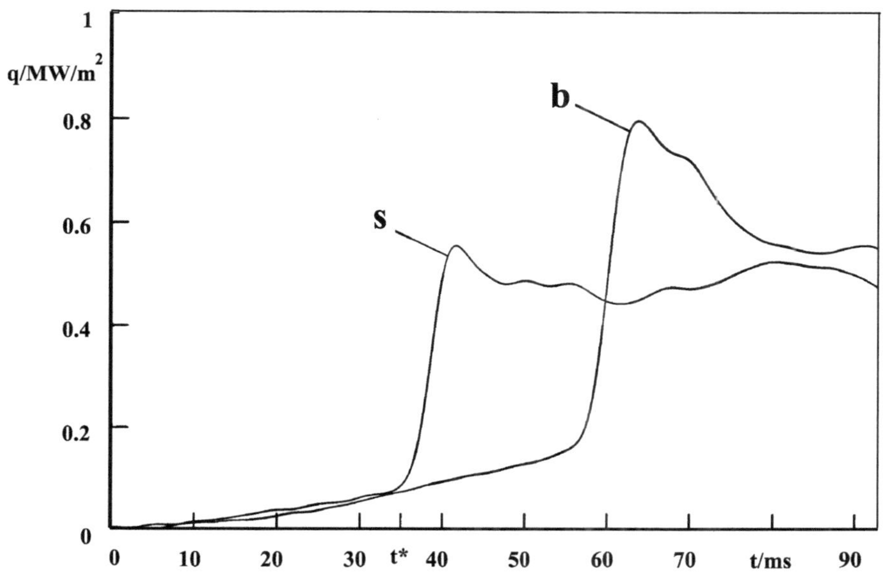

Fig. 3.9. Heat flux profiles in mode T deduced from the data of Fig. 3.3

The heat transfer profiles for each mode exhibit a self-similar pattern. Its characteristic features are illustrated by Fig. 3.10.

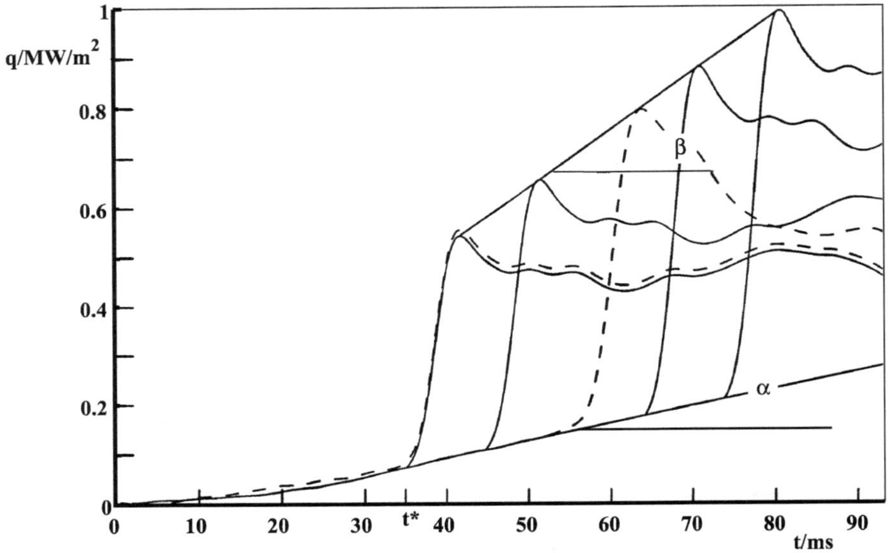

Fig. 3.10. Self-similarity features of heat flux profiles

This observation furnished a rationale for self-similarity model, on the basis of which the heat flux profiles can be integrated to evaluate the total amount of energy lost by heat transfer to the walls. This was accomplished as follows.

The evaluation of heat transferred to the walls involves double integration

$$q_w(t) = \int_o^t [\int_o^A \dot{q}''(A,t_i)\, \mathrm{d}A]\mathrm{d}t_i \tag{3.2}$$

where subscript f denotes the final state of the combustion event when the exothermic zone extends over the total wall area of the enclosure.

The evolution of heat transfer from a combustion event to the walls of the enclosure consists of two stages.

Up to t = t*, when the exothermic zone gets first "in touch" with the wall at A = A*, the walls are heated only by radiation from the products. Hence, at t ≤ t*,

$$q_{wR_o}(t) = A_f \int_o^{t^*} \dot{q}''_{wR}(t_i)\,\mathrm{d}t_i \tag{3.3}$$

At $t \geq t^*$, both the reactants and the products are "in touch" with the walls. For the part of the walls that is not yet "in touch" with the products

$$q_{wR}(t) = \int_{t^*}^{t} [A_f - A(\tau)]\,\dot{q}''_R(\tau)\,\mathrm{d}\tau \tag{3.4}$$

while heat transfer to the part of the walls "in touch" with the products is expressed in terms of a full Duhamel integral

$$q_{wP}(t) = A_f \int_{t^*}^{t} [\int_0^{\tau} \dot{q}''_P(t_j;t_i - t_j)\frac{\mathrm{d}\Lambda(t_j)}{\mathrm{d}t_j}\mathrm{d}t_j]\,\mathrm{d}\tau \tag{3.5}$$

where $\Lambda(t_j) \equiv A(t_j)/A_f$, while, with reference to the geometry of the self-similar profiles displayed in Fig. 3.3,

$$\dot{q}''_P(t_j;t_i - t_j) = (t_j - t^*)\Lambda(t_j)\,\dot{q}''_P(t^*;t_i - t^*) + [\dot{q}''_R(t_j) - \dot{q}''_R(t^*)] \tag{3.6}$$

The deformation of the self-similar heat flux profile, $\dot{q}''_P(t^*;t_i - t^*)$, is prescribed by slopes α and β, noted in Fig. 3.10, that delineate the growth of its amplitude. To evaluate the kernel of the integral presented by (3.5), $\dot{q}''_P(t_j;t_i - t_j)$, knowledge of function $\Lambda(tj)$, expressing the growth of the combustion front, is required. This can be inferred from the schlieren photographs, from which, especially t early times, the wall area in contact with the combustion products can be discerned. It was found thereby that the profile of $\Lambda(tj)$ is similar to the pressure profile, $P(tj)$. Hence it is postulated that

$$\Lambda(t_j) = [P(t_j) - P^*]/[P_f - P^*] \tag{3.7}$$

where $P^*=P(t^*)$.

The total amount of heat transferred to the walls of the enclosure is given by the sum of the integrals expressed by (3.3), (3.4) and (3.5)

$$q_w(t) = q_{wR_o}(t < t^*) + q_{wR}(t > t^*) + q_{wP}(t > t^*) \tag{3.8}$$

The results of calculations carried out for the three modes of combustion according to the above heat transfer model, are presented by Fig.3.11.

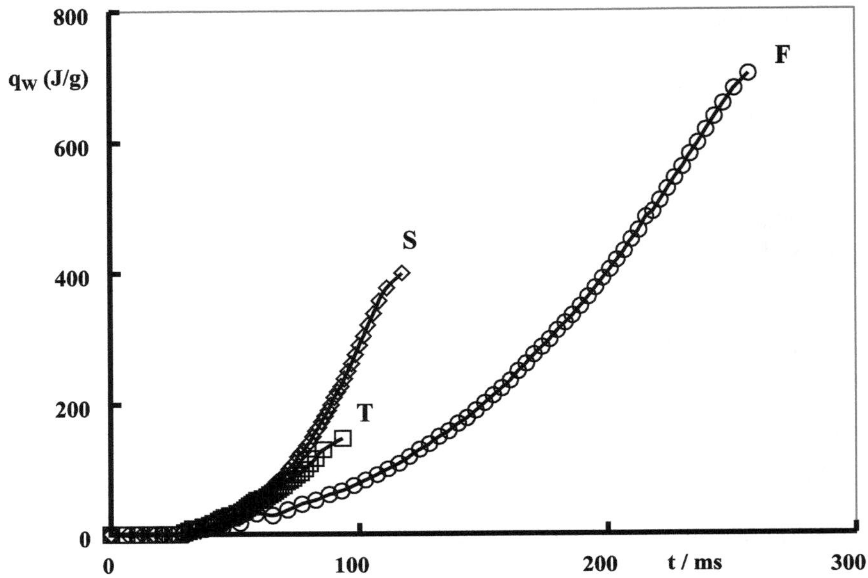

Fig. 3.11. Profiles of energy lost by heat transfer to the walls

3.4 Pressure Diagnostics

3.4.1 Dynamic Aspects

The measured pressure profiles, depicted by Figs. 3.1, 3.2, and 3.3, normalized with respect to the initial pressure of 5 atm, are presented by Fig. 3.12. Profiles of their life functions are displayed by Fig. 3.13 in comparison to the data.

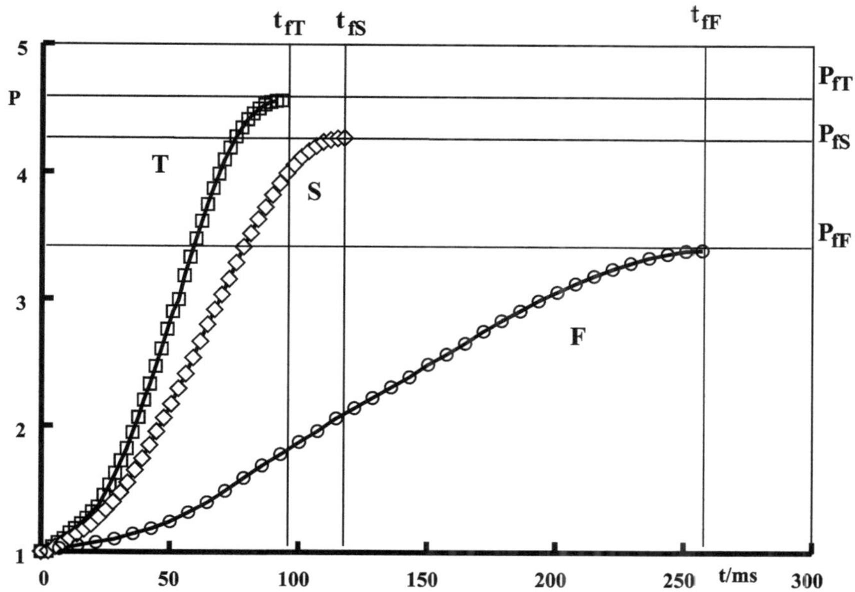

Fig. 3.12. Measured pressure profiles normalized with respect to state **i**

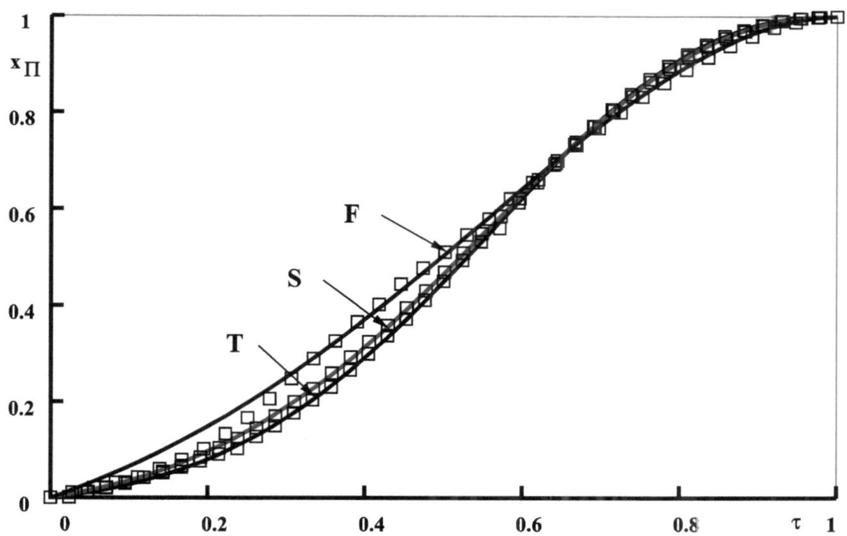

Fig. 3.13. Life functions of pressure profiles with their data marked with circles for mode F, diamonds for mode S and squares for mode T

3.4.2 Thermal Aspects

State diagrams of the three modes of combustion are displayed in Figs. 3.14, 3.15 and 3.16. Profiles of the mass fractions of fuel and products, evaluated by the use of (1.32), with the data of Fig. 3.13, are displayed in Figs. 3.17, 3.18 and 3.19. The corresponding phase diagrams of the mass fractions fuel and products are presented by Figs. 3.20, 3.21 and 3.1.22. The temperature profiles are depicted by Fig. 3.23.

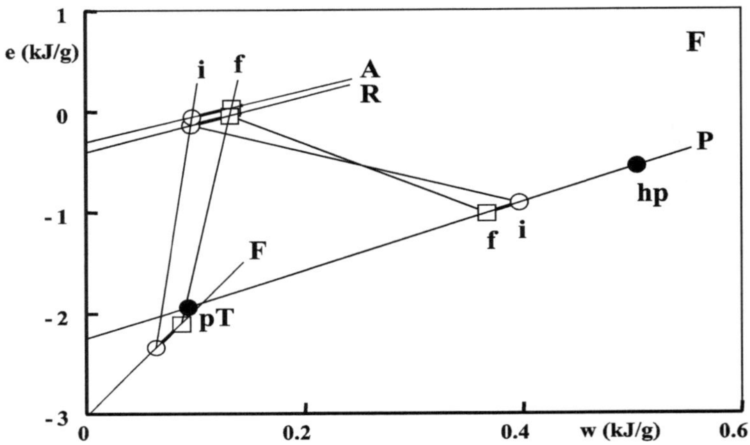

Fig. 3.14. State diagram for mode F

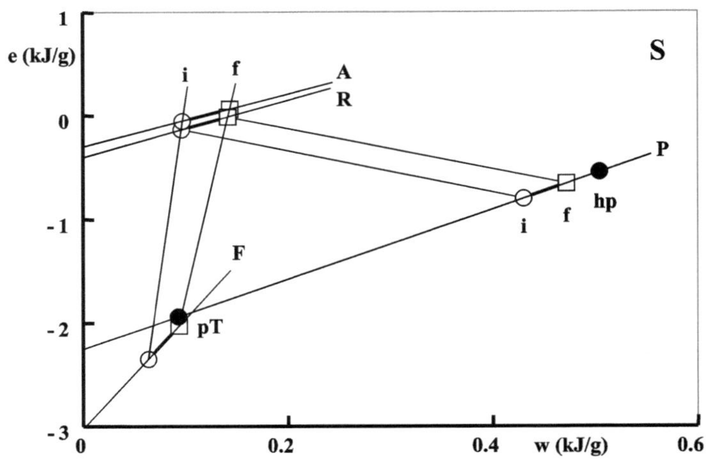

Fig. 3.15. State diagram for mode S

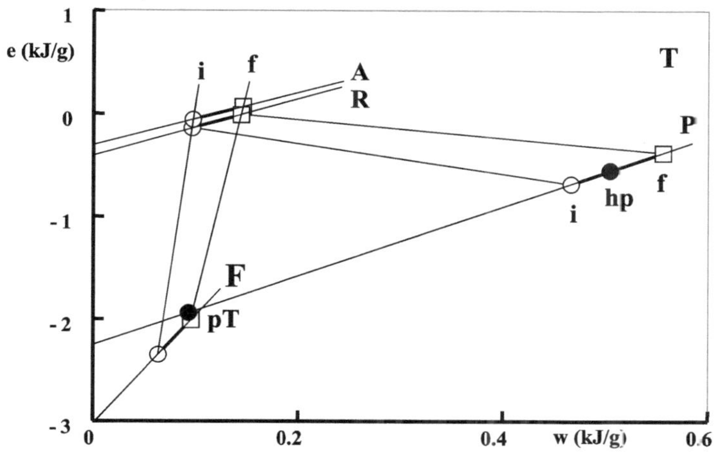

Fig. 3.16. State diagram for mode T

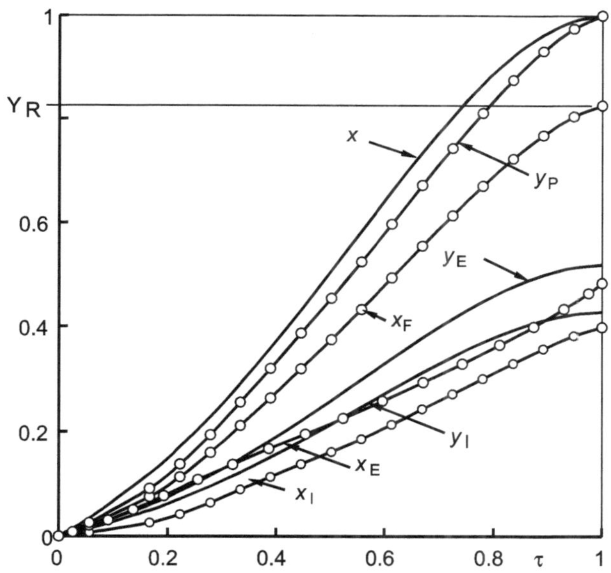

Fig. 3.17. Profiles of mass fractions of fuel and products in mode F

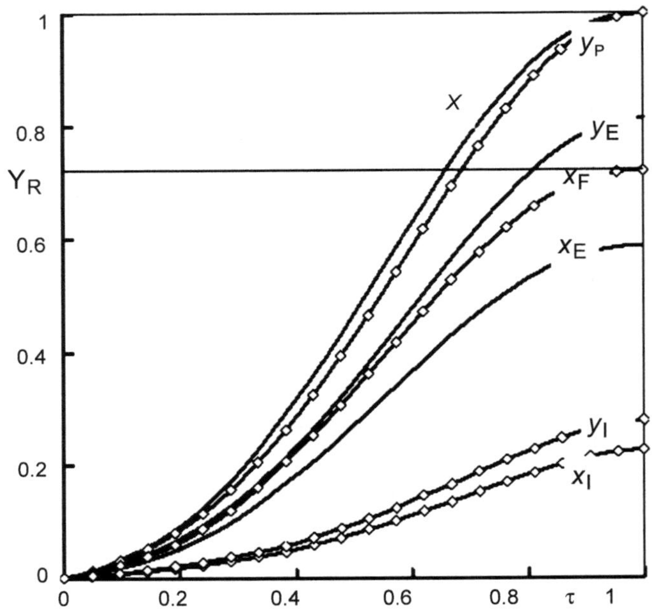

Fig. 3.18. Profiles of mass fractions of fuel and products in mode S

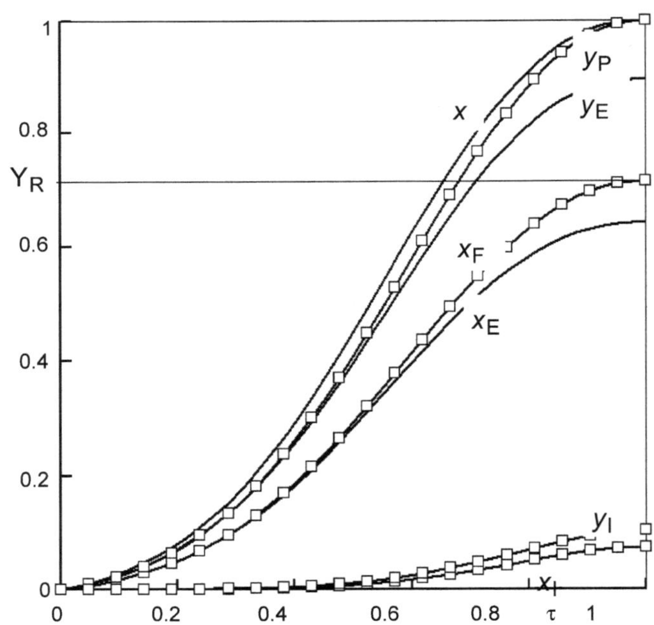

Fig. 3.19. Profiles of mass fractions of fuel and products in mode T

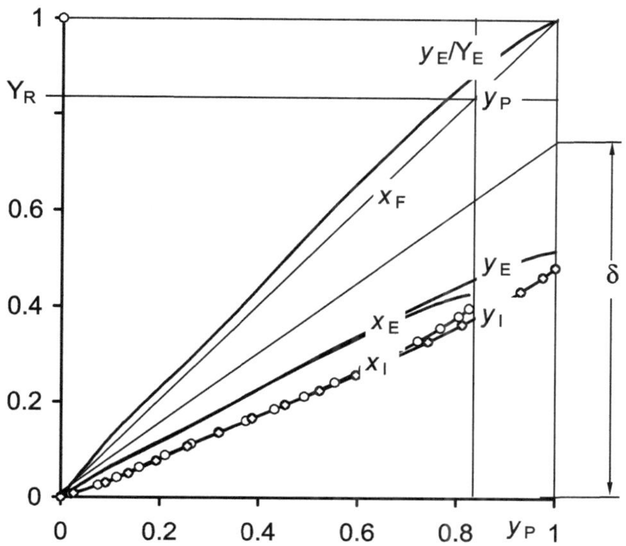

Fig. 3.20. Phase diagram of mass fractions of fuel and products in mode S

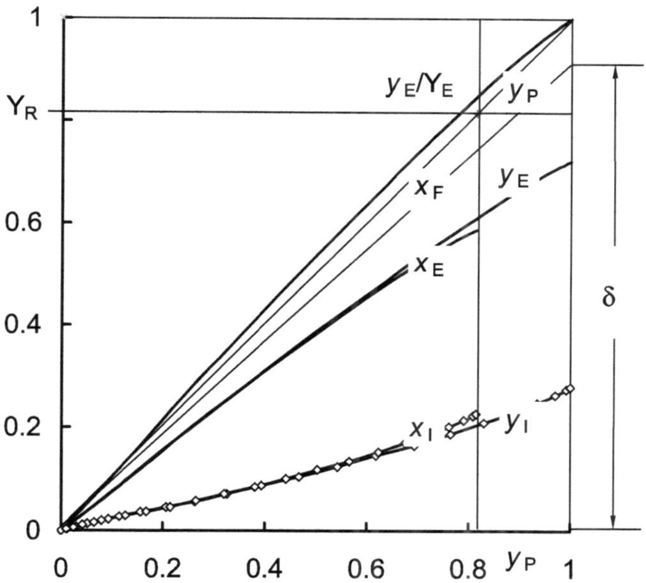

Fig. 3.21. Phase diagram of the mass fractions of fuel and products in mode S

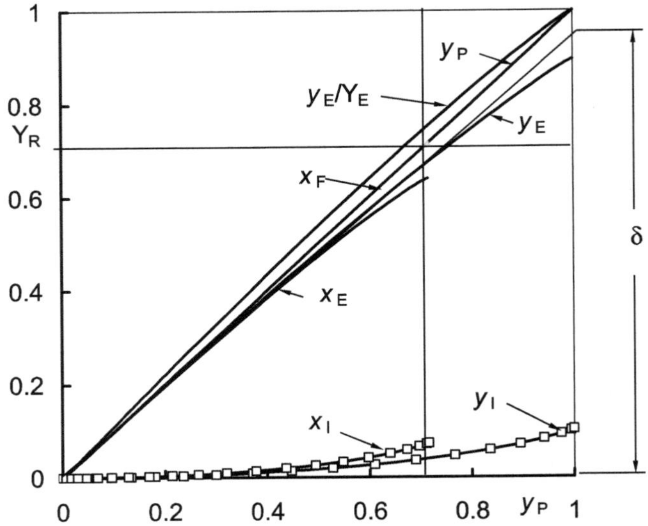

Fig. 3.22. Phase diagram of the mass fractions of fuel and products in mode T

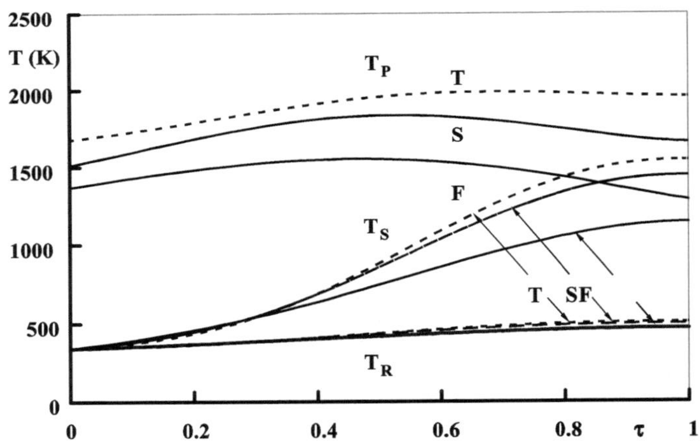

Fig. 3.23. Temperature profiles

The lifetimes, maximum pressure levels, and the parameters of the state diagrams and life functions are presented by Table 3.1.

The mass fraction of reactants, Y_R, and of the effectiveness, Y_E, provided by the data of Figs. 3.17-19, are listed in Table 3.2.

Table 3.1. Parameters of thermal aspects

Mode	**F**	**S**	**T**
T/ms	257	118	93
p_t/bar	16.95	21.35	22.80
P_t	3.39	4.27	4.56
C_R	3.86	3.86	3.86
C_P	2.74	2.74	2.74
Q	6.98	6.98	6.98
α	3.76	7.46	9.00
χ	0.85	1.22	1.32

Table 3.2. Mass fraction of reactants and the effectiveness of fuel utilization

Mode	F	S	T
Y_R	0.8203	0.8147	0.7155
Y_E	0.5206	0.7212	0.8962

As brought out in the above, higher values of effectiveness, Y_E, associated with higher pressure peaks displayed in Fig. 3.12, are produced by smaller mass fraction of reactants, Y_R.

It is of interest to note that, in the phase diagrams presented by Figs. 3.20, 3.21 and 3.22, the data of $y_E(y_P)$, can be expressed in terms of the functional relationship

$$y_E / Y_E = 1 - (1 - y_P)^{\delta/Y_E} \tag{3.9}$$

where δ is the slope of $y_E(y_P)$ at $y_P = 0$. In its form, this function is identical to that of exponent, $\zeta(\tau)$, defined by (2.46) for the life function.

The mass fraction of products is prescribed then by the inverse of (3.9), i.e.

$$y_P = 1 - (1 - y_E / Y_E)^{Y_E/\delta} \tag{3.10}$$

where $\delta = Y_E^{1/4}$ within an uncertainty range below 2.5% for all the three modes of combustion under study.

4 Chemical Kinetic Aspects

4.1 Introduction

The conversion of reactants into products is of essential significance to the dynamics of exothermic systems. Its analysis is based on thermo-chemical principles, subject to gasdynamic conditions of constraint. The first is on the microscopic scale of molecular interactions, treated by chemical kinetics. The second is on the macroscopic scale of exothermic centers in the flow field. Their salient features are presented here in turn.

The role of chemical kinetics is prominently displayed in the thermochemical phase space. This has been recognized by Semenov (1944 & 1958/59) right from the outset of his chain reaction theory, and its mechanism in this space was formulated in terms of a set of non-linear ordinary differential equations (ODE's) that are *autonomous* with respect to time, the sole independent variable. As typical of non-linear dynamics, the time was then eliminated by a simple expedient of dividing the constitutive equations by each other. Obtained thus were the fundamental relationships governing the kinetic behavior of the system, expressed eventually in terms of integral curves, or trajectories, in the thermophysical phase space.

The co-ordinates of the phase space are formed by the dependent variables of the problem. The thermochemical phase space is, therefore, multidimensional in nature, the number of its co-ordinates being equal to the number of degrees of freedom. The essential properties of the solution are expressed there by integral curves, or trajectories, whose manifolds are delineated by separatrices and attractors whose intersections specify the coordinates of singular points providing the boundary conditions for these curves. The intrinsic nature of these curves is revealed by their projections on the planes of any two co-ordinates selected for this purpose.

In particular, the projection of the trajectories of the thermochemical phase space upon the plane of the temperature and the concentration of a chain carrier played an important role in physical chemistry of combustion. It is on its basis that Gray and Yang (1965) [with its sequel by Yang and

Gray, 1967], developed the concept of the "unification of the thermal and chain theories of explosion limits." They demonstrated that the effect of chain branching, referred to as the "chemical kinetic explosion," is manifested there by a saddle-point singularity located at the intersection of a separator between the trajectories of the initial stages of combustion and the attractor for their final stages. Thereupon, Peter Gray and his collaborators applied this method of approach with great success to a number of chemical systems (vid. e.g. Gray & Scott, 1990 and Gray & Lee, 1967). Particularly noteworthy is the explanation provided by Griffiths (1990) of the roles played by singularities in the course of thermo-kinetic interactions, encompassing a variety of nodal and saddle, as well as spiral points, the latter expressing the oscillatory behavior of cool flames.

At later stages of the exothermic process of combustion, the integral curves tend to bunch together, forming manifolds attracted by the coordinates of thermodynamic equilibrium, as demonstrated by Maas and Pope (1992). This can be interpreted as an evidence that the reacting system is then so much impressed by the final state of thermodynamic equilibrium, that, in its asymptotic approach, it becomes virtually independent of the events associated with its initiation (ignition) and evolution (chain branching).

4.2 Principles

4.2.1 Formulation

The evolution of an exothermic center, introduced in Chapter 1, is intimately associated with the progress of chemical reaction, subject to heat transfer across its boundaries. In order to reveal its dynamic features, the influence of molecular diffusion, which tends to obscure the chemical kinetic mechanism, is excluded from the analysis, while that of conductivity is expressed in terms of a relaxation time.

The process under study is, then, that of transformation of a chemical system from the state of the reactants into that of the products. The former are at a given composition and temperature, $T_R = T_i$, and the latter are at T^*, which may be identified with either that of the surroundings, so that $T^* < T$, or with the thermodynamic equilibrium, reached at T_f, of the products, in which case $T^* > T$. Constitutive equations of chemical kinetics consist of the *chemical source* and the *thermal source*.

The *chemical source* is formulated in terms of the so-called law of mass action, which, in the form of the popular computer algorithm, CHEMKIN, is expressed by the species conservation equation in the following manner.

For K chemical reactant species, A_k (k = 1,2,...K), reacting in I elementary steps (i = 1,2,...I), each of the form

$$\sum_{k=1}^{K} a'_{ki} A_k \Leftrightarrow \sum_{k=1}^{K} a''_{ki} A_k$$

where a'_{ki} and a''_{ki} denote stoichiometric coefficients of, respectively, the reactants and the products, the rate of gain in the mass fraction of the kth component

$$\frac{dy_k}{dt} = v_S M_k \sum_{i=1}^{I} (a''_{ki} - a'_{ki}) \{k_i^+ \prod_{i=1}^{I} [A_k]^{a'_{ki}} - k_i^- \prod_{i=1}^{I} [A_k]^{a''_{ki}}\} \tag{4.1}$$

for which the reaction rate constant

$$k_i = A_i T^{n_i} \exp(-E_i / RT)$$

while v_s is the specific volume of the system and M_k is the molar mass of the kth component.

The *thermal source* is based on the elementary energy balance for an exothermic center

$$dq = dh - v dp \tag{4.2}$$

where

$$dh = \sum_{k=1}^{K} y_k dh_k + \sum_{k=1}^{K} h_k dy_k \tag{4.3}$$

Then, since, according to the JANAF data base, the reaction constituents are perfect gases, $dh_k = c_{p,k} dT$ and the rate of the temperature rise

$$\frac{dT}{dt} = -\frac{1}{c_p}\sum_{k=1}^{K} h_k \frac{dy_k}{dt} + \frac{v}{c_p}\frac{dp}{dt} + \frac{1}{c_p}\frac{dq}{dt} \tag{4.4}$$

where $c_p \equiv \sum_{k=1}^{K} c_{p,k} y_k$.

The last term in these equations is due to diffusion - a gradient in the flow field expressed in the form of a partial differential. In an ordinary differential form of (4.4), it is expressed in terms of a thermal relaxation time, τ_T, so that

$$\frac{1}{c_p}\frac{dq}{dt} = \frac{T_P - T}{\tau_T} \tag{4.5}$$

The integration of (4.1) and (4.3) is carried out along a linear path of constant pressure between the corresponding points on R of $x_w = 0$ and on P of $x_w = 1$, delineating the evolution of consecutive exothermic centers. By making sure that the ratio of rate constants $k^+/k^- = K$ - the equilibrium constant - the integration is, in effect, a solution of a double boundary value problem, for which the end state is specified by the condition of thermodynamic equilibrium.

The solution is associated with an essential difficulty due to the fact that these equations are essentially stiff, as pointed out originally by Hirschfelder et al (1960), demanding a special treatment to assure convergence. This task nowadays significantly aided by the CHEMKIN method of Kee et al (1980, 1989 & 1993), and Miller et al (1990). Its application involves the LSODE (Linear Solver of Ordinary Equations) procedure of Hindmarsh (1971), using the multi-step integration method of Gear (1971). Salient features of most numerical methods developed for treating such equations, in comparison to that of Gear, were reviewed by Bui et al (1984).

4.2.2 Illustration

An implementation of the theoretical background provided above to the evolution of an exothermic center is illustrated by ignition of a stoichiometric hydrogen–oxygen system, $H_2 + 0.5O_2 + 1.88N_2$, contained in a constant volume vessel. The kinetic mechanism of this system is provided by Table 4.1 - a source of data for all the parameters invoked in (4.1).

Table 4.1. Kinetic mechanism and parameters of the H_2-O_2 system

	Elementary steps	A	n	E	Notes
1.	$OH + O \rightarrow O_2 + H$	$0.18\ 10^{14}$	0	0	—
2.	$O + H_2 \rightarrow OH + H$	$0.15\ 10^{8}$	2	7.55	—
3.	$OH + H_2 \rightarrow H_2O + H$	$0.18\ 10^{9}$	1.6	3.3	—
4.	$OH + OH \rightarrow H_2O + O$	$0.15\ 10^{10}$	1.14	0	—
5.	$H + H + M \rightarrow H_2 + M$	$0.64\ 10^{18}$	–1	0	M = Ar, H_2
6.	$H + H + H_2 \rightarrow H_2 + H_2$	$0.97\ 10^{17}$	–0.6	0	—
7.	$H + OH + M \rightarrow H_2O + M$	$0.14\ 10^{24}$	–2	0	—
8.	$H + O_2 + M \rightarrow HO_2 + M$	$0.7\ 10^{18}$	0.8	0	M = Ar
9.	$O + O + M \rightarrow O_2 + M$	$0.1\ 10^{18}$	–1	0	M = Ar
10.	$H + HO_2 \rightarrow OH + OH$	$0.15\ 10^{15}$	0	1	—
11.	$H + HO_2 \rightarrow H_2 + O_2$	$0.25\ 10^{14}$	0	0.69	—
12.	$O + HO_2 \rightarrow OH + O_2$	$0.2\ 10^{14}$	0	0	—
13.	$OH + HO_2 \rightarrow H_2O + O_2$	$0.2\ 10^{14}$	0	0	—
14.	$HO_2 + HO_2 \rightarrow H_2O_2 + O_2$	$0.2\ 10^{13}$	0	0	—
15.	$OH + OH + M \rightarrow H_2O_2 + M$	$0.13\ 10^{23}$	–2	0	M = O_2
16.	$H + H_2O_2 \rightarrow H_2 + HO_2$	$0.17\ 10^{13}$	0	3.75	—
17.	$O + H_2O_2 \rightarrow OH + HO_2$	$0.28\ 10^{14}$	0	6.4	—
18.	$OH + H_2O_2 \rightarrow H_2O + HO_2$	$0.7\ 10^{13}$	0	1.43	—
19.	$H + H_2O_2 \rightarrow H_2O + OH$	$0.1\ 10^{14}$	0	3.58	—

[$k = AT^n \exp(-E/RT)$, where k is in moles/cm^{3}sec, T in K, and E in kcal/mole]

For a specific example, let us consider the case when the exothermic reaction takes place at a pressure is $p_i = 6$ atm, while the thermal relaxation time $\tau = 1$sec, whereas the surroundings are at a temperature of $T_a = 300$K, so that $T_z = T_a < T$ and $\dot{Z}$ is negative.

Concentration histories of active radicals and the temperature profiles for the thermal case, i.e. when the components are devoid of active radicals, are displayed in Fig. 4.1. The two initial temperatures of 774 and 775 K, adopted for this example, provide quite an accurate identification of what is known as 'auto-ignition' for the case under study. The influence of energy loss brought about by heat transfer to the surroundings is then of particular significance. Without it, i.e. for $\tau_T = \infty$, extinction cannot take place, because, at constant pressure, the rate of the temperature rise an exothermic reaction is always positive.

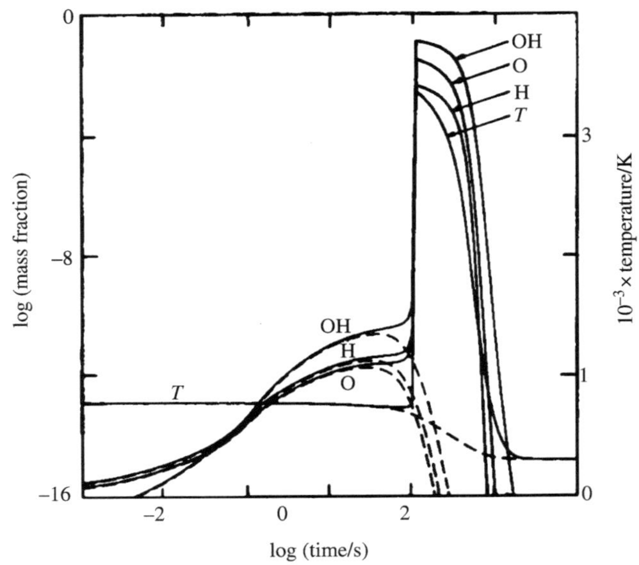

Fig. 4.1. Temperature and concentration profiles of active radicals upon thermal ignition of an exothermic center for a stoichiometric hydrogen-oxygen mixture at 6 atm. and 775 K (continuous lines) and 774 K (broken lines) in surroundings at 288 K, with the thermal relaxation time $\tau_T = 1$ sec

4.2.3 Phase space

The most informative way to display the solutions of (4.1) and (4.4) for a system of K chemical species is by means of integral curves (trajectories) on an (K + 2) dimensional phase space. Thus, projections of such integral curves for our example are presented in Fig. 4.2 on the plane of the mass fraction of hydrogen atom - the most effective chain carrier - and the temperature.

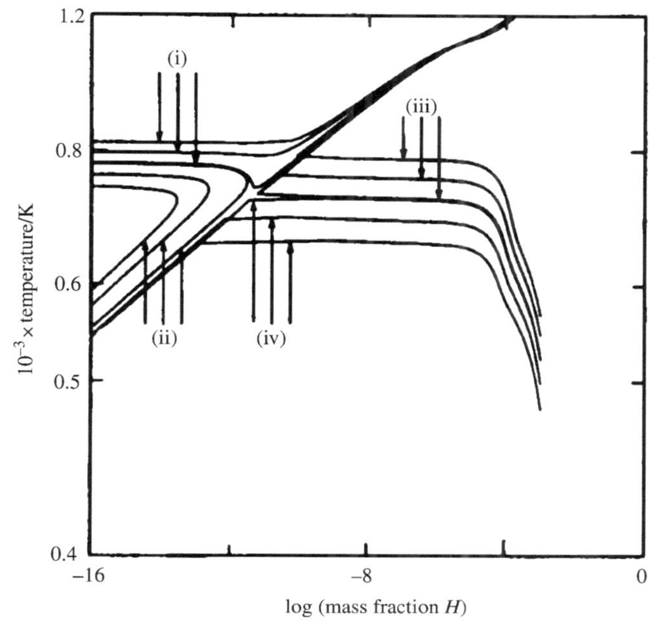

Fig. 4.2. Integral curves for ignition of a stoichiometric hydrogen-oxygen mixture at 6 atm. in surroundings at 288 K, with the thermal relaxation time $\tau_T = 1$ sec

There are four families of curves evident on this diagram. They correspond to two sets of initial temperatures for two cases of active radical concentration, i.e. either zero, $(Y_H)_o = (Y_O)_o = (Y_{OH})_o = 0$ - the thermal case - or finite, i.e. $(Y_H)_o = 10^{-3}$,while $(Y_O)_o = (Y_{OH})_o = 0$ - the thermochemical case. These families correspond to the following processes: (i) ignition for the thermal case (a), (ii) extinction for the thermal case (a), (iii) ignition for the thermochemical case, and (iv) extinction for the thermochemical case. The borderline between ignition and extinction is a separator, and that between the thermal and thermochemical cases is an attractor. The point of their intersection is a saddle point singularity.

Initial states of the former specify the thresholds for ignition – referred to as the self-ignition, or auto-ignition, temperature - while the latter defines the critical states of the radical pool into which all the trajectories tend to merge. From the way the integral curves get bunched together on this plane, it is evident that the chemical system is so strongly attracted by the final equilibrium state, that its influence becomes prevalent long before it is attained.

For a chemical kinetic mechanism involving just one chain carrier, the phase space is, of course, expressed in terms of a single phase trajectory. The thermochemical phase space is reduced them to a plane. It is, in fact, such a phase plane that has been elucidated in the seminal paper of Gray and Young (1965) on the thermochemistry of photolysis.

The most remarkable feature of Fig. 4.3 is that the temperature threshold for ignition is significantly lower in the thermochemical case than in the thermal. Initial concentration of active radicals acts evidently as an effective substitute for the temperature. Thus, the attractor represents, on one hand, the envelope of all the integral curves for thermal ignition, and, on the other, it specifies the minimum concentration of active radicals having a significant effect upon the temperature for ignition in the thermochemical case.

4.2.4 Ignition limits

Critical temperatures of thermal ignition, evaluated in this manner for τ_T = 1 sec and τ_T = 30 sec, over a range of pressures from 10^{-3} to 10 atm, are depicted on the pressure – temperature plane of Fig. 4.4. Shown there for comparison is the classical, experimentally established, thermal ignition limit, referred to conventionally as the "auto-ignition". or "kinetic explosion" limits (Lewis and Von Elbe, 1961). The computed curves are in good agreement with the classical plot over all its three segments: the 'first explosion limit' at low pressures, ascribed in the classical literature as one due to the quenching effects of the wall, the 'second explosion limit' at intermediate pressures, credited to the escalating influence of chemical kinetic chain branching, and the 'third explosion limit' at high pressure, ascribed to thermal conduction loss. The ability to account for all three comprehensively provides a noteworthy unification of the theoretical framework. It is of interest then to note that, although the thermal ignition limit is nonexistent unless the relaxation time is finite, the magnitude of this parameter has a relatively small influence upon its coordinates on the plane of initial pressures and temperatures.

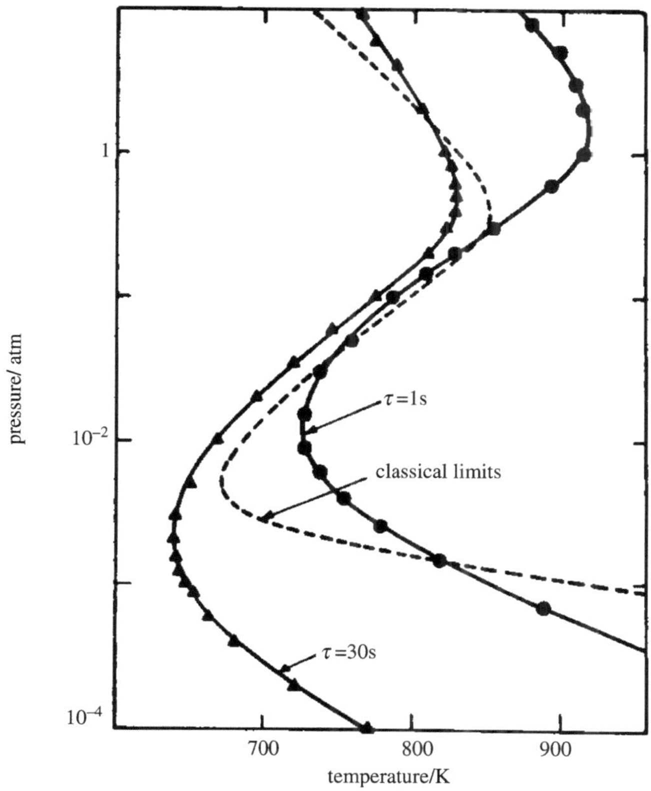

Fig. 4.3. Thermal ignition ('kinetic explosion') limits in the temperature-pressure diagram

4.3 Exothermic Center

4.3.1 Formation

The runaway tendency of an exothermic center is usually modulated by molecular transport phenomena, leading to the formation of a flame kernel. In a turbulent field, however, there are, as a rule, a number of centers created at about the same time. Under such circumstances, the process of ignition may be *weak* or *strong*. The line of demarcation between their regimes on the plane of initial temperatures and pressures is referred to as the *strong ignition limit*. The case of strong ignition is known as the onset of knock – the phenomenon which played a crucial role in the evolution of internal combustion engines and attracted, therefore, a good deal of scientific attention. See, for example, Sokolik 1960; Saytzev & Soloukhin 1962; Voyevodsky & Soloukhin 1965; Meyer & Oppenheim 1971 *a, b*. There is also ample background available for the theory of exothermic centers (e.g. Van Tiggelen 1969, Borisov 1974, Oppenheim et al 1974,1975). In particular, numerical analyses of strong exothermic centers were provided by Zajac & Oppenheim (1971) and by Cohen et al. (1975a), as well as by Cohen & Oppenheim (1975).

The dynamic effects of ignition depend primarily on the amplitude of the specific power pulse - the rate of the exothermic process – that can be controlled by modifying the initial state of the reactants, their pressure, temperature and composition. The most informative insight into such effects is provided by the reflected wave technique in shock tubes.

Examples of the results obtained by such experiments are provided by cinematographic laser-illuminated schlieren records presented on the next two figures.

Figure 4.4 illustrates the case of weak ignition manifested at its outset by the formation of flame kernels that occur at lower pressures and temperatures. Such kernels are developed as a rule in corner eddies generated by the interaction of the reflected wave with the boundary layer created by flow behind the incident shock. The record of Fig. 4.4 was obtained by the use of an iso-octane-oxygen mixture at a pressure of 6 atm and a temperature of 1300K attained behind the reflected wave, taken at a frequency of 0.5 MHz.

Figure 4.5 illustrates the case of strong ignition, manifested by the appearance of a shock front of a blast wave without any prior evidence of flame kernels that occurs at higher temperatures or pressures. The record of Fig. 4.5 was obtained under the same conditions as that of Fig. 4.4, except for the temperature of 1400K, 100°C higher than before.

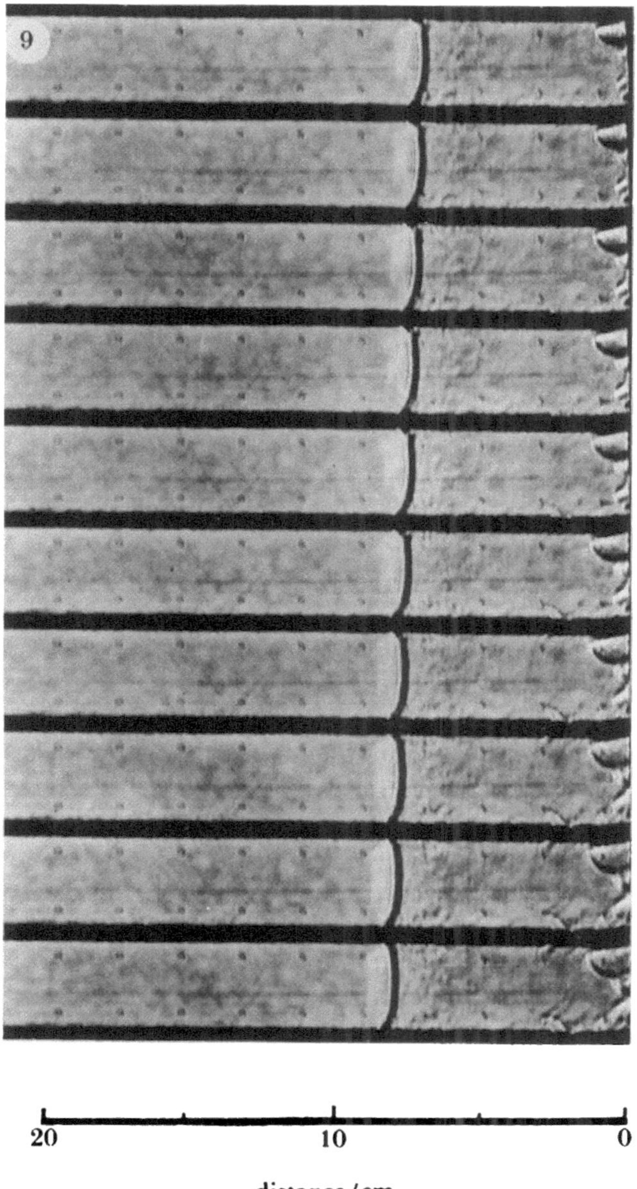

Fig. 4.4. Cinematographic schlieren record of weak ignition behind a reflected shock front in a stoichiometric iso-octane-oxygen mixture diluted in 70% argon at p = 6 atm and T = 1300 K (Vermeer et al. 1972)
End wall of the shock tube is at the right edge of the photographs. Sequence of frames is from top to bottom at time intervals of 2 μs.

Fig. 4.5. Cinematographic schlieren record of strong ignition at the same conditions as Fig. 4.8, except for T = 1400 K, rather than 1300 K (Vermeer et al. 1972)

4.3.2 Strong ignition limit

The line of demarcation between the regimes of mild and strong ignition on the plane of initial temperatures and pressures defines then, as pointed here at the outset, the strong ignition limit. The existence of such a limit has been pointed out first by Saytzev & Soloukhin (1962) and, according to an early theory proposed by initially by Downs et al (1951), corroborated later by Voyevodsky & Soloukhin (1965), it has been interpreted as an extension of the 'second explosion limit' into the regime of 'explosion.'

An experimentally determined strong ignition limit for a stoichiometrichydrogen-oxygen mixture is presented in Fig. 4.6 (Meyer & Oppenheim 1971a)

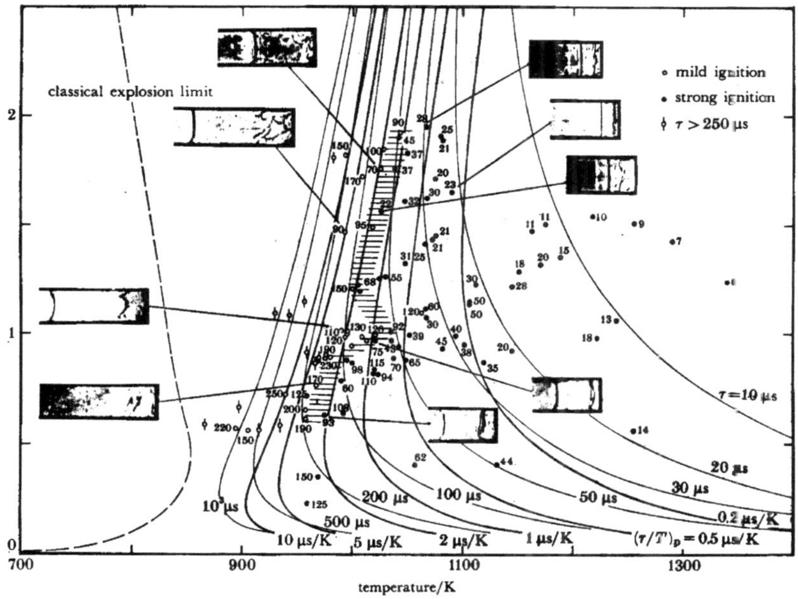

Fig. 4.6. Strong ignition limit - a line of demarcation between weak and strong ignitions - for a stoichiometric hydrogen-oxygen mixture

As indicated by the horizontally shaded zone marking the demarcation between the weak and strong ignitions, instead of appearing as an extension of the second limit, it coincides most closely with a locus of a partial derivative of induction time, τ_i, with respect to temperature at constant pressure, i.e. $(\partial\tau_i/\partial T)_p = \text{const}$. This empirical observation can be rationalized by the randomness in temperature distribution (Meyer & Oppenheim 1971 *b*).

Consider a set of N exothermic kernels per unit mass of the reacting mixture, developing under the influence of a Gaussian temperature distribution, so that

$$\frac{d\ln N}{dT} = \frac{1}{\sqrt{2\pi}\sigma} \exp[-\frac{1}{2}(\frac{T - T_m}{\sigma})^2] \tag{4.6}$$

where σ is the standard deviation and subscript m denotes the mean.

By virtue of its exponential dependence on the reciprocal of the temperature, expressed by the Arrhenius relation, the induction time can be expressed by a Taylor series truncated at the linear term, i.e.

$$\tau_T = \tau_m - (\partial\tau/\partial T)_p (T - T_m) \tag{4.7}$$

and (3.6) becomes

$$\frac{d\ln N}{d\tau_T} = \frac{1}{\sqrt{2\pi}\varpi} \exp[-\frac{1}{2}(\frac{T - T_m}{\sigma})^2] \tag{4.8}$$

where $\varpi \equiv -(\partial\tau_T / \partial T)_p \sigma$

In order to establish the rule for the limit, consider the decay of a power pulse from its initial form of a square wave extending over a time interval, δ, with amplitude equal to a uniform rate of specific exothermic energy deposition, $\dot{q}(\delta)$. In turbulent field, where the temperature distribution is random, the specific exothermic power at a given instant of time, t, is obtained from the contribution of centers whose induction times are within the interval $\tau_T' = t - \delta$ and $\tau_T'' = t$. Thus, by integrating (3.8) from τ_t' to τ_t'', with normalized amplitude of exothermic energy and time, the former with respect to its initially uniform value, $Q \equiv q(t)/q(\delta)$, and the latter with respect to the pulse width, $\Theta \equiv t/\delta$, one obtains the following expression for the specific exothermic power

$$\dot{Q} \equiv \frac{dQ}{d\Theta} = \frac{N(\delta)}{N} = \frac{1}{2}[\text{erf}(\frac{N}{\sqrt{2\pi}}\frac{\delta}{\varpi}) - \text{erf}(\frac{\Theta - \Theta_m - 1}{\sqrt{2\pi}}\frac{\delta}{\varpi})] \tag{4.9}$$

Expressed thus, in effect, is the dependence of the specific exothermic power pulse on the standard deviation in induction times, ϖ.

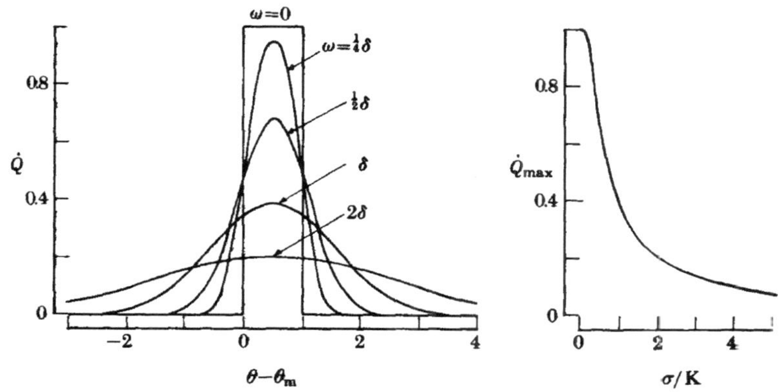

Fig. 4.7. Decay of a power pulse for a set of exothermic centers in terms of standard deviation in induction times, and the decay of their maxima in terms of standard deviation in the temperature distribution for a stoichiometric hydrogen-oxygen mixture at T = 1000K and p = 6 atm (Meyer & Oppenheim 1971b)

The variation of $\dot{Q}$ as a function of Θ-Θ_m is presented in Fig. 4.7, where the value of ϖ is expressed in terms of multiples of the exothermic pulse width, δ. As expected, the decay of the power pulse is manifested by spreading out, as typical of diffusion. According to (3.9) the decrease of their peak at the center, where $\Theta - \Theta_m = 0.5$, depends on the value of δ/ϖ, and, hence, by virtue of the definition of ϖ, on σ, so that

$$\dot{Q}_{max} = \mathrm{erf}[\frac{\delta}{2\sqrt{2\pi}(\partial\tau_T/\partial T)_p\sigma}] \tag{4.10}$$

For the particular case of the stoichiometric hydrogen oxygen system of Fig. 4.6, where the strong ignition limit corresponds evidently to the locus of

$$(\partial\tau_i/\partial T)_p = -2\mu s/K$$

while, at the level of 1000K, $\delta = 2$ μs, independently of any local temperature variations. The decay of the power pulse peaks, obtained from (4.8), is displayed on the right side of Fig. 4.7. As apparent from it, the strength of the power pulse, expressed in terms of its peak, is remarkably insensitive to the standard deviation in the temperature, its value of only four degrees at a level of 1000 K being associated with a tenfold decrease in peak specific power, exposing thus the critical nature of the strong ignition limit.

The concept of the strong ignition limit is of general significance to the dynamic features of combustion. Its specific expression, $(\partial\tau_\iota/\partial T)_p$, may differ somewhat from this straightforward derivative. For instance, in the case of hydrocarbons, it was found to be rather $(\partial \ln\tau_\iota/\partial T)_\mathrm{p}$, as revealed by Vermeer et al. (1972). According to (4.10), this discrepancy is ascribable to the fact that, for hydrocarbons, the width of the exothermic power pulse, δ, is proportional to the induction time, τ_ι, whereas for hydrogen it is independent of it.

Revealed thus is the mechanism of such well known intrinsic instability phenomena in combustion as knock in internal combustion engines (see, for example, Sokolik 1960) and the 'explosion in explosion' at the onset of detonation (Urtiew & Oppenheim 1966).

4.4 Engine Combustion

4.4.1 Status Quo

The chemical kinetic processes taking place in the course of combustion in a piston engine are illustrated here by combustion in the Renault engine whose dynamic and thermodynamic properties were treated in Section 1.4.

The relaxation time, τ_T, required for this purpose according to (4.5), was determined by calibration. Concentrations of NO were, first calculated for a set of relaxation times. Then, as displayed by Fig. 4.8, the values of τ_T that matches their concentrations of 1459 PPM and 357 PPM, measured` at full and part loads, were established at, respectively, 20 μsec, and 4 μsec, as indicated by broken lines.

Chemical kinetic computations for auto-ignition were carried out for a sequence of discrete exothermic centers, over intervals of crank angles adjusted to rapid changes of the temperature by small steps, and to its gradual changes by large steps.

The temperature profiles of representative exothermic centers are presented by Fig. 4.9. The concentration profiles of NO and CO are displayed, respectively, in Fig. 4.10 and Fig. 4.11. The mass averages of concentrations, $c_k = \int_i^f y_k dx_P$, for k = NO, CO, are presented by Fig. 4.12, where their measured engine-out values are indicated by broken line segments next to the appropriate coordinate axes. The latter are evidently in quite a satisfactory agreement with the results of calculations.

To provide an insight into the thermochemical mechanism of chemical transformations, displayed by Fig. 4.13 are the projections of integral curves in the multidimensional phase space for k = NO and k = CO into a (T-logc_k) plane. Time progresses from the initial state at -∞ in the directions indicated by arrows. Particularly noteworthy in these diagrams are the vertical segments that indicate concentration freezing as the temperature drops.

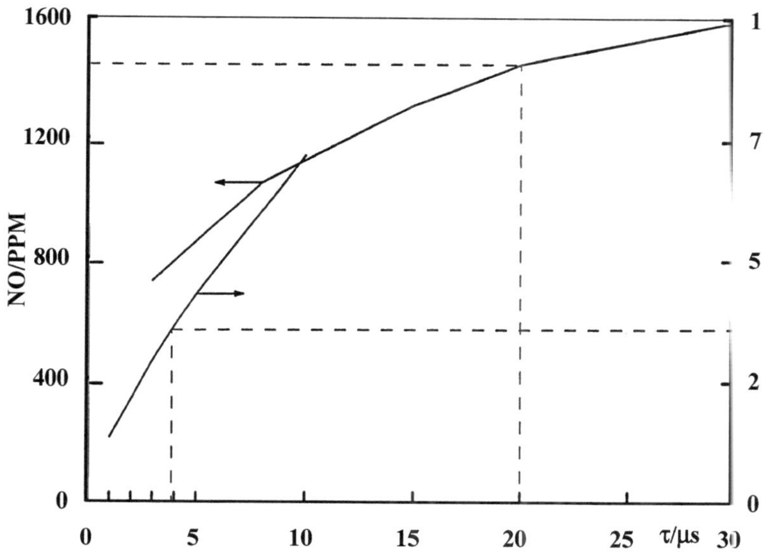

Fig. 4.8. Calibration of relaxation times

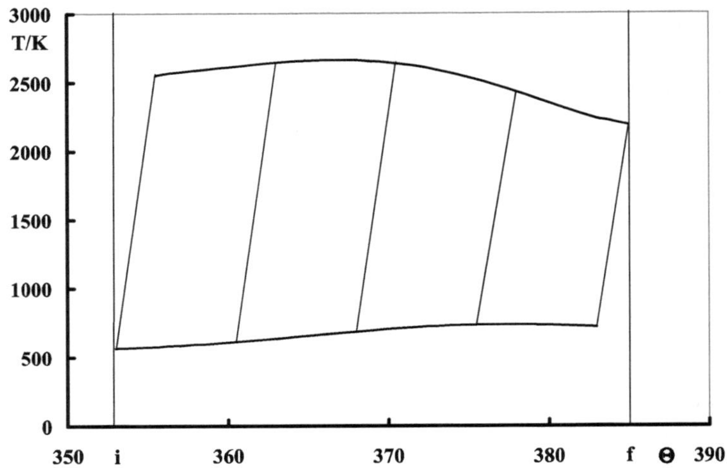

FULL LOAD

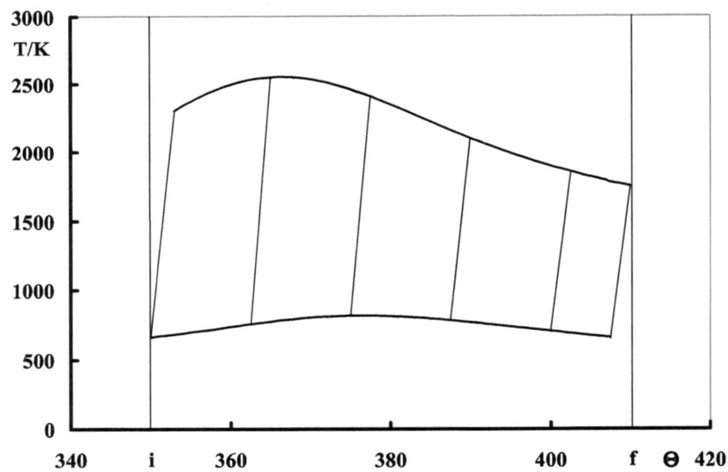

PART LOAD

Fig. 4.9. Temperature profiles

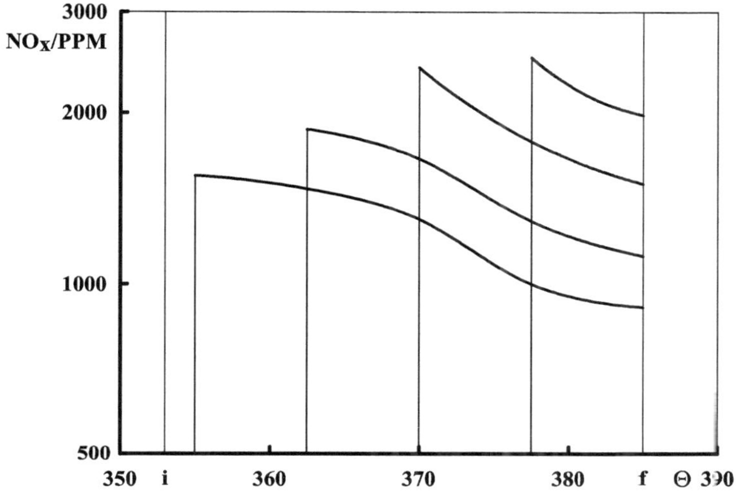

FULL LOAD

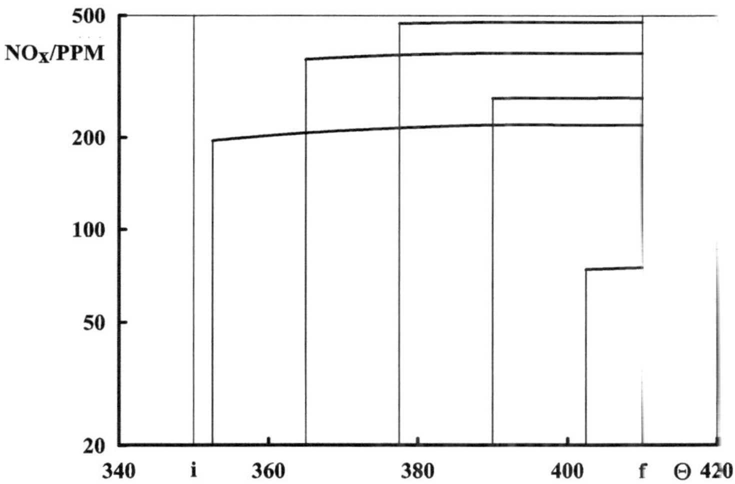

PART LOAD

Fig. 4.10. Profiles of (NO)

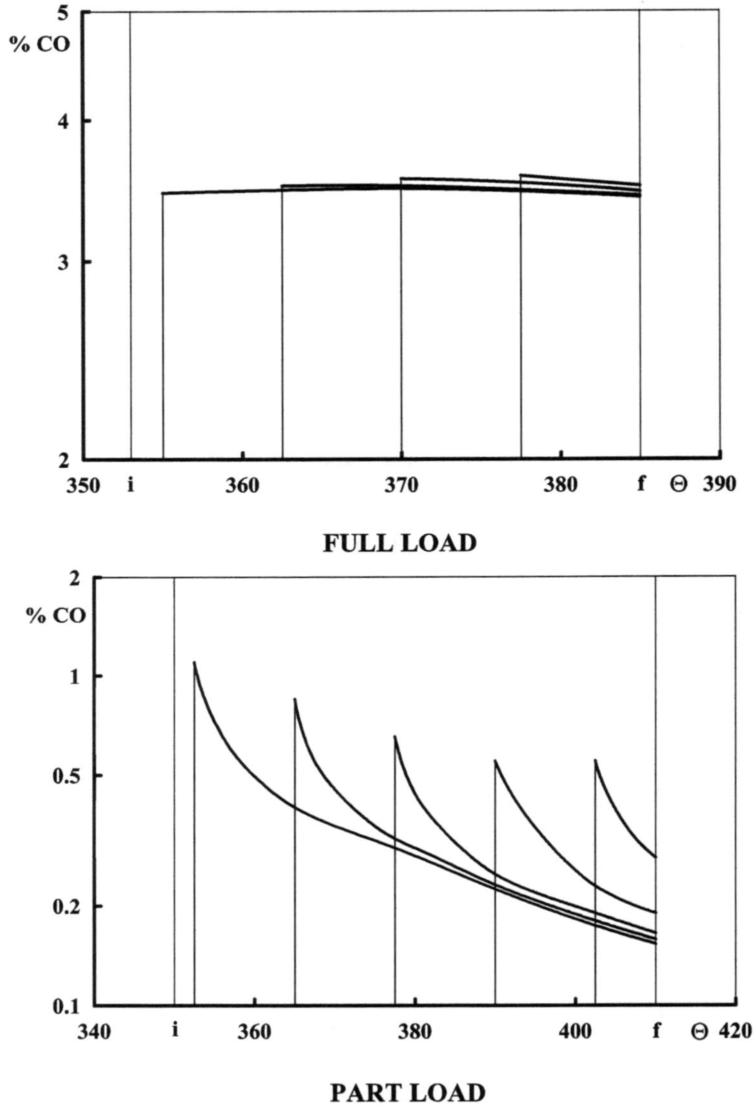

Fig. 4.11. Profiles of (CO)

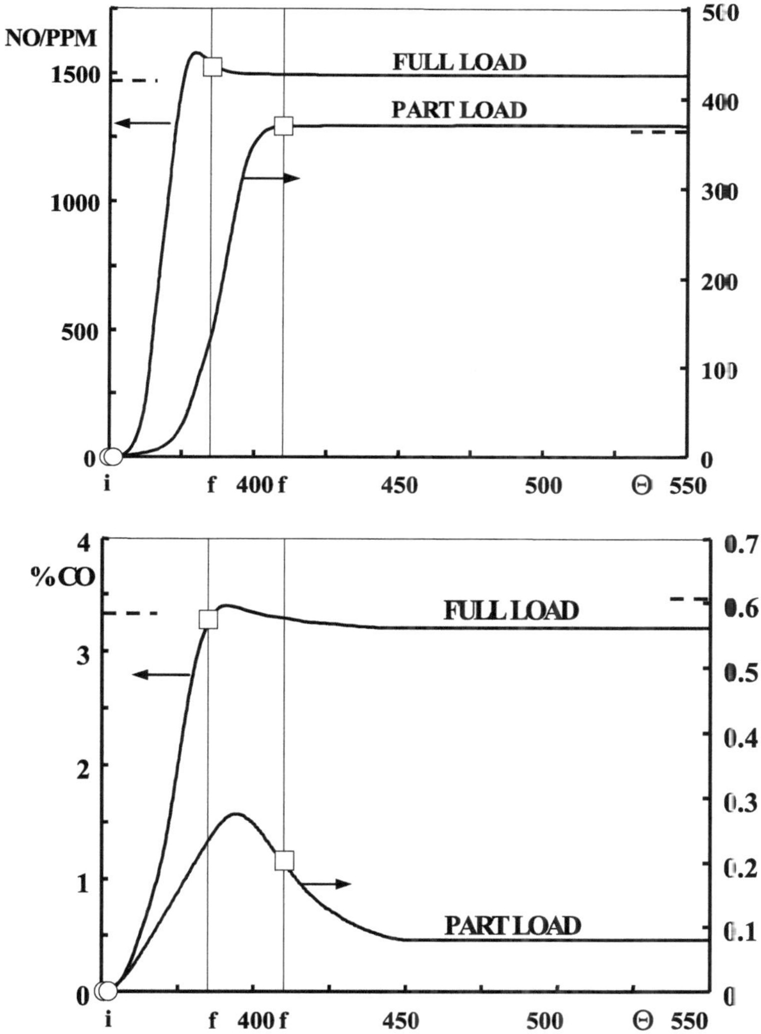

Fig. 4.12. Profiles of mass-average (NO) and (CO)

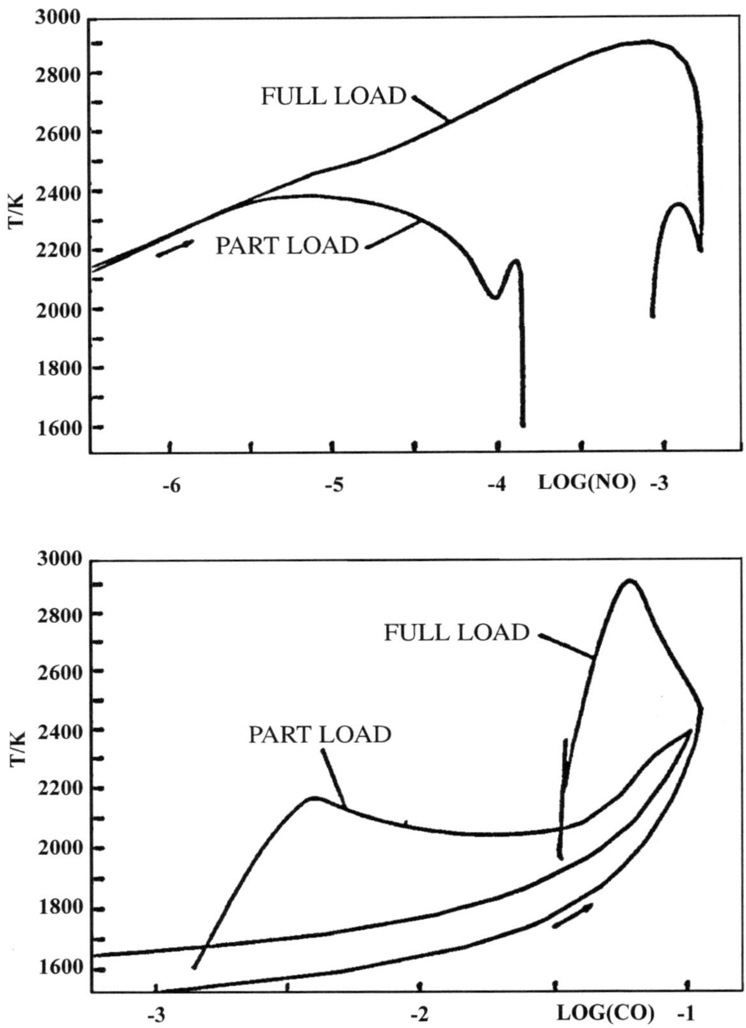

Fig. 4.13. Integral curves of (NO) and (CO)

4.4.2 Prognosis

The chemical kinetic analysis presented above provides the required background for evaluation of improvements that can be brought about by modifications of the exothermic process of combustion in an engine. The means used for this purpose are referred to as internal (or in-cylinder) treatment, in contrast to the external treatment that today is in universal use for reduction of pollutant emissions by employing a chemical processing plant, such as the catalytic converter, in the exhaust pipe. The technology of internal treatment is based on the advantages attained by executing the exothermic process at a minimum allowable temperature – a condition achievable by reduction of heat transfer to the walls in its course. As a consequence of this action the unavailable energy is minimized(the exergy is maximized), leading to significant reduction in the formation of chemically generated pollutants, NO_X and CO.

Specifically, the gains attainable by the Renault F7P engine if, instead of running it in a conventional, throttled, Flame Traversing the Charge (FTC) manner, it is operated in a stratified charge, wide open throttle (WOT), Fireball Mode of Combustion (FMC) are as follows (Oppenheim et al 1994).

Table 4.2. Operating conditions (Oppenheim et al 1994)

	θ_i°	θ_f°	Q	λ_o	λ_r	CASE
FTC	335	410	Q_H	1	1	0
`FMC	335	410	Q_H	4.44	1.05	1
			$Q_H/2$	8.00	1.33	2
			0	9.09	2.00	3
	335	392.5	$Q_H/2$	5.41	1.67	4
			$Q_H/4$	6.90	2.00	5
			0	9.52	2.50	6

The operating conditions examined for this purpose are listed in Table 4.2, where Q expresses the net energy loss incurred by heat transfer to the walls, while λ_o and λ_r are the air excess coefficients in the cylinder charge and in the chemically reacting mixture, respectively. Identified thus are six cases of FMC, besides the reference case of FTC, denoted by 0, for which $\lambda_o = 1$.. Cases 1, 2, 3, are concerned with the effects of diminished heat transfer loss achievable by reducing the contact of the reacting medium with the walls of the cylinder-piston enclosure, while the time interval within which fuel is consumed remains unchanged. Cases 4, 5, 6, take into account, moreover, the consequences of having the lifetime of the dynamic process reduced by a factor of two – an outcome of increased of combustion rate that is attainable by turbulent mixing induced by jet injection and jet ignition.

In all the cases, the indicated power was maintained carefully at the same level, as displayed by the indicator diagrams of the FTC and FMC modes of combustion portrayed in Fig. 4.14, for which the initial and final states of the dynamic process, θ_i and θ_f were adjusted for maximum IMEP, corresponding to MBT (Maximum Brake Torque) condition.

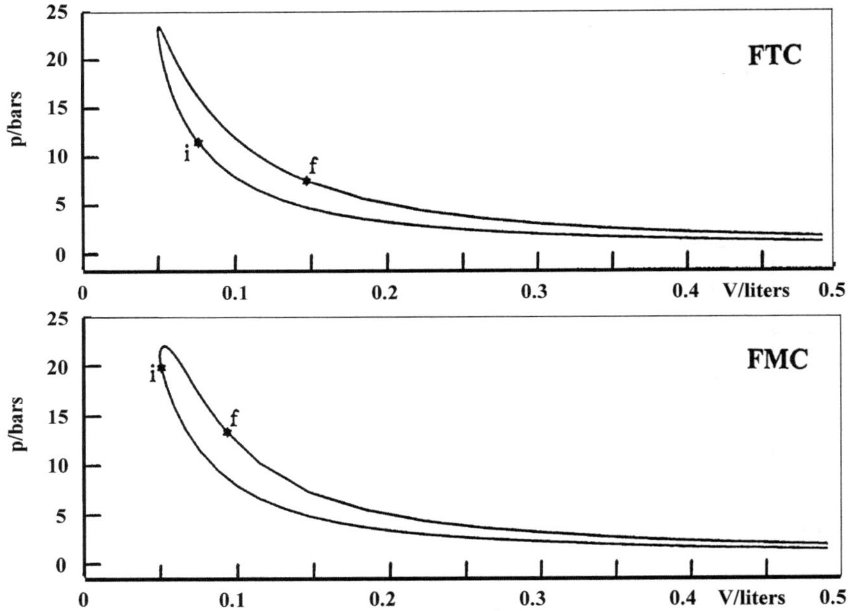

Fig. 4.14. Indicator diagrams for part load operation of the Renault F7P-700 engine at 2000 rpm, operated either by spark-ignited FTC (Flame Traversing the Charge) or in a jet generated FMC (Fireball Mode of Combustion). (Oppenheim et al 1994)

The thermal relaxation time was fixed at $\tau = 5\ \mu s$, appropriate, according to Fig. 4.8, for part-load operation of the Renault engine. The computations were based on the detailed kinetic reaction data for oxidation of propane provided by Westbrook and Pitz [1984].

For stratified charge of FMC, presented by cases 1-7, the operating conditions listed in Table 4.2 were established as follows.

With $Q_H \equiv Q_I / Q_I^o$, the superscript "o" referring to the reference cycle of Case 0, the postulated magnitude of $Q_H < 1$ was satisfied by decreasing the value of Q_P so that the effective mass fraction of products, y_E, is appropriately increased – an iterative procedure yielding a higher air equivalence ratio, λ_o, of the charge, specified in the fifth column of the table.

Chemical kinetic calculations reveal that the corresponding temperature of the reactants at state **i** is then lower than that of the separator for auto-ignition, whereas ignition by infusion of active radicals is ruled out as improbable. In order to reach it, the charge has to be stratified, so that the air-equivalence ratio in the reaction zone, of is sufficiently low to satisfy this criterion. Its low limit, λ_r, provided in column 6 of Table 4.2, has been evaluated for this purpose by an iterative procedure on λ for a fixed initial temperature of the charge at state **i**.

Upon the evaluation of the overall air excess coefficient, λ_o, to provide the required IMEP, the chemical kinetic calculations of combustion in the fireball were performed for a set of postulated air excess coefficients in the reactants, λ_r.

Established thus was a critical threshold, λ_r*. For smaller values of λ_r, the chemical induction time remained at a reasonably low level, while, for $\lambda_r > \lambda_r$*, computation of the oxidation mechanism at the same temperature ceased to proceed. Marked thus was a sharply defined critical limit of extinction, reached as a consequence of dilution by excess air at a fixed temperature, rather than, as is usually the case, by too low temperature at a fixed composition specified by λ_r. Critical values of the air excess coefficient, λ_r, are listed in Table 4.2, and it is for them that all the results of calculations for combustion in fireballs are presented here.

Profiles of the temperatures, as well as the concentrations of NO and CO for all the seven cases are presented by Figs. 4.15-4.21 (Oppenheim et al 1994).

Principal performance parameters, thus established, are displayed in Fig. 4.22 and listed in Table 4.3.

They present the ISFC[1], ISNO[2] and ISCO[3], obtained for all the cases specified in Table 4.2.

---

1 Indicated Specific Fuel Consumption
2 Indicated Specific NO
3 Indicated Specific CO

As apparent from them, appreciable gains can be obtained by executing the exothermic process of combustion in an engine cylinder in such a manner that the energy lost by heat transfer to the walls is significantly diminished. Primary means to attain these gains are provided by turbulent mixing produced by turbulent jets. The peak temperature can be thereby brought down to well below 2000K, at which, with the same power output, the consumption of fuel is reduced by 50%, the formation of NO is practically annihilated and the concentration of CO is decreased by an order of magnitude.

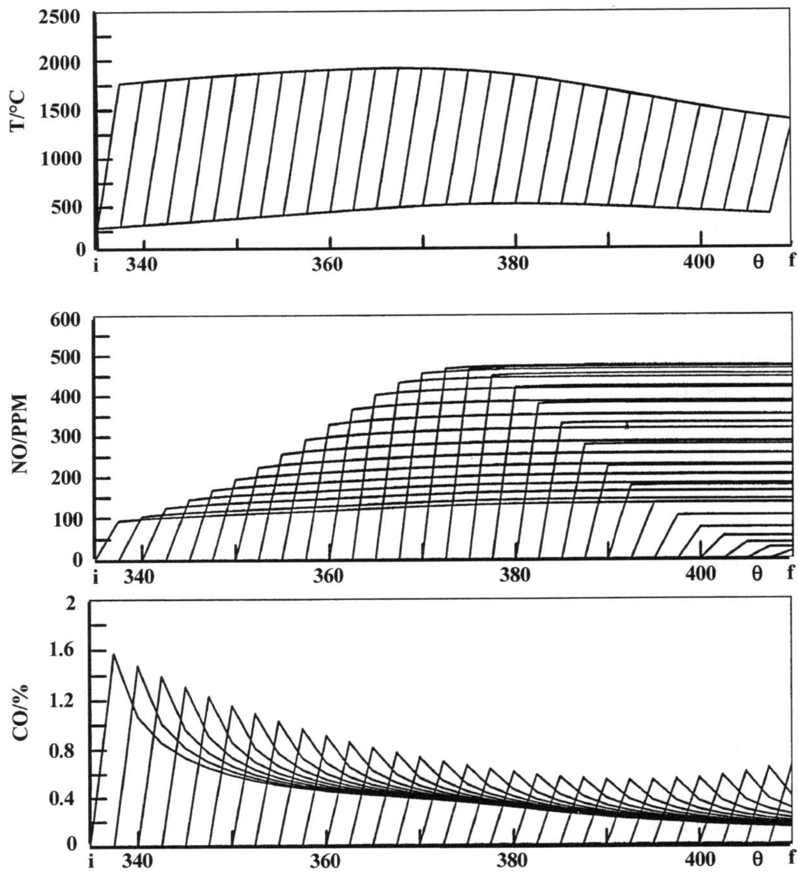

Fig. 4.15. Profiles of the temperature and mole fractions of NO and CO for FTC in case 0 (Oppenheim et al 1994)

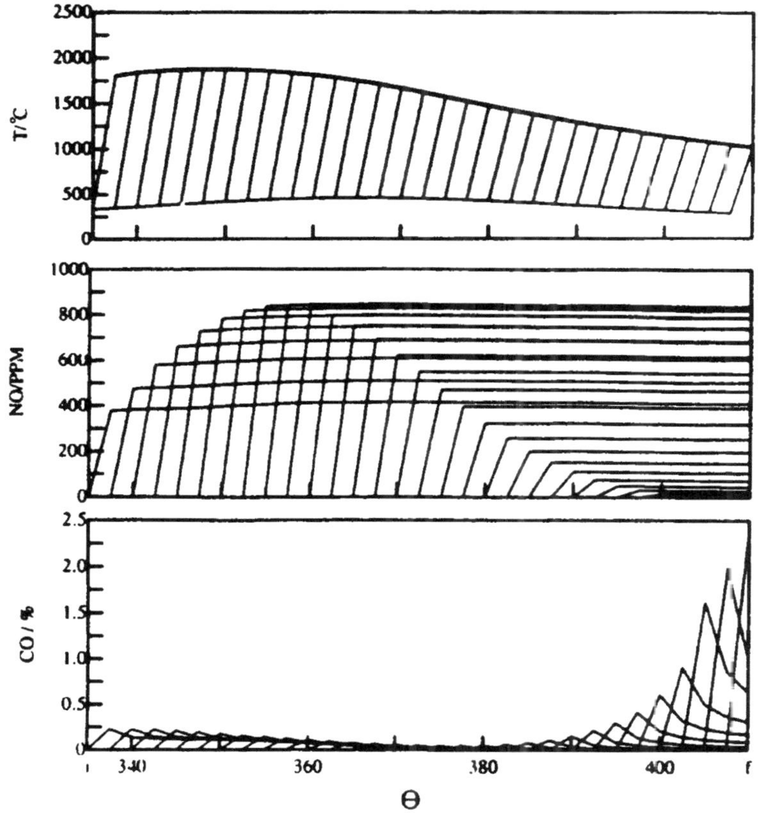

Fig. 4.16. Profiles of the temperature and mole fractions of NO and CO for FMC in case 1 (Oppenheim et al 1994)

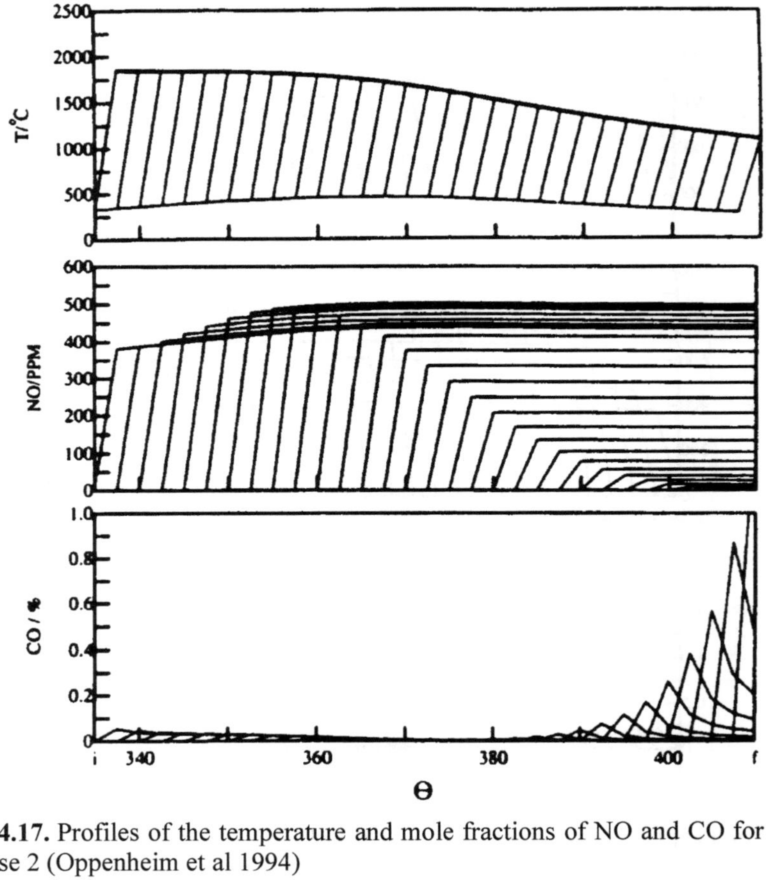

Fig. 4.17. Profiles of the temperature and mole fractions of NO and CO for FMC in case 2 (Oppenheim et al 1994)

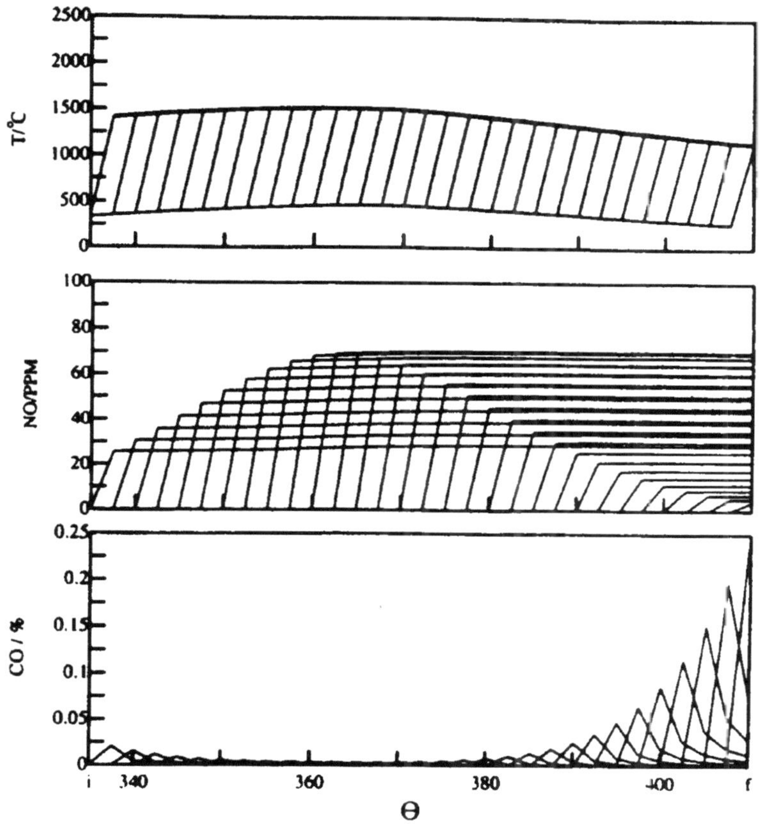

Fig. 4.18. Profiles of the temperature and mole fractions of NO and CO for FMC in case 3 (Oppenheim et al 1994)

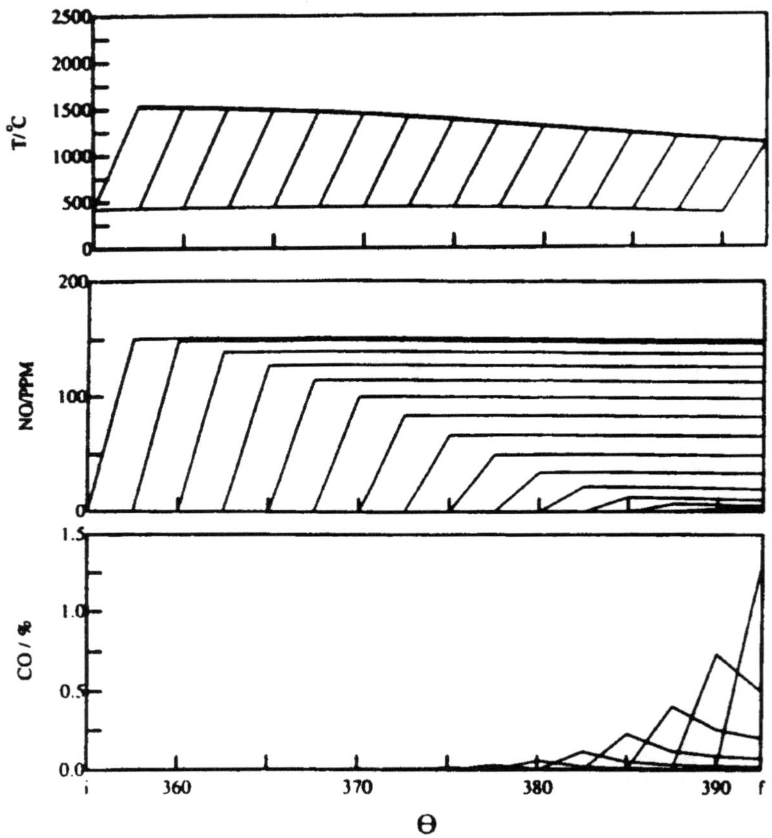

Fig. 4.19. Profiles of the temperature and mole fractions of NO and CO for FMC in case 4 (Oppenheim et al 1994)

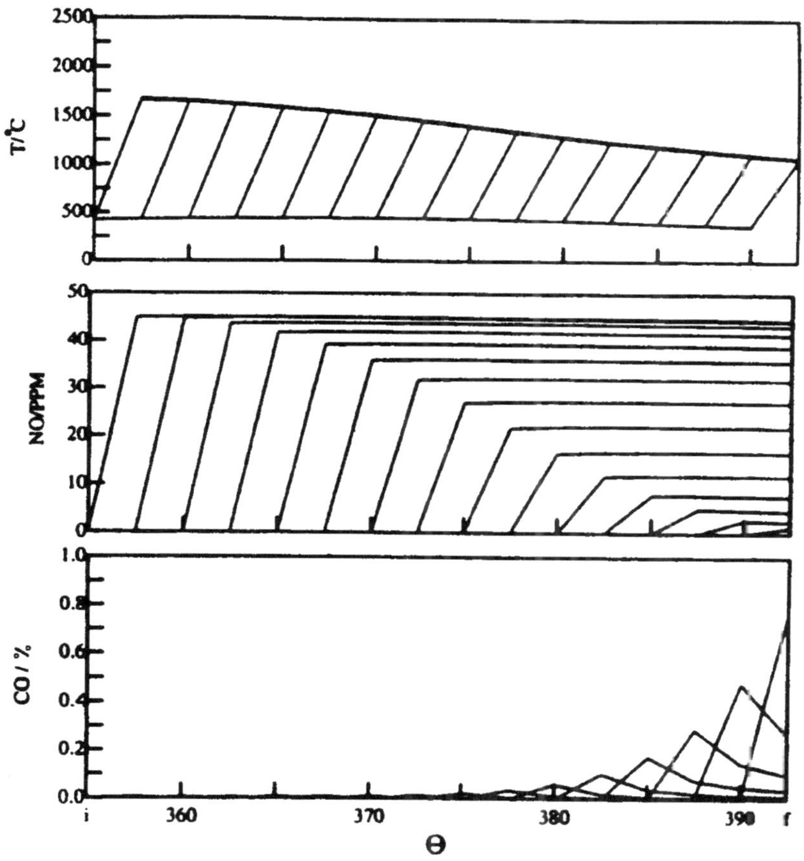

Fig. 4.20. Profiles of the temperature and mole fractions of NO and CO for FMC in case 5 (Oppenheim et al 1994)

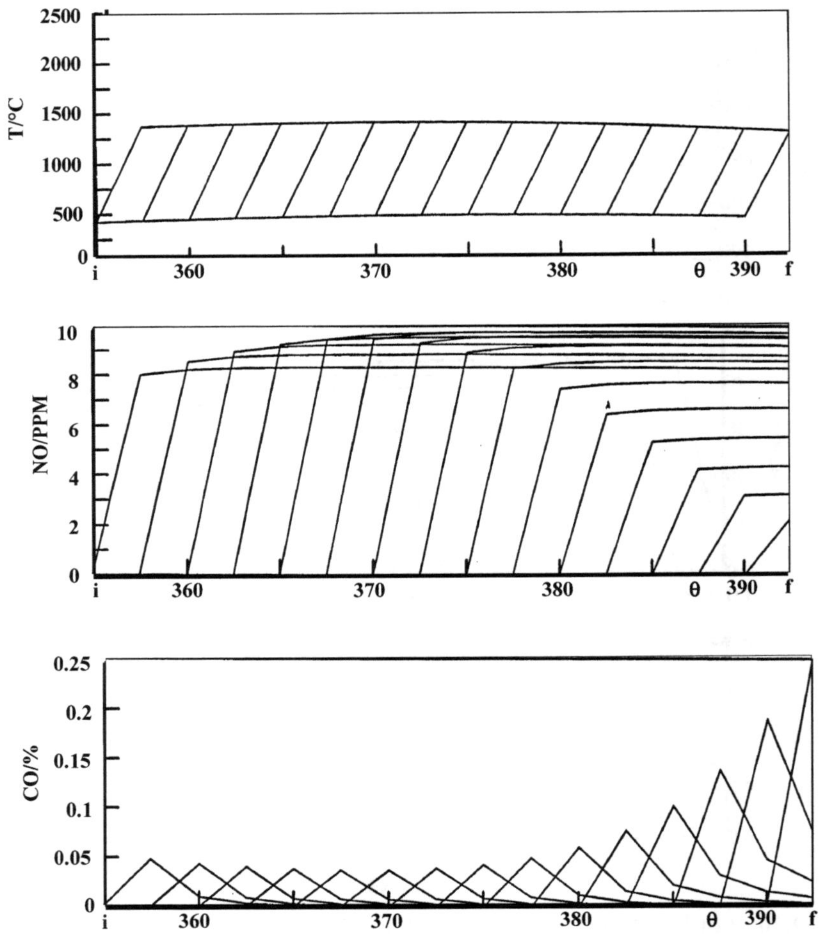

Fig. 4.21. Profiles of the temperature and mole fractions of NO and CO for FMC in case 6 (Oppenheim et al 1994)

Table 4.3. Performance parameters (Oppenheim et al 1994)

CASE	0	1	2	3	4	5	6
ISFC(g/kWh)	388	365	274	170	284	226	165
PPM	368	88	70	8	14	5	1
Mg/g_{fuel}	6.5	5.7	6.3	1.1	1.2	0.6	0.2
ISNO(g/kWh)	2.5	2.1	1.7	0.19	0.35	0.13	0.03
PPM	1600	150	51	10	134	51	19
Mg/g_{fuel}	26.5	9.1	2.5	1.3	11.1	5.4	2.8
ISNO(g/kWh)	5.9	3.3	0.67	0.22	3.2	1.2	0.46

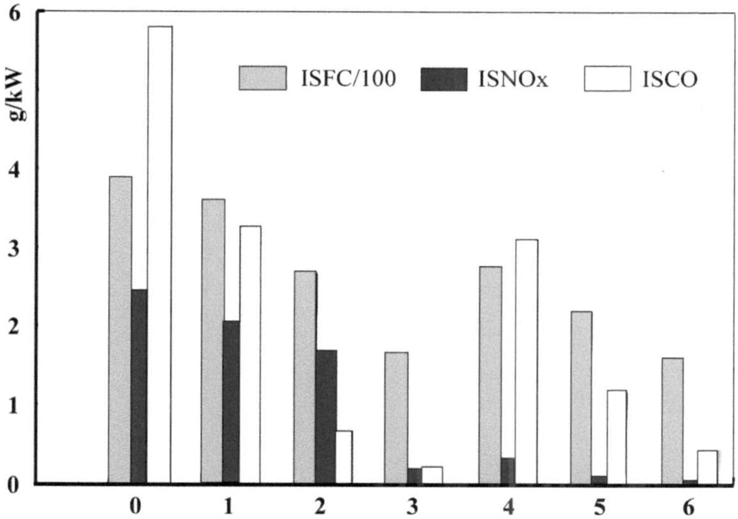

Fig. 4.22. Prognosis of engine performance parameters (Oppenheim et al 1994)

PART 2

FIELD

5 Aerodynamic Aspects

5.1 Introduction

Aerodynamics deals with a class of problems that can be resolved on the basis of solely the momentum and continuity equations. In its application to the dynamics of combustion, the field is contained in an enclosure, S, whose volume is prescribed (like an engine cylinder equipped by a piston activated by a crankshaft mechanism to convert its reciprocating motion into rotary) as a function of time, i.e. $V_S = V(t)$. The working substance is expressed in terms of a multi-fluid system, made out of fuel, F, and oxidizer, usually air, A, forming the reactants, R, that undergo the exothermic process in the course of which they are converted into products, P. In recognition of the fact that ignition - the initiation of the exothermic process of combustion - is an essential singularity of the system (since the specific volume of products at time t = 0, $v_P \equiv V/M = 0/0$ (!), its intrinsic mechanism is taken out of scope.

The reactants, R, are at frozen (fixed) composition of the system that consists of the substance initially in the enclosure and that exchanged, by intake and exhaust, with the surroundings, are distinguished from each other by impermeable interfaces.

The products, P, are at thermodynamic equilibrium and, as a consequence of turbulent mixing, their composition is determined by the condition of maximum entropy or minimum Gibbs function. In contrast to interfaces, the boundaries between products and reactants are permeable and are therefore referred to as fronts.

5.2 Transport

Under such circumstances, the behavior of the turbulent field is governed by the transport of momentum described by the Navier-Stokes equation of motion:

$$\frac{Du}{Dt} = -\nu\nabla p + \mathrm{Re}^{-1}[\nabla^2 u + \beta\nabla(\nabla\cdot u)] \tag{5.1}$$

where $D/Dt \equiv \partial/\partial t + u\cdot\nabla$, u is the velocity vector, ν - the specific volume, p - pressure, Re - the Reynolds number, while $\beta \equiv \kappa/\mu + 3^{-1}$, κ denoting the bulk viscosity and μ the shear viscosity.

By expressing it, respectively, in terms of its rotational (or curl) and divergent components, it provides expressions for the transport of vorticity, $\omega \equiv \nabla\times u$:

$$\frac{D\omega}{Dt} = -\nabla\nu\times\nabla p + \mathrm{Re}^{-1}\nabla^2\omega + (\omega\cdot\nabla)u - \omega(\nabla\cdot u) \tag{5.2}$$

and for the transport of dilatation, $\Delta \equiv \nabla\cdot u$:

$$\frac{\mathrm{D}\Delta}{\mathrm{Dt}} = -\nabla\nu\cdot\nabla p - \nu\nabla^2 p + \omega\cdot\omega + \frac{1+\beta}{\mathrm{Re}}\nabla^2\Delta \tag{5.3}$$

The problem is fully formulated in terms of (6.1), combined with the global equation of continuity and appropriate initial and boundary conditions. Equations (5.2) and 5.3) provide information on the two fundamental components of the flow field: vorticity and dilatation.

5.3 Velocity Field

The mechanism of momentum transport, thus expressed, is revealed by a synthetic method of approach, based on the Helmholz (1858) velocity decomposition theorem (known also as the Hodge decomposition), that plays a principal role in modern computational fluid mechanics. According to it, the velocity vector is decomposed into a divergence-free, rotational and curl-free, irrotational components, i.e.

$$u = u_\omega + u_\Delta \tag{5.4}$$

where $\nabla\cdot u_\omega \equiv 0$, while $\nabla\times u_\Delta \equiv 0$.

The rotational component is expressed in terms of the vector potential, B, so that

$$u_\omega \equiv \nabla \times B \tag{5.5}$$

whence

$$\nabla \times u_\omega = \nabla \times (\nabla \times B) \equiv \nabla(\nabla \cdot B) - \nabla^2 B = \omega(x') \tag{5.6}$$

The irrotational component is expressed in terms of the scalar potential, Φ, according to which

$$u_\Delta \equiv \nabla \Phi \tag{5.7}$$

whence, according to the global continuity equation,

$$\nabla \cdot u_\Delta = \frac{\partial}{\partial t}\ln v + (u \cdot \nabla)\ln v = \nabla^2 \Phi = \Delta(x') \tag{5.8}$$

so that the dilatation is expressed in terms of an eikonal equation for lnv.

The fluid dynamic consequences of the Helmholz decomposition are succinctly presented by Batchelor (1967) and what follows is, in effect, an implementation of his exposition.

5.4 Rotational Component

In the absence of sources for the vector potential, **B**, its gradient $\nabla \cdot \mathbf{B} = 0$, refining, under such circumstances, its definition without any loss in generality. It follows therefore from (5.6) that

$$\nabla^2 B(x) = -\omega(x') \tag{5.9}$$

- a Poisson equation for the vector potential, *B*. Its solution is expressed in terms of the convolution integral

$$B(x) = -G * \omega(x') \equiv \frac{1}{4\pi}\int \frac{\omega(x')}{|x - x'|}\mathrm{d}V(x') \tag{5.10}$$

where $G = -\frac{1}{4\pi}\frac{1}{|x - x'|}$ is the Green's function.

Then, in view of the definition of **B,** invoked by (5.5),

$$u_{\omega}(x) = \frac{1}{4\pi}\int \nabla \times \frac{\omega(x')}{|x-x'|} \mathrm{d}V(x')$$
$$= -\frac{1}{4\pi}\int \frac{(x-x')\times\omega(x')}{|x-x'|^3} \mathrm{d}V(x') \tag{5.11}$$

which represents the Biot-Savart law for vorticity poles.

Expressed thus is the formalism of vortex dynamics and, in particular, of the random vortex method of Chorin (1973, 1978, 1989, 1994), which was applied to turbulent combustion by Ghoniem et al (1982), as recounted by Oppenheim (1985).

5.5 Irrotational Component

According to (5.8), the Poisson equation for the scalar potential is straightforwardly

$$\nabla^2 \Phi(x) = \Delta(x') \tag{5.12}$$

Then, as before, the scalar potential is expressed in terms of the convolution integral, based on the same Green's function, so that

$$\Phi(x) = G * \Delta(x') \equiv -\frac{1}{4\pi}\int \frac{\Delta(x')}{|x-x'|} \mathrm{d}V(x') \tag{5.13}$$

whence, by the definition of Φ, invoked in (5.7),

$$u_{\Delta}(x) = -\frac{1}{4\pi}\int \nabla \frac{\Delta(x')}{|x-x'|} \mathrm{d}V(x')$$
$$= \frac{1}{4\pi}\int \frac{(x-x')\Delta(x')}{|x-x'|^3} \mathrm{d}V(x') \tag{5.14}$$

which represents the Biot-Savart law for dilatation poles.

Similarly to the rotational component, the above expresses the formalism of the fluid dynamic effects of dilatation due to exothermic centers.

5.6 Multi-Fluid Systems

The field where combustion takes place is a multi-fluid system whose thermodynamic properties are exposed in Chapter 1. As displayed in Fig. 1.1, the system is made out initially of the charge C, of which part, the reactants R, is transformed by tye exothermic process of combustion into the products P. The predominant fluid dynamic feature of the system is turbulent mixing. A typical lifetime of the chemical transformation is orders of magnitude shorter than the characteristic time of the fluid dynamic events – a feature that lends itself to clear exposition of the field by treating it at the limit of the Damköhler number Da = ∞.

Under such circumstances, the identity of each components of the multi-fluid system is maintained by having its bounds delineated by interfaces that, in the absence of combustion, are impermeable contact surfaces and, in its course, are semi-permeable combustion fronts made out of exothermic centers due to one-directional transformation of R into P.

The impermeable interfaces between the components of a multi-fluid system cannot exert any influence upon the flow-field. In the Eulerian grid-based computations, their motion is traced by Level Set Methods (Sethian 1996). In the Lagrangian grid-less computations, like the vortex method where the identity of each elementary particle, like the vortex blob, can be tagged, the interfaces are delineated by distinction between the tags.

The permeable combustion front acting, in effect, as exothermic centers, affect the flow field by the Biot-Savart effect of dilatation. For the case of Da = ∞, a finite change in specific volume, v, takes place across then at constant pressure, invoking a change in dynamic potential, w, and internal energy, e.

According to (5.8), the dilatation at an exothermic center at ξ_k,

$$\Delta_k(\xi_k) \equiv (\nabla \cdot u_\Delta)_k = \frac{\partial}{\partial t} \ln v_k \tag{5.15}$$

where $\xi_k \equiv x_R - u_R t$, while, in terms of $\alpha \equiv v_P/v_R$, whose value is prescribed by (5.9),

$$v_k(\mathbf{x}') = v_R [1 + (\alpha - 1) \, f(t)] \tag{5.16}$$

Hence

$$\Delta_k(\boldsymbol{\xi}) = \frac{v_R}{v_k}(\alpha - 1)\dot{f}(t) \tag{5.17}$$

At the limit of Da = ∞, the transformation of ν_R into ν_P takes place in a step. Consequently, f(t) is expressed by a Heavyside function and its derivative is a Dirac δ. Since, at at ξ_k, $\nu_k = \nu_R$, (6.19) becomes

$$\Delta_k(\xi) = (\alpha - 1)\delta(t - t_i) \tag{5.18}$$

Upon their formation, the exothermic centers form exothermic fronts, whose location and direction are determined by the intersection between two velocity vectors, one of the reactant field, $\mathbf{u}_R$, and the other of the product field, $\mathbf{u}_P$ The hodograph of these velocities defines a plane depicted by Fig. 5.1.

As displayed there, the vector difference between them, $U_R - U_P$, identifies the direction of the vector $\mathbf{n}_F$, normal to the front. The paths of particles and the front in time-space coordinate system are presented by Fig. 5.2.

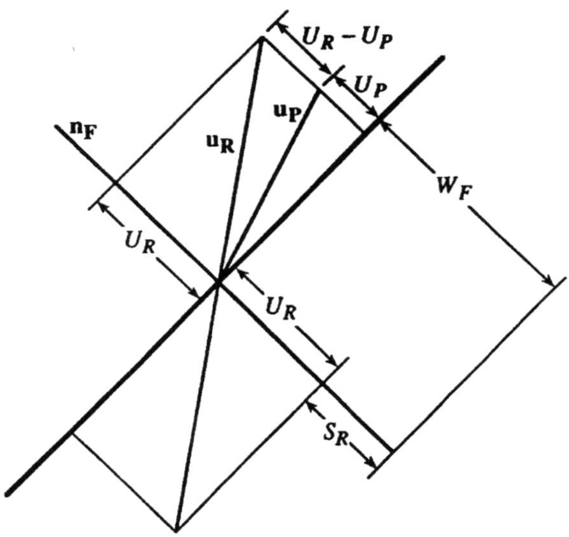

Fig. 5.1. Velocity hodograph at an exothermic center

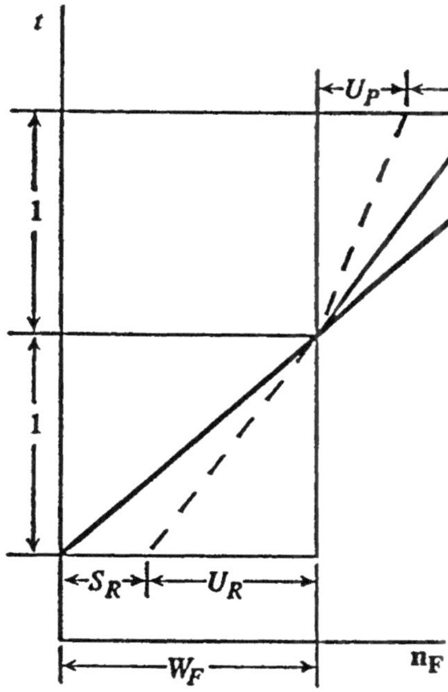

Fig. 5.2. Time-space diagram of particle and front paths

The exothermic front moves at a velocity, $\mathbf{w}_F$, whose component, normal to the front, is

$$W_F = S_P + U_P = S_R + U_R \tag{5.19}$$

where $W_F \equiv w_F \cdot n_F$, $S_K \equiv s_K \cdot n_F$ and $U_K \equiv u_K \cdot n_F$ (K = P,R), while $\mathbf{n}_F$ is a unit vector normal to the front.

To conserve mass across the front

$$\frac{S_P}{S_R} = \frac{\nu_P}{\nu_R} \equiv \alpha \tag{5.20}$$

whence (5.19) yields

$$W_F = \frac{\alpha U_R - U_P}{\alpha - 1} \tag{5.21}$$

Then, as illustrated by Fig. 5.2, the transition within it can be expressed in terms of a front co-ordinate, $\xi \equiv x \cdot n_{\mathrm{F}} - W_{\mathrm{F}} t$, so that, in terms of an S curve of a 'diffusion wave front' delineated by **f** (ξ),

$$\frac{U(\xi)}{U_{\mathrm{R}}} = \frac{\nu(\xi)}{\nu_{\mathrm{R}}} = 1 + (\alpha - 1) f(\xi) \tag{5.22}$$

whereas

$$\frac{\partial}{\partial t} = -W_{\mathrm{F}} \frac{d}{d\xi} \text{ while } (u_{\mathrm{F}} \cdot \nabla) = U_{\mathrm{F}} \frac{d}{d\xi} \tag{5.23}$$

According to (5.8),

$$\Delta_{\mathrm{F}} = \Delta(x_{\mathrm{F}}) = \frac{\partial}{\partial t} \ln \nu_{\mathrm{F}} + (u_{\mathrm{F}} \cdot \nabla) \ln \nu_{\mathrm{F}} \tag{5.24}$$

on the basis of which, by virtue of (5.25), and noting that, in view of (5.19), $U(\xi) + S(\xi) = W_F$,

$$\begin{aligned} \Delta x_{\mathrm{F}} &= \frac{-W_{\mathrm{F}} + U_{\mathrm{F}}(\xi)}{S_F(\xi)} \frac{\mathrm{d}}{\mathrm{d}\xi} S_{\mathrm{F}}(\xi) \\ &= -S_{\mathrm{R}} (\alpha - 1) \dot{f}(\xi) \end{aligned} \tag{5.25}$$

For the sake of sharpness, the transition across the front is considered here again to take place in a step. Thus, as before, $f(\xi)$ is expressed by a Heavyside function, its derivative in (5.25) becoming a Dirac δ, whence, in view of (5.20) and (5.21),

$$\Delta_{\mathrm{F}} = (U_{\mathrm{R}} - U_{\mathrm{P}})\, \delta(\xi - \xi_{\mathrm{F}}) = (\alpha - 1) S_R \delta(\xi - \xi_{\mathrm{F}}) \tag{5.26}$$

The propagation velocity of the exothermic front, as well as the strength of its dilatation, are thus functions of the turbulent velocity vector field, rather than being prescribed by the structure of the flame, as it is in a reaction-diffusion system of a principally laminar flame.

5.7 Front

The front is propagated by virtue of three mechanisms:

(1) advection, due to the velocity vector field in which it resides

(2) propagation, driven by dilatation due to its exothermic nature

(3) baroclinicity, providing an additional source of vorticity

The motion of the front, generated by the first two factors, is determined by implementing a level set method (Sethian, 1996). Its position is, according to it, given by the zero level set specified by

$$\partial\phi/\partial t + F|\nabla\phi| = 0 \tag{5.27}$$

symbol F denoting the rate at which ϕ is advanced, referred to as the speed function. If it depends only on the position, i.e. F(**x**) only, (5.27) is the well-known eikonal equation. The dependent variable

$$\phi \equiv V_P / V_c \tag{5.28}$$

where the subscript P denotes the volume of the products, while c is that of the cell. If the value of ϕ is known, (5.27) provides an expression for the evaluation of F(**x**). This is, indeed, the case in an unmixed system, where, according to (5.18),

$$\frac{1}{\phi}\frac{d\phi}{dt} = \frac{1}{V_P}\frac{dV_P}{dt} = \Delta_k(\xi) = (\alpha - 1)\delta(t - t_i) \tag{5.29}$$

whence, in a finite difference form, noting that the computational time step $k = \Delta t\, \delta(t - t_i)$,

$$\Delta\phi = (\alpha - 1)\phi k \tag{5.30}$$

It should be noted that a useful technique for tracking the motion of an exothermic front was was developed by Chorin (I980) and implemented by Ghoniem et all (1982), using the algorithm of Noh & Woodard (1976), known as SLIC (Simple Line Interface Calculation. Today, this technique should be considered as a simplified version of the level set method. It was used, as a matter of fact, in all the examples presented in the next chapter

The third factor arises as a consequence of the turbulent nature of the field, where the exothermic front is intrinsically curved. It generates thus vorticity due to the baroclinic effect expressed by the first term in (6.2). Its principal action is evident on the plane of the front depicted in Fig. 6.2, while its vector, normal to this plane, is located in the field of the products. Pragmatically, the jump in vorticity across the front, due to the baroclinic effect, can be described (vid. Rotman et al 1988) in terms of a finite difference form as follows

$$\Delta\omega = -\frac{\Delta\mathrm{v}}{\Delta n}\frac{\Delta\mathrm{p}}{\Delta s}\Delta t \tag{5.31}$$

for which

$$\frac{\Delta\mathrm{v}}{\Delta n} = \frac{\mathrm{v_P} - \mathrm{v}_R}{S_\mathrm{P}\Delta t} \tag{5.32}$$

while

$$\frac{\Delta\mathrm{p}}{\Delta s} = -\frac{U_\mathrm{S}}{\mathrm{v_R}}\frac{\Delta U_\mathrm{S}}{\Delta s} \tag{5.33}$$

Thus, since the stretch factor

$$\mathrm{K} \equiv \frac{1}{\mathrm{A}}\frac{\Delta\mathrm{A}}{\Delta t} = \frac{\Delta U_\mathrm{S}}{\Delta s} \tag{5.34}$$

then, taking into account (5.17) and (5.24), the vorticity increment due to the baroclinic effect

$$\Delta\omega = \frac{\alpha - 1}{\alpha}\frac{U_\mathrm{S}}{S_\mathrm{R}}\mathrm{K} \tag{5.35}$$

6 Random Vortex Method

6.1 Background

A direct way to implement the fluid mechanical equations for turbulent flow fields, with the dynamic features of combustion taken into account, is by a grid-free algorithm of the Random Vortex Method (RVM), introduced by Chorin (1973, 1978). For this purpose, instead of integrating the Navier-Stokes equation directly, its differential, the vortex transport equation is considered constitutive with the stipulation that its dependent variable behaves as a set of discrete Lagrangian particles in random motion, putting it into the class of Fokker-Planck equations describing the time evolution of probability density functions for stochastic processes. The consequences of diffusion are then modeled by expressing the Laplacian in the vortex transport equation by random walks of discrete vortex elements, referred to as vortex blobs and sheets.

In general, vortex methods are applicable to three-dimensional flows, as demonstrated by Ghoniem and Gharakhani (1997). Here, the implementation of RVM is presented just for two-dimensional planar flow fields.

6.2 Formulation

Presented here is RVM addressed to the solution of problems with following idealizations.

1. the flow field is planar
2. the working substance consists of only two components: the reactants and the products
3. the combustion front acts as a line of demarcation between the two components that behaves as a constant pressure deflagration, propagating at a prescribed normal burning velocity

4. the exothermicity of combustion is manifested by an increase in specific volume associated with the transformation of reactants into products across the front.

The problem thus prescribed is consistent with the well known model of thin flames used universally in the theory of mixing-controlled turbulent combustion.

As a consequence of the two-dimensional formulation, the effects of vortex stretching are neglected, whereas the action of the most significant effect of turbulence – the vorticity – is treated emphatically.

The Navier-Stokes and continuity equations are then, respectively, reduced to

$$\mathrm{D}\boldsymbol{u}/\mathrm{D}t = \mathrm{R}^{-1}\nabla^2\boldsymbol{u} - \nabla p \tag{6.1}$$

and

$$\nabla \cdot \boldsymbol{u} = \varepsilon(\boldsymbol{r}_\mathrm{F}) \tag{6.2}$$

where $\boldsymbol{u} = (u,v)$ is the normalized velocity vector, R is the Reynolds number at the inlet to the system, p is the normalized pressure, ε is the rate of dilatation at $\boldsymbol{r}_\mathrm{F}$, subscript F denoting the front, while $\boldsymbol{\Delta}$ expresses the gradient operator, Δ^2 represents the Laplacian, and $\mathrm{D}\boldsymbol{u}/\mathrm{D}t = \partial/\partial t + \boldsymbol{u}\cdot\nabla$ is the substantial derivative.

The flow field is specified by the solution of these equations, subject to the boundary conditions

(1) at inlet: $$\boldsymbol{u} = (1, 0) \tag{6.3}$$

(2) along the walls: $$\boldsymbol{u} = 0 \tag{6.4}$$

The position of the front, specifying the location of the rate of dilatation, ε ($\boldsymbol{r}_\mathrm{f}$), is established by the eikonal flame propagation equation

$$\frac{\partial \boldsymbol{r}_\mathrm{F}}{\partial t} = \boldsymbol{u} + \mathrm{S}_\mathrm{u}\boldsymbol{n}_\mathrm{t} \tag{6.5}$$

where S_u is the velocity of advection, referred in the literature as the normal burning speed, and $\boldsymbol{n}_\mathrm{t}$ is the unit vector normal to the front.

Here, the elementary component of the flow field is the vorticity

$$\boldsymbol{\omega} \equiv \nabla \times \boldsymbol{u} \tag{6.6}$$

whose propagation is specified by the vortex transport equation, the curl of (6.2),

$$\mathrm{D}\boldsymbol{\omega}/\mathrm{D}t = \mathrm{R}^{-1}\nabla^2\boldsymbol{\omega} \tag{6.7}$$

The velocity field, $\boldsymbol{u}(r)$, is determined by (6.1) and (6.6), while (6.7) is used to update the vorticity field, $\boldsymbol{\omega}\,(x, y)$. The rate of dilatation, $\varepsilon(x, y)$ is established by the front propagation algorithm for the solution of (6.5).

The velocity vector, $\boldsymbol{u}$, is decomposed, according to (5.4), into a divergence-free component, $\boldsymbol{u}_\omega$, and a curl-free component, $\boldsymbol{u}_\Delta$, where

$$\nabla \cdot \boldsymbol{u}_\omega \equiv 0 \tag{6.8}$$

while

$$\nabla \times \boldsymbol{u}_\Delta \equiv 0 \tag{6.9}$$

Both the velocity components have to satisfy the zero normal velocity boundary conditions,

$$\boldsymbol{u}_\omega \cdot \boldsymbol{n} = 0 \quad \text{and} \quad \boldsymbol{u}_\Delta \cdot \boldsymbol{n} = 0 \tag{6.10}$$

where $\boldsymbol{n}$ is the unit vector normal to the walls, whereas only the total velocity, $\boldsymbol{u}$, is required to satisfy the no-slip condition

$$\boldsymbol{u} \cdot \boldsymbol{s} = 0 \tag{6.11}$$

where $\boldsymbol{s}$ is the unit vector tangential to the walls.

6.3 Vortex dynamics

6.3.1 Vortex blobs

A vortex blob is a discrete elementary vorticity, $\boldsymbol{\omega}_j$, that acts within an elementary volume, ΔV_j, located at $\boldsymbol{r}_j$. Its magnitude is expressed in terms of the Dirac delta function, i.e.

$$\boldsymbol{\omega}_j = \Gamma_j \, \delta(\mathbf{r} - \mathbf{r}_j) \tag{6.12}$$

where

$$\boldsymbol{\Gamma}_j = \lim_{\Delta V_j \to 0} \int_{\Delta V_j} \boldsymbol{\omega}_j \mathrm{dV} \tag{6.13}$$

is its local circulation.

For a planar field, the relevant component of the vector potential, **B**, is the stream function, $\boldsymbol{\psi}$, so that (5.9) is reduced to

$$\nabla^2 \boldsymbol{\psi} = -\omega \tag{6.14}$$

As brought up by (5.10), its solution, for a given distribution of discrete vortex blobs, $\boldsymbol{\omega}(\mathbf{r}_j)$, is

$$\boldsymbol{\psi}(\mathbf{x}) = \int_{\mathbf{A}} \mathrm{G}(\mathbf{r} - \mathbf{r}_j)\, \omega(\mathbf{r}_j)\, \mathrm{d}A \tag{6.15}$$

where

$$\mathrm{G}(\mathbf{r}, \mathbf{r}_j) = (\boldsymbol{\Gamma}_j / 2\pi) \ln |\, \mathbf{r} - \mathbf{r}_j \,| \tag{6.16}$$

is the Green's function, while A is the area of the blob.

The elementary vorticity, $\boldsymbol{\omega}(\boldsymbol{r}_j)$, is a function of small support that tends to a Dirac delta as the area where it exists, ΔAj, approaches zero. This process requires smoothing of the Green's function to eliminate the singularity at its center (Chorin 1973). As in a finite difference scheme, the integration of (6.4) is executed by summation, so that

$$\psi(\mathbf{x}) = \sum_j G_j \boldsymbol{\Gamma}_j \tag{6.17}$$

for which

$$\boldsymbol{\Gamma}_j = \int_{\Delta A_j} \omega_j dA \tag{6.18}$$

where ΔA_j is finite, while G_j is a smoothed Green function at $\mathbf{r}_j$, (Chorin 1973; Hald 1979).

The velocity field it produces is determined by the stream function, whence, by definition,

$$u \equiv \partial\Psi / \partial y \qquad v \equiv -\partial\Psi / \partial x \tag{6.19}$$

so that, in terms of complex variables,

$$w_{\Psi}(z, z_j) = \frac{-i\Gamma_j |z-z_j|}{2\pi \max(|z-z_j|, r_o)} \frac{1}{z-z_j} \tag{6.20}$$

where $w = u - iv$ ($i = \sqrt{-1}$), $z = x + iy$, while r_o is the cut-off radius, i.e. the radius of the core within which $|u|$ is constant. The velocity distribution of such a blob is displayed in Fig. 6.1.

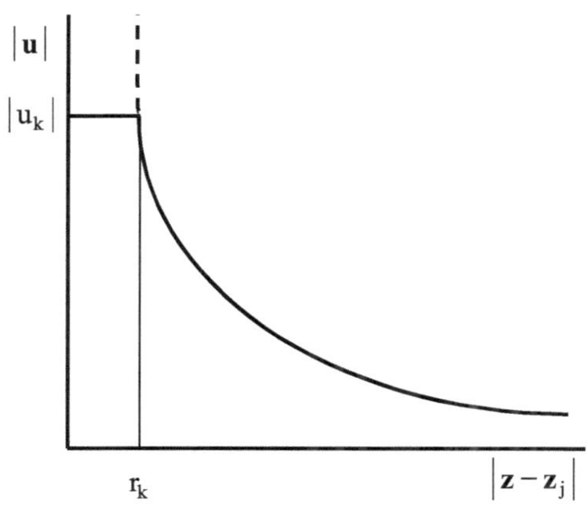

Fig. 6.1. Velocity distribution induced by a vortex blob

Velocity vector fields satisfying zero normal boundary conditions, $\mathbf{u}_{\omega} \cdot \mathbf{n} = 0$ and $\mathbf{u}_{\Delta} \cdot \mathbf{n} = 0$, can be determined by the use of any Poisson solver. Employed here is the classical method of conformal mapping, whereby the vector field is transformed into the upper-half of the ζ-plane.

The vector field produced in this plane by a vortex blob at ζ_j is prescribed by

$$w_{\omega}(\zeta, \zeta_j) = w_{\Psi}(\zeta, \zeta_j) - w_{\Psi}(\zeta, \zeta_j^*) \tag{6.21}$$

where $w(\zeta, \zeta_j)$ is given by (6.6), while the asterisk denotes a complex conjugate. The total velocity produced by a set of J_b vortex blobs

$$w_{\omega}(\zeta) = w_p(\zeta) + \sum_{j=1}^{J_b} w_{\omega}(\zeta, \zeta_j) \tag{6.22}$$

The solution in the physical domain, the z-plane, can be, thereupon, obtained by virtue of the Schwarz-Christoffel theorem, according to which, for a given geometry of the field, the differential of the transform function is

$$F(\zeta) \equiv d\zeta / dz \tag{6.23}$$

whence

$$w(z) = w(\zeta) F(\zeta) \tag{6.24}$$

expressing the velocity vector, u_{ω}, of (5.4) in terms of its complex variable.

The vorticity field, $\boldsymbol{\omega}(\mathbf{x})$, is updated at every computational time step, k, by a solution of the vortex transport (5.2) in fractional steps, made out of the contribution of the convection operator

$$D\boldsymbol{\omega} / Dt = 0 \tag{6.25}$$

and that of the diffusion operator

$$\partial \boldsymbol{\omega} / \partial t = \mathrm{R}^{-1} \nabla^2 \boldsymbol{\omega} \tag{6.26}$$

The key to the random vortex method of Chorin is the observation that the solution corresponding to a time step, k, of a one-dimensional component of the above diffusion equation, when the initial condition is provided by the Dirac delta function, δ(0), is the Green function

$$G(x, \mathrm{k}) = (4\pi \mathrm{k} / \mathrm{R})^{-1/2} \exp(-\mathrm{R}x^2 / 4\mathrm{k}) \tag{6.27}$$

This is the probability density function of a Gaussian random variable with zero mean and standard deviation of $\sigma = (2\mathrm{k/R})^{1/2}$!

Thus, if the initial vorticity field is distributed over a set of discrete vortex elements and each is given a displacement from the origin by an amount drawn from a set of Gaussian random numbers of an appropriate variance, it provides a sample of the distribution specified by (6.27). According to its exact solution for a given distribution of vorticity, $\omega(\mathbf{x})$, after a time interval k, the circulation per unit length

$$\gamma(\mathbf{x}) = \int_A \mathrm{G}(\mathbf{x} - \mathbf{x}_\mathrm{j}, \mathrm{k}) \omega(\mathbf{x}_\mathrm{j}) d\mathbf{x}_\mathrm{j} \tag{6.28}$$

for which G is prescribed by(6.27). The probabilistic counterpart of this solution is obtained by displacing each vortex element from its position **x**' through a distance η_j. The random walk is then constructed by repeating this procedure at each time step. Two-dimensional random walk is treated in essentially the same way, the vortex elements being moved in two mutually perpendicular directions x and y, by two independent Gaussian random variables, η_j, with zero mean and standard deviation of $\sigma = (2\mathrm{k/R})^{1/2}$.

The convection and diffusion contributions in the **z**-plane are combined, according to equation (1.2), by summation

$$z_j(t+k) = z_j(t) + w^*(z_j)k + \eta_j \tag{6.29}$$

where $w = w_\omega + w_\Delta$ and :$\eta_j = \eta_x + i\eta_y$, or, in the ζ-plane, in terms of its transform

$$\zeta_j\,(t+k) = \zeta_j\,(t) + w^*(\zeta_j)\,F^*(\zeta_j)\,F(\zeta_j)\,k + (\eta_j)\,F(\zeta_j) \tag{6.30}$$

Since the velocity is calculated in the ζ-plane by means of (6.22), it is more direct and, hence, simpler to use for this purpose (6.30), rather than (6.29).

To satisfy the no-slip boundary condition expressed by (6.11), the velocity, **w,** has to be calculated at points along the wall. The points are selected to be a distance h apart along each wall. Wherever the tangential component of velocity ***u*** at wall is not zero, a vortex with a circulation hu_w is created and included in the computations at the next time step, according to equation (6.30) or equation (6.29). However, this procedure of vorticity creation is not accurate since on the average one-half of the newly created blobs are lost through diffusion across the wall. This implies that Kelvin's theorem is not satisfied exactly and the accuracy near the wall is poor. Furthermore, vortex blobs do not provide a good description of the flow near solid walls where velocity gradients are very high, because inside the core of a blob the velocity is considered to be constant. This motivates the introduction of vortex sheets to take up the role of blobs in shear layers at the walls.

6.3.2 Vortex sheets

To satisfy the no-slip boundary condition, $\boldsymbol{u} \cdot \boldsymbol{s} = 0$, the velocity, **w**, has to be calculated at points along the wall. For this purpose Chorin (1978) introduced the concept of a numerical shear layer – a relatively thin slice of the field where the role of blobs is taken over by vortex sheets: vortex elements satisfying two conditions

(i) $\partial v / \partial x << \partial u / \partial y$

(ii) diffusion in the x-direction is negligible in comparison to convection.

For sheets, the expression for vorticity, (1.6), is reduced to

$$\boldsymbol{\omega}_\delta = -\partial u / \partial y \tag{6.31}$$

The velocity vector, $\boldsymbol{u}_{\omega}$ (***r***), within the sheer layer is calculated then as follows.

A definite integral of (6.31) from the outer edge of the shear layer, δ_S, to an internal level, y_i, is

$$u_\delta(x_i) - u(x_i, y_i) = -\int_{y_i}^{\delta_s} \omega dy \tag{6.32}$$

where $u_\delta = u$ at $y = \delta$, is transformed into a summation by partitioning the value of ω along discrete intervals Δy, so that the circulation per unit length of a vortex sheet is expressed as

$$\gamma_j = \lim_{\nabla y \to 0} \int_{y_i}^{y_i + \nabla y} \omega dy \tag{6.33}$$

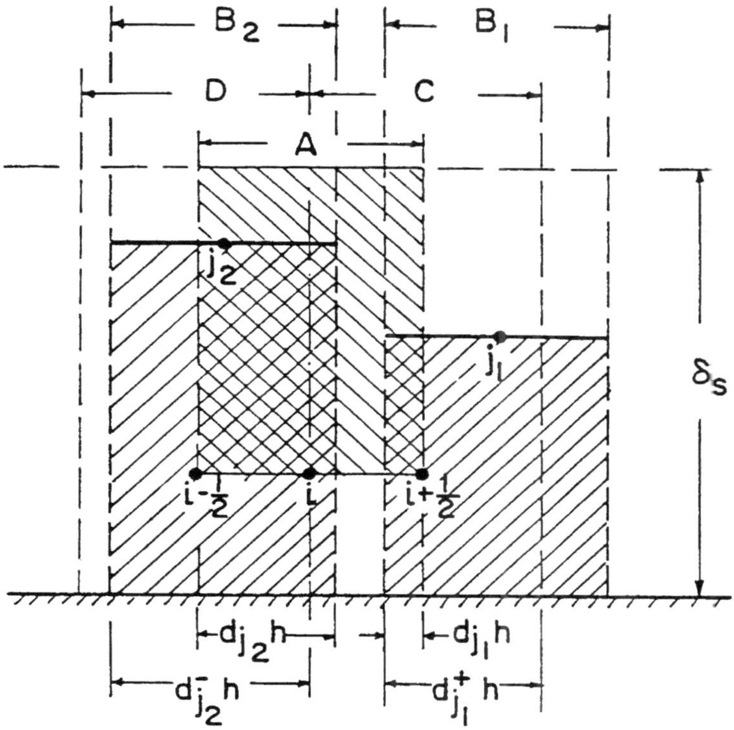

Fig. 6.2. Geometry of interdependence in the numerical shear layer
A: *zone of dependence over point* i; B: *zone of influence under sheet* j;
C: *zone of dependence around point* i+½; D: *zone of dependence around point* i–½.

The circulation induced by a vortex sheet of length h is then

$$\Gamma_j = h\gamma_j \tag{6.34}$$

so that, the velocity jump per unit length across it is

$$\Delta u_j = \gamma_j \tag{6.35}$$

Unlike the elliptic flow field modeled by vortex blobs, where their effects extend throughout the field, the zone of influence of a vortex sheet is, as a consequence of (6.32), restricted to a 'shadow' below it, as shown in Fig. 6.2 by vertically hatched areas.

Thus, the flow velocity at a point (x_i, y_i), where $y_i < y_j$, is determined by the summation equivalent of the integral relation expressed by (6.32), with (6.33) and (6.34) taken into account, whence

$$u(x_i, y_i) = u_\delta(x_i) - \sum_j \gamma_j d_j \tag{6.36}$$

where

$$d_j = 1 - |x_i - x_j|/h \tag{6.37}$$

is the influence factor of sheet j on point i, expressing the fraction of the length of the sheet extending over the zone of dependence over point i, marked by horizontally hatched area Fig. 6.2.

The normal velocity component, v, is determined from the condition of $\nabla \cdot u_\omega = 0$, invoked in the Helmholz decomposition theorem, combined with (6.57), whence

$$v = -\partial I / \partial x \tag{6.38}$$

where

$$I \equiv \int_0^{y_i} u dy = u(x_i) y_i - \int_0^{y_i} y du = u(x_i) y_i - \sum_j \gamma_j d_j y_j \tag{6.39}$$

In a finite-difference form, the above is expressed as

$$v(x_i, y_i) = -\{I^+ - I^-\}/h \tag{6.40}$$

where, according to (6.39),

$$I^{\pm} = u_\delta(x_i \pm ½h)y_i - \sum_j y_j^o \gamma_j d_j^{\pm} \tag{6.41}$$

while, as indicated in Fig. 6.2,

$$d_j^{\pm} = 1 - (x_i \pm ½h - x_j)/h \quad \text{and} \quad y^o = \min(y_i, y_j\} \tag{6.42}$$

The motion of a sheet is governed by the random displacement with u evaluated by the use of (6.41) and (6.42), while $\eta_i = 0 + i\,\eta_y$, in accordance with condition (ii), formulated at the outset of this section. To make sure that the motion of a vortex sheet is matched with the vortex blob into which it transforms, a correction term of $-½\gamma_j$ is added to the expression for $u(x_i, y_i)$ specified by (6.41) to account for the effect of the image of the blob. The paper introducing the vortex sheet method (Chorin 1978) contains information on techniques to reduce the statistical error and speed up the convergence of the algorithm.

6.3.3 Algorithm

Computations of turbulent flows made out of vortex blobs and sheets are accomplished by adopting the strength of the sheet, h, - a feature specifying their spatial resolution - and fixing, thereupon, the corresponding time step, k, in accordance with the Courant stability condition, $k \le h/\max|\mathbf{u}|$ (Chorin 1980a). For a given Reynolds number, identified thereby is the standard deviation, σ. The thickness of the numerical shear layer δ_s is then taken as a multiple of σ whereby, as shown in Fig. 6.3, the loss of vortex blobs due to their random walk is minimized. Finally, the number of sheets initially in the stack is chosen, limiting the maximum allowable value for γ. These decisions are equivalent to those made in choosing a grid size and the corresponding step in a finite difference algorithm.

For initial conditions specified in terms of inlet flow velocity, the velocity along the wall is first evaluated by the potential flow solution of (6.14).

The core radius, r_o, is fixed to abide with the no-slip boundary condition – a requirement satisfied with a minimum error by setting $r_O > \delta_S$. The potential velocity at the wall produced by a vortex blob is, according to (6.34),

$$u_o = \Gamma_j / \pi r_o \tag{6.43}$$

whence, by virtue of (6.35) with $\Delta u_j = u_O$,

$$r_O = h / \pi \tag{6.44}$$

providing an explicit relation between the length of the vortex sheet and the core radius of a vortex blob.

The displacement of the sheets in the numerical shear layer is calculated by the use (6.29) for velocities specified by (6.36) and (6.40). The various consequences of random displacement of vortex sheet are presented by Fig. 6.3. When a sheet jumps out of the boundary layer into the flow field, it becomes a blob. If a sheet jumps out on the other side of the wall, it becomes restored by its mirror image either in the shear layer as a sheet or in the flow field as a blob. If a vortex blob jumps from the flow field into the numerical sheer layer it is transformed into a sheet. The possibility of losing a vortex blob depicted in Fig. 6.3 is minimized by having the layer thickness $\delta_S < r_O$ fixed by the size of the core radius, as pointed out at the outset.

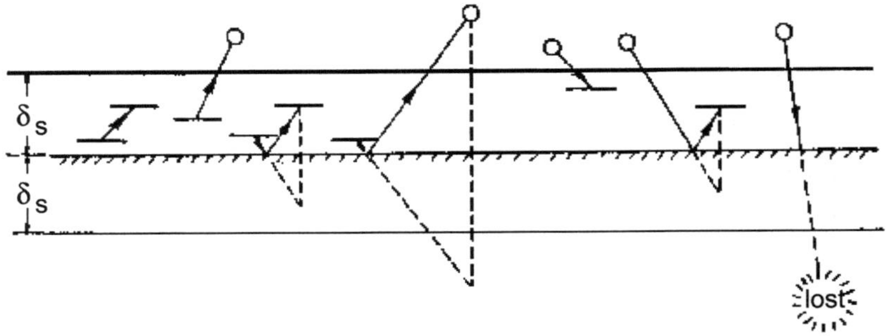

Fig. 6.3. Transformation of vortex sheets into blobs at the numerical shear layer

Once the position and strength of both the sheets and blobs are established, the flow field at a given time step is fully determined. It should be noted that vortex blobs are generated only as a consequence of the displacement of vortex sheets outside the boundary layer, modeling the physical mechanism of turbulence generation in actual flow conditions..

6.4 Source blobs

Source blobs manifest the dynamic effects of exothermic centers identified by (1.15). Their intrinsic feature is dilatation. Their fluid mechanical action is analogous to vortex blobs. Whereas the former are elements of the rotational field, the latter are essentially irrotational. Hence, albeit the velocity profile produced by a source blob is the same as that of a vortex blob depicted in Fig. 6.1, it expresses the effects of the velocity potential, rather than of the stream function.

Thus, according to the definition of the velocity potential, the components of the velocity vector induced by a source blob are

$$u \equiv \frac{\partial \Phi}{\partial x}, \quad v \equiv \frac{\partial \Phi}{\partial y} \tag{6.45}$$

for which, according to (6.2),

$$\nabla^2 \Phi = \varepsilon(r_\mathrm{F}) \tag{6.46}$$

Similarly as before, its solution is given by

$$\Phi(x) = \int_\mathrm{A} G(x, \mathbf{x}_\mathrm{j}) \boldsymbol{\Delta}(\mathbf{x}_\mathrm{j}) \mathrm{d}A \tag{6.47}$$

where

$$G(x, \mathbf{x}_\mathrm{j}) = (\boldsymbol{\Gamma}_\mathrm{j}/2\pi 2 \ln |\, \mathbf{x} - \mathbf{x}_\mathrm{j} \,| \tag{6.48}$$

is the Green function, while A is the area of the blob.

The solution of (6.47) is approximated by the summation

$$\Phi(x) = \sum_{\mathrm{j}} \mathrm{G}(\boldsymbol{x}, \mathbf{x}_{\mathrm{j}})\Delta_{\mathrm{j}} \tag{6.49}$$

where

$$\Delta_{\mathrm{j}} = \int_{A_{\mathrm{j}}} \varepsilon_{\mathrm{j}} dA \tag{6.50}$$

Then, the total velocity produced by a source blob in the ζ-plane is

$$\mathrm{w}_{\Delta}(\zeta, \zeta_{\mathrm{j}}) = \mathrm{w}_{\Phi}(\zeta, \zeta_{\mathrm{j}}) + \mathrm{w}_{\Phi}(\zeta, \zeta_{\mathrm{j}}^{*}) \tag{6.51}$$

and, for $\mathrm{J_S}$ source blobs,

$$\mathrm{w}_{\Delta}(\zeta) = \sum_{\mathrm{j=1}}^{\mathrm{J_S}} \mathrm{w}_{\Delta}(\zeta, \zeta_{\mathrm{j}}) \tag{6.52}$$

6.5 Implementation

In calculations by the RVM, source blobs in the combustion zone are coincident with vortex blobs, forming a compound blob.

Away from the center, irrespectively of its nature, such a blob induces an inviscid flow velocity. Its far field is, therefore, potential.

Thus, according to the Poisson equation

$$\nabla^2 \Phi = \delta \cdot u_0 \tag{6.53}$$

where Φ is the velocity potential, δ the Dirac delta function and u_0 the inlet flow velocity.

To examine the flow field induced by the action of such a blob, its effects in a channel simulating a stream tube are determined by solutions of (6.51), subject to boundary conditions of

$$\nabla \Phi \cdot \mathbf{n} = 0 \tag{6.54}$$

where **n** is the vector normal to the channel walls.

The motion of the blob's front is formulated as an initial value problem of a Stefan-like interface at the cross section of a channel where the reactants flow from right to left across it. According to the three elementary components of the front identified in section 5.7, its effects are investigated in three consecutive steps

(1) advection induced by the vortex, plus

(2) self-advancement at an appropriate normal speed, plus

(3) dilatation due to the exothermicity

The solutions are obtained by means of conformal mapping using the Schwarz-Christoffel transformation, according to which, for a channel flow

$$\frac{d\zeta}{dz} = \pi\zeta \tag{6.55}$$

so that the position vector in the transformed plane

$$\zeta = \exp(\pi\zeta) \tag{6.56}$$

where z is the position vector in the physical plane.

The initial velocity vector field is given by the complex potential of a vortex;

$$w = \Phi + \mathrm{i}\Psi = \mathrm{i}\Gamma \ln z \tag{6.57}$$

where the circulation is normalized so that $\Gamma = \pi^{-1}$.

The motion of the front is determined on the basis the Huygens principle, whose application to was presented by Chorin (1980). The results derived for a particular case of the normal speed, $S_{SR} = 0.4|\nabla\Phi|_{wall}$, after 80 time steps of $1.785 \cdot 10^{-3}\,h|\nabla\Phi|_{wall}$, where h is the width of the channel, are displayed in Fig. 6.4.

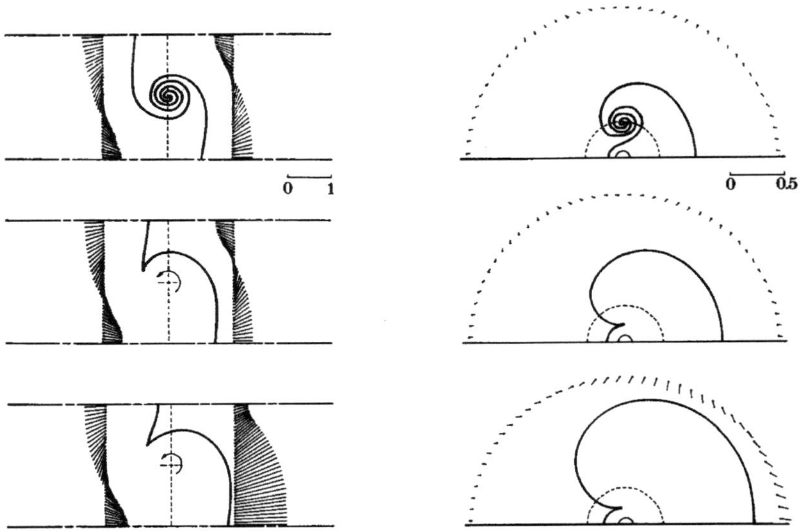

Fig. 6.4. Compound blob in a channel stream

On the left are the solutions in the physical z-plane, and on the right, their counterparts in the transformed ζ-plane. The initial position of the front is marked a straight broken line passing through the blob's center. The effect of each elementary component of the propagation mechanism is expressed by the successive deformation of the front in the three consecutive diagrams of each column. Shown on both sides of the deformed front are sets of velocity vectors displayed as straight line segments starting from a row of arbitrarily selected cross sections to which they refer.

The first two cases are entirely passive and therefore exert no influence upon the velocity vector field. However, the dilatation due to the exothermic center in third case manifests its effect upon the flow field. Surprisingly enough, the only side thereby affected is that of the reactants, the flow field of the products remaining unchanged. The reason for this is displayed in the transformed plane, where the singularity at the center of the blob is anchored at its initial position and acts as the pivot for all the stream lines.

6.6 Applications

The results of applications of the RVM to representative problems are illustrated here by the following examples:

(1) turbulent combustion behind a step in a channel;

(2) wakes in channels with sudden expansion;

(3) turbulent jets in closed channels;

(4) turbulent combustion fronts in rectangular channels.

Initiation of a combustion zone is simulated by the action of compound blobs situated at prescribed points of ignition starting at a prescribed instant of time.

6.6.1 Turbulent Combustion behind a Step

A turbulent combustion field behind a step is presented by a sequence of cinematographic schlieren records displayed by Fig. 6.5.

The height of the test section is 2.54 cm; its width is 3.81 cm. The reactants, made out of a propane/air mixture at fuel-equivalence ratio, ϕ =0.57, are flown into the test section at atmospheric pressure and room temperature. Their velocity is U_o = 13.6 m/sec, corresponding to Reynolds number Re = 22000. The time interval between frames is 1.22 msec.

For application of the RVM, conformal transformation is used as a Poisson solver with its z-plane and the ς-plane, evaluated according to the Schwarz-Christopher theorem.

The velocity functions of ζ are

$$w_p(\zeta) = H / \pi\zeta \tag{6.58}$$

and

$$F(\zeta) = \frac{\pi\zeta}{H}\left(\frac{\zeta - 4}{\zeta - 1}\right)^{1/2} \tag{6.59}$$

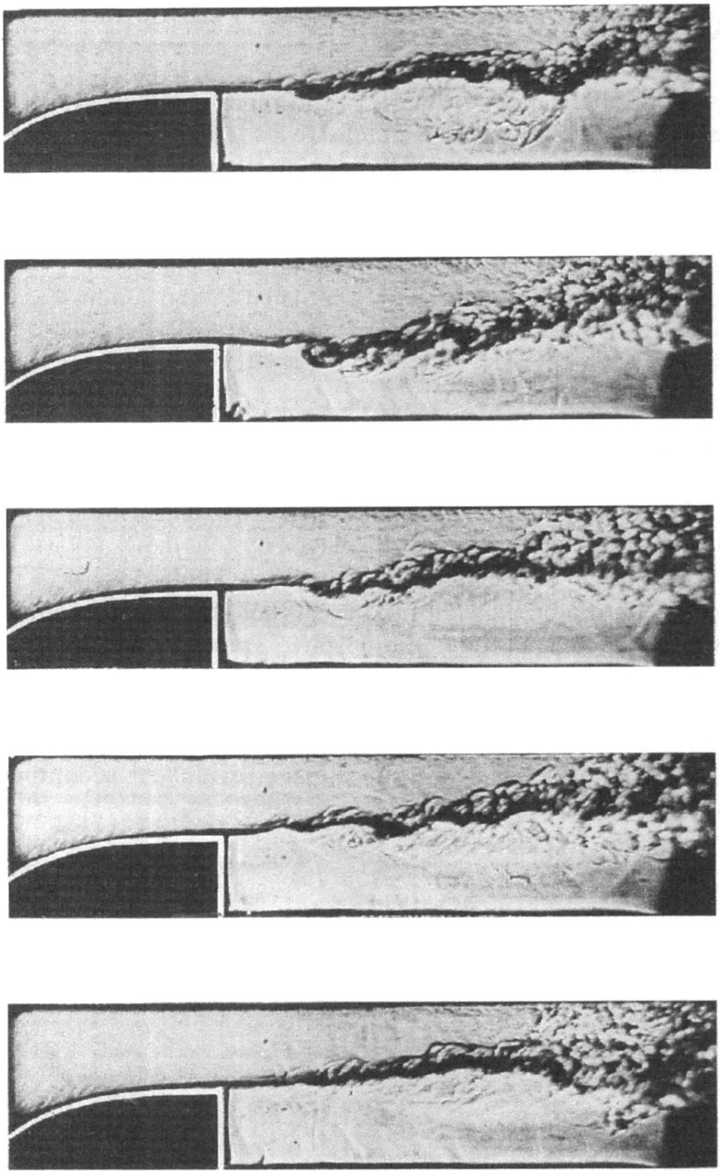

Fig. 6.5. Cinematographic Schlieren record of turbulent combustion field behind a step

For the case at hand, the z and ς planes are displayed in Fig. 6.6.

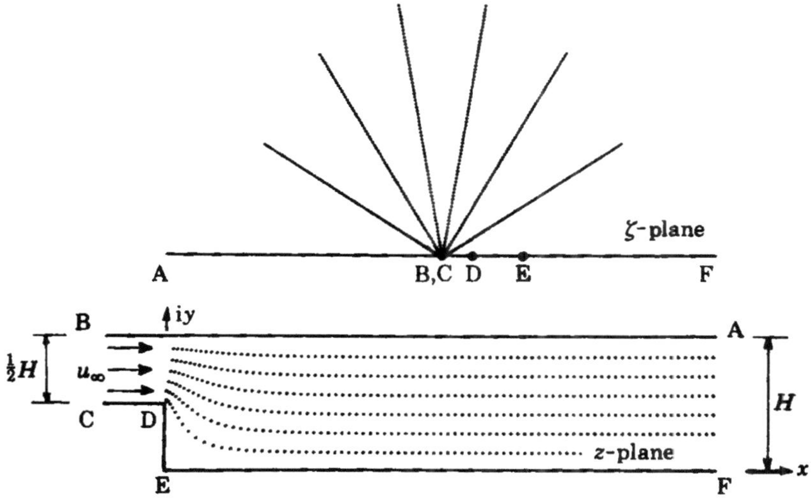

Fig. 6.6. Streamlines flow behind a step in transformed and physical planes

The evolution of a turbulent velocity vector field of a non-reacting fluid, evaluated by the RVM at computational steps of $\Delta t = H/2u_\infty$ sec, is presented by Fig. 6.7. The vectors display the velocities of all the vortex blobs used in the computations. It should be noted that vectors in all the displays of this kind presented in this book are depicted by line segments used conventionally for them, but, instead of arrowheads at the end, they are provided by small circles situated at the blobs to which they pertain.

The turbulent velocity vector field of a non-reacting flow portrayed by Fig. 6.5, is presented by Fig. 6.7. The turbulent velocity vector field of a reacting flow is displayed by Fig. 6.8, for which the conditions of calculations, as well as the time interval between frames are the same as those for Fig. 6.7. In computations for the reaction zone, exothermic blobs are superimposed upon vortex blobs. The contour of the exothermic front - the undulating interface between the reaction zone and the unreacted fluid referred to popularly as the flame that is recorded by schlieren photography – is delineated by thick lines.

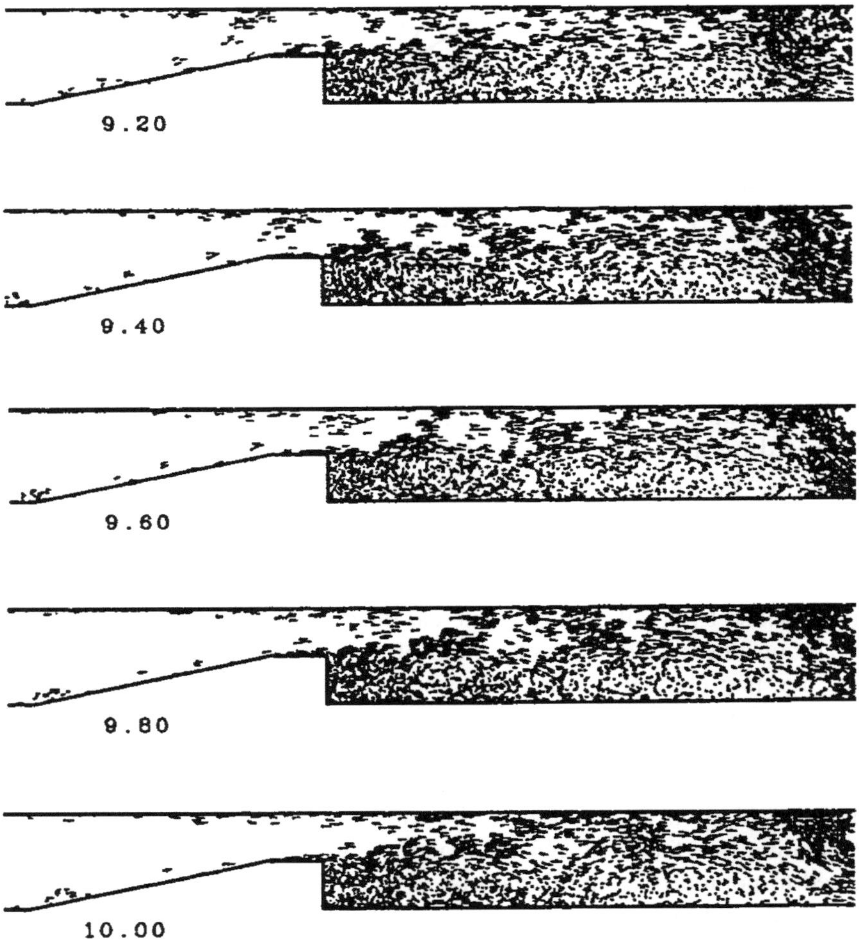

Fig. 6.7. Velocity vector field of a non-reacting turbulent flow at inlet Re = 22,000 displayed at time intervals of $0.2\Delta t$

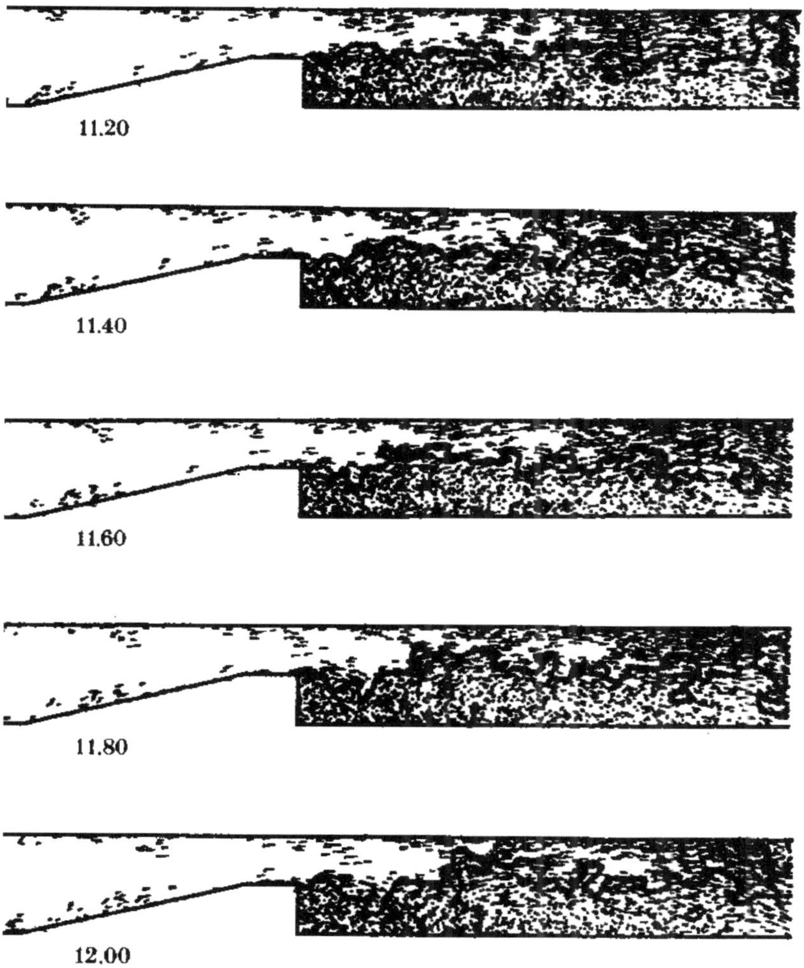

Fig. 6.8. Evolution of the velocity vector field of an exothermically reacting fluid portrayed by Fig. 6.4

Velocity profiles at a number of cross sections downstream of the step are calculated by averaging the results of 20 consecutive computational time steps. Since the flow has reached a stationary state and the inlet flow is steady, time-averaging is identical to ensemble averaging. While the flow field is continuously perturbed by the random samples used in the random walk modeling the effects of diffusion, the growth and decay of instabilities is governed by the nonlinear interaction due to convection.

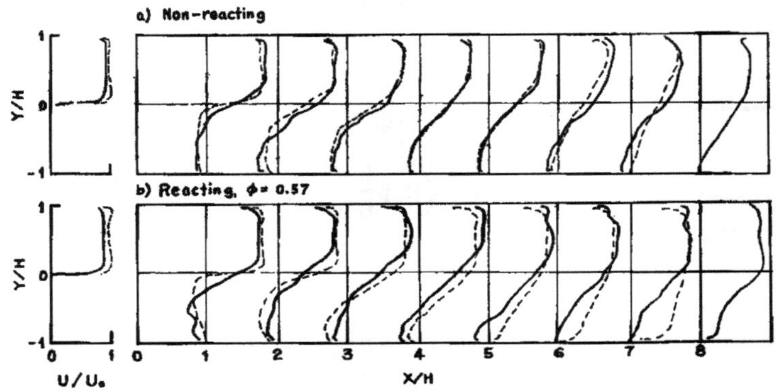

Fig. 6.9. Average velocity profiles for non-reacting and reacting flow fields at Re = 22,000
Symbol U_o *denotes the inlet velocity at the step, while H is the step height. Numerical results are presented by solid lines, while experimental data of Pitz and Daily 1983 are displayed by broken lines.*

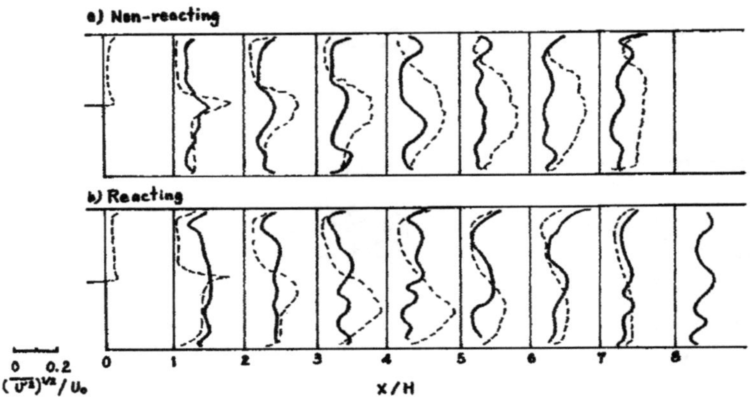

Fig. 6.10. Streamwise turbulence intensity profiles $(u'^2)^{1/2}/u_o$, for non-reacting and reacting flow fields at Re = 22,000
Solid lines represent numerical solution, while broken lines display experimental data.

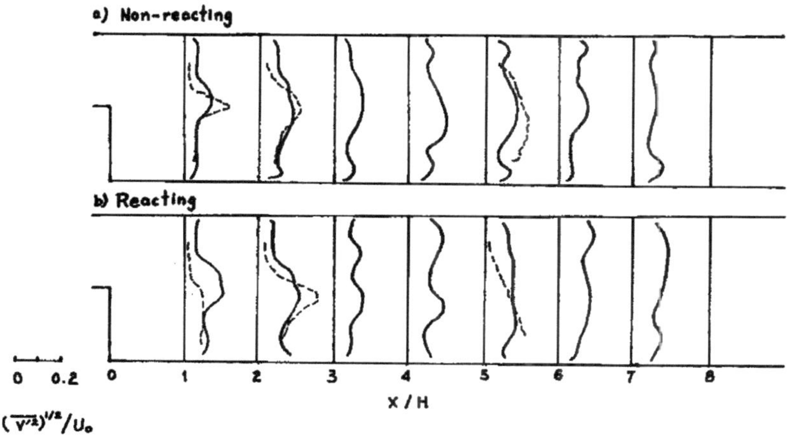

Fig. 6.11. Transverse turbulence intensity profiles $(v'^2)^{1/2}/u_o$, for non-reacting and reacting flow fields at Re = 22,000
Solid lines represent numerical solution, while broken lines display experimental data.

Figure 6.9 displays the average velocity profiles evaluated by the RMV and the experimental data of Pitz and Daily 1983 obtained by laser Dopler velocimetry for the non-reacting and reacting flow fields at Re = 22,000. Figures 6.10 and 6.11 provide a comparison between the RMV computations and the experimentally measured turbulence intensities in streamwise and transverse directions, $(u'^2)^{1/2}/U_o$ and $(v'^2)^{1/2}/U_o$, for the non-reacting and reacting flow fields. As expected, they are at their maxima in the immediate vicinity of the walls and at the start of the mixing layer. Turbulence due to high shear is there continuously generated, giving rise to peak levels followed by their gradual decay downstream.

From a similarity between Figs. 6.5 and 6.8, it appears that the results of computations made by the RVM for a two-dimensional flow field, are in quite a satisfactory agreement with the cinematographic schlieren record. This conclusion is enforced by the similarity between numerical results of RVM calculations and experimental measurements of velocity profiles presented by Fig. 6.9, and the turbulence intensities displayed by Figs. 6.10 and 6.11, in spite of the relatively small samples of computational outputs taken into account.

Upon the study of turbulent flow fields behind a step, similar fields taking place in channels with sudden expansion in their cross-section become of interest. Numerical solutions obtained by the RVM of such fields are provided in the next section.

6.6.2 Turbulent flow fields in channels with sudden expansion

In order to examine the effects of the singularity at the corner of the step, the non-slip condition at the walls are here relaxed. Presented in this section are three cases whose geometrical configuration differs only by the expansion ratios of the channels, 2:1 in Fig. 6.12, 4:1 in Fig. 6.13, and 20:1 in Fig. 6.14. At inlet in all of them $R_e = 10^4$.

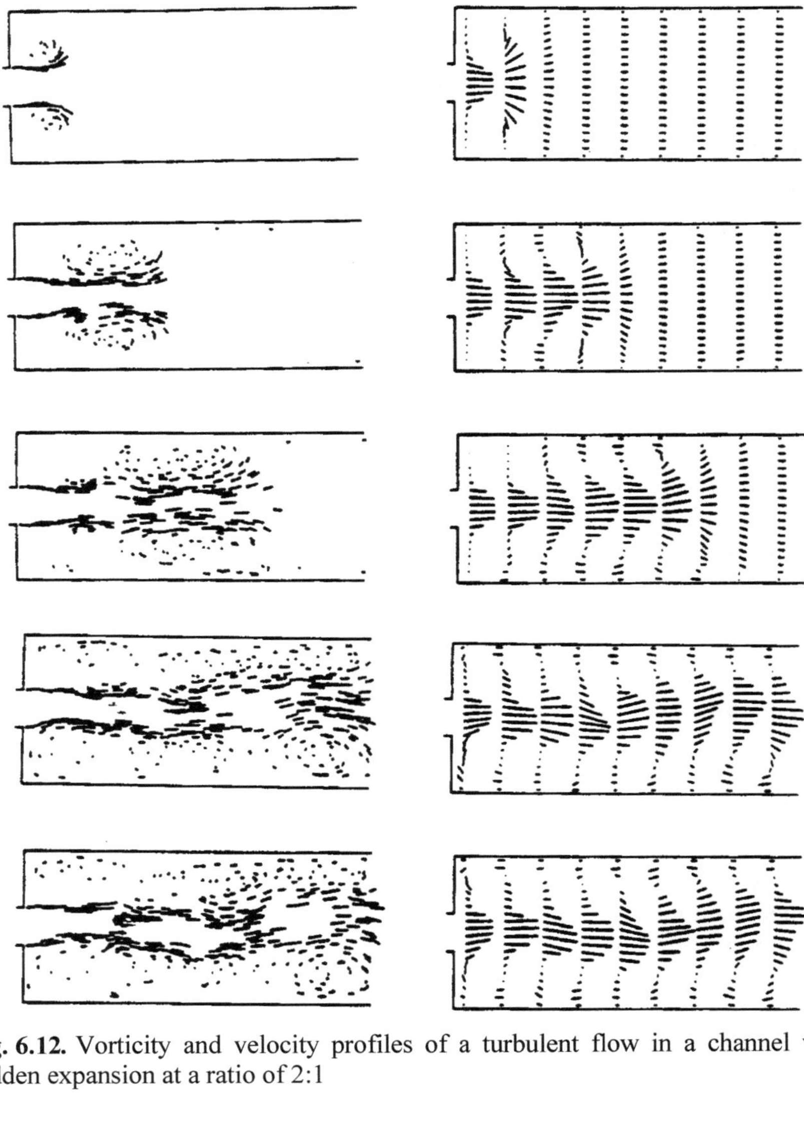

Fig. 6.12. Vorticity and velocity profiles of a turbulent flow in a channel with sudden expansion at a ratio of 2:1

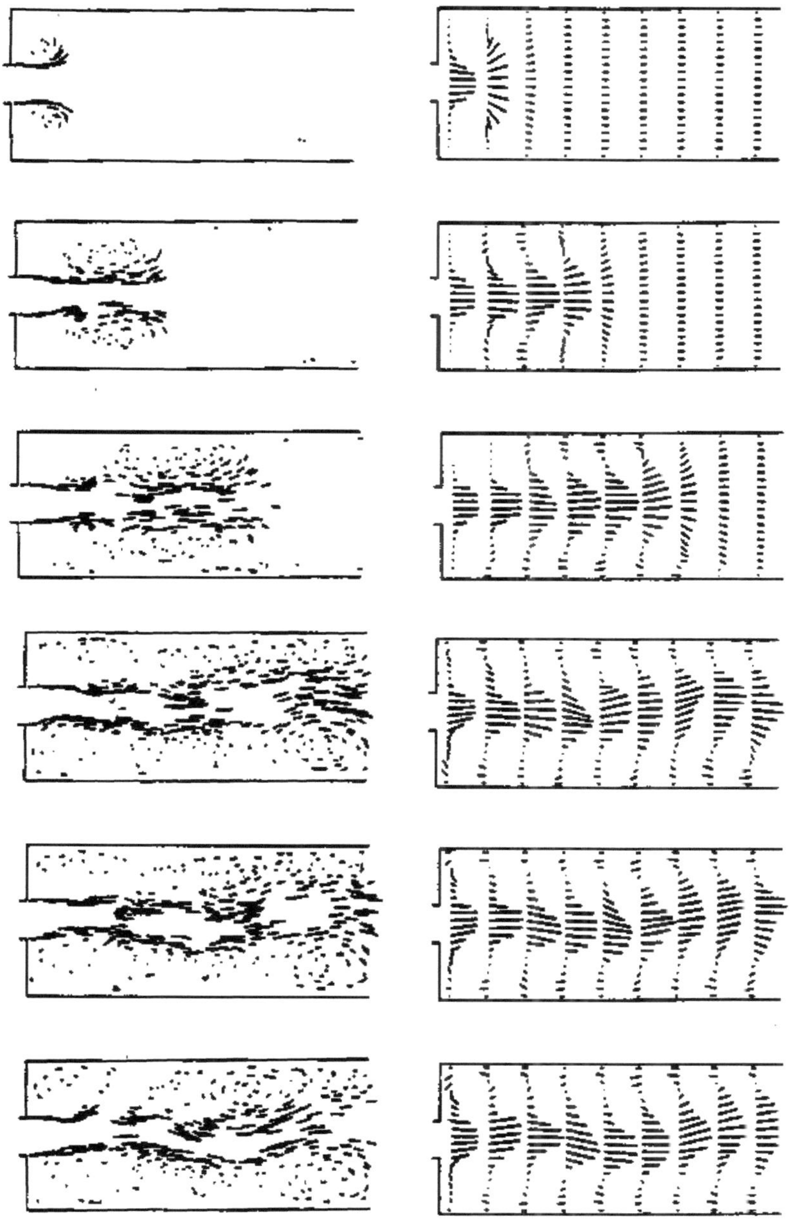

Fig. 6.13. Vorticity and velocity profiles of a turbulent flow in a channel with sudden expansion at a ratio of 4:1

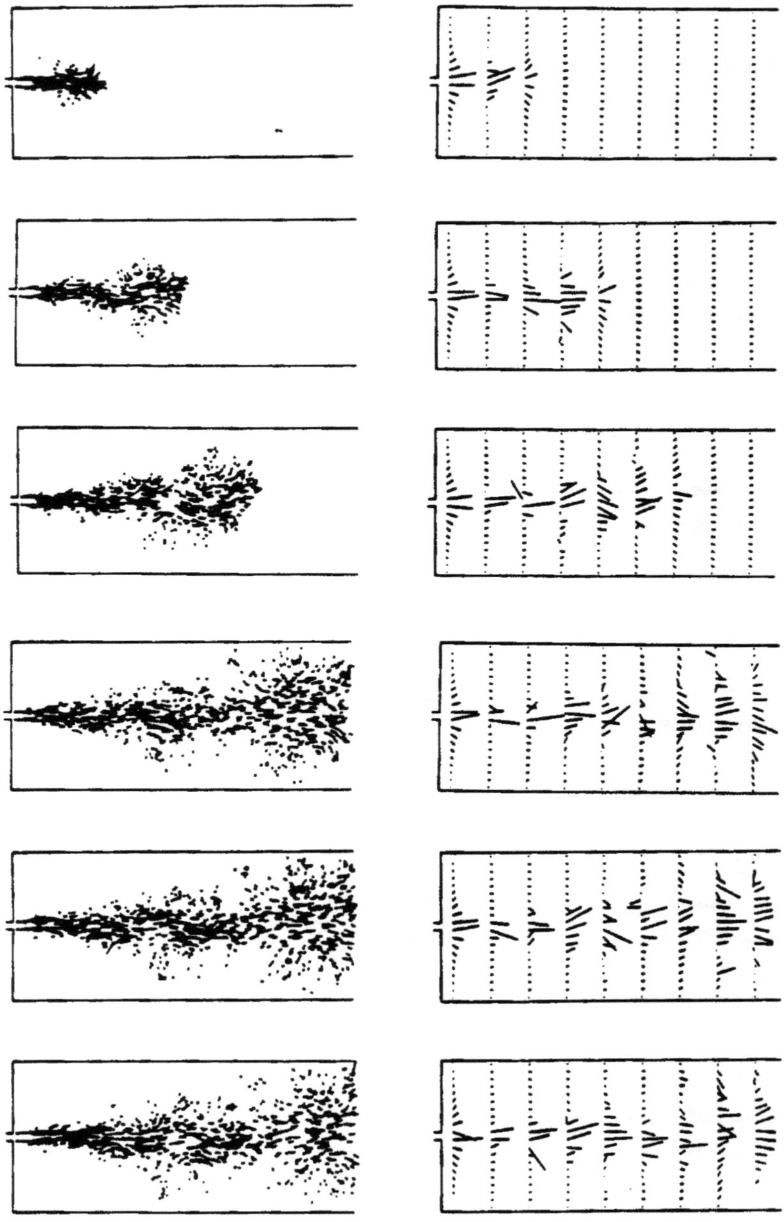

Fig. 6.14. Vorticity and velocity profiles of a turbulent flow in a channel with sudden expansion at a ratio of 20:1

The difference between the first two cases and the third is quite distinct. Whereas the former display turbulent wakes associated with recirculation manifested by reverse direction of velocity vectors at the corners of the channel, the third portrays the formation of a turbulent jet unencumbered by the walls of the enclosure.

6.6.3 Turbulent jets in a piston-compressed channel

Turbulent jets injected transversely into a piston compressed channel are influenced by the flow field induced by the piston. This effect is displayed in Figs. 6.15-18 by velocity vector fields and interfaces between the jet fluid and that in the channel initially at rest. The jet velocity at inlet is 10 times higher than that of the piston. Its Reynolds number Re = 10^4.

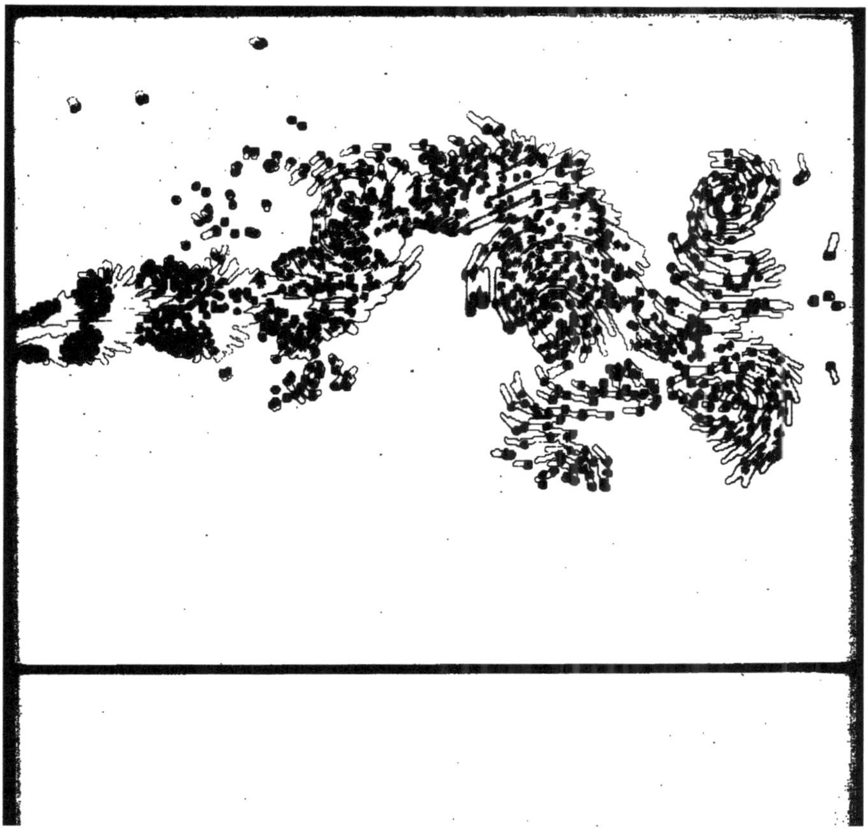

Fig. 6.15. Velocity vectors of a single jet injected into a piston compressed channel

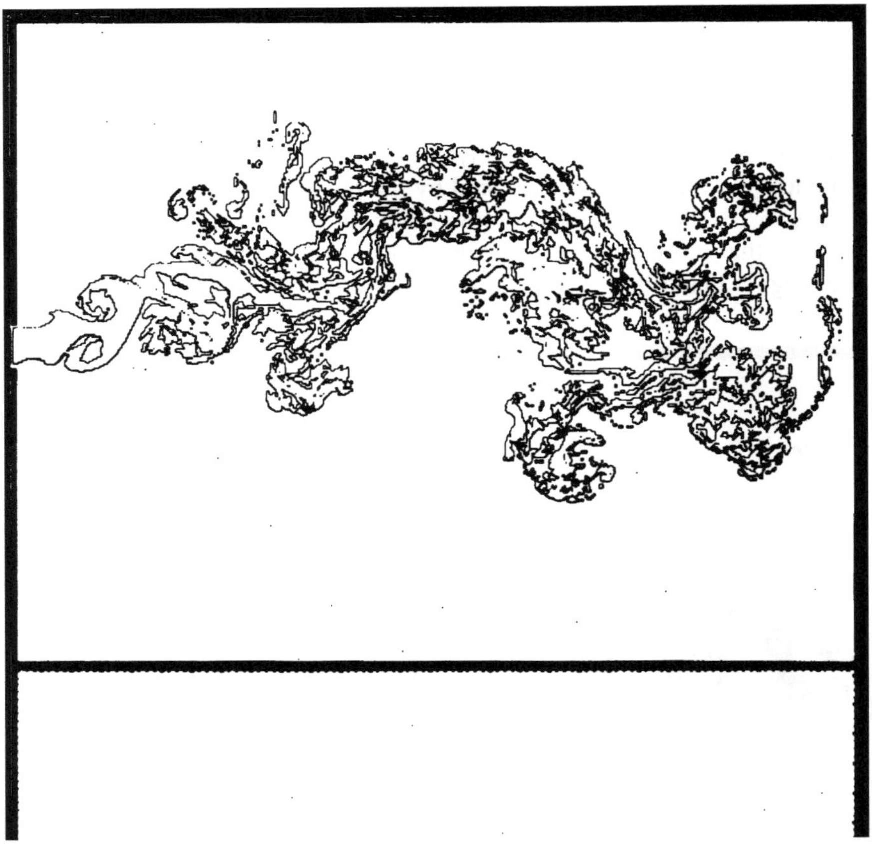

Fig. 6.16. Interfaces between the fluid of a single jet and that in the channel

These figures simulate the formation of the turbulent jet plume displayed by cinematographic schlieren records in Fig. 3.3.

Figures 6.17 and 6.18 depict the formation of a turbulent jet plume generated by opposed injection, demonstrating the advantage accruable by multi injection systems in keeping the combustible zone away from the walls

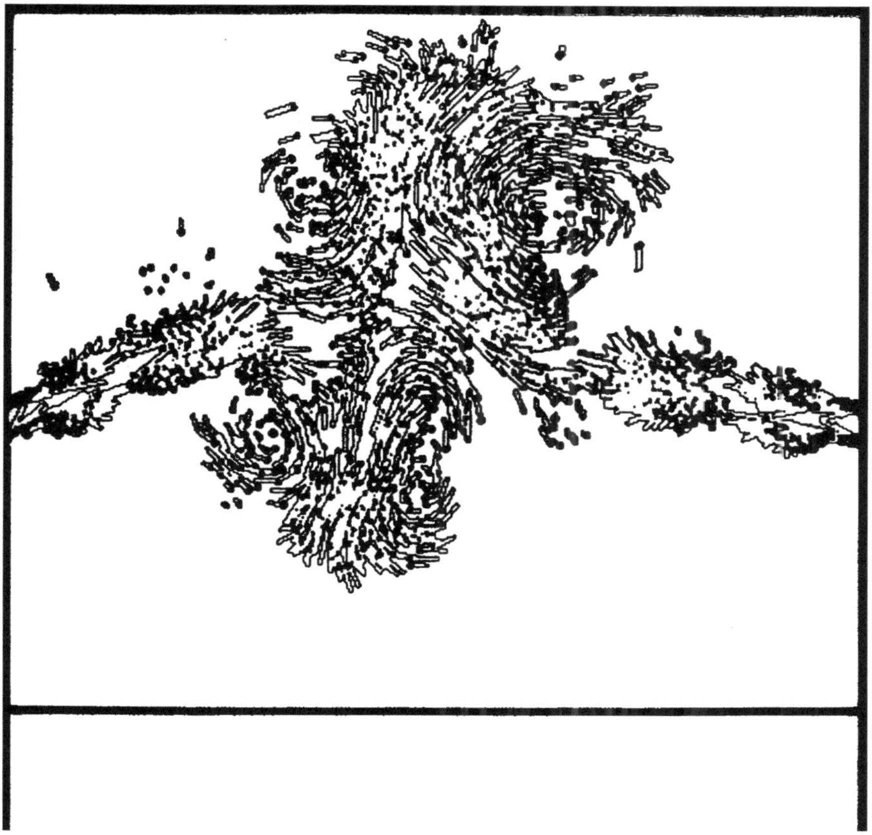

Fig. 6.17. Velocity vectors of opposed jets injected into a piston compressed channel

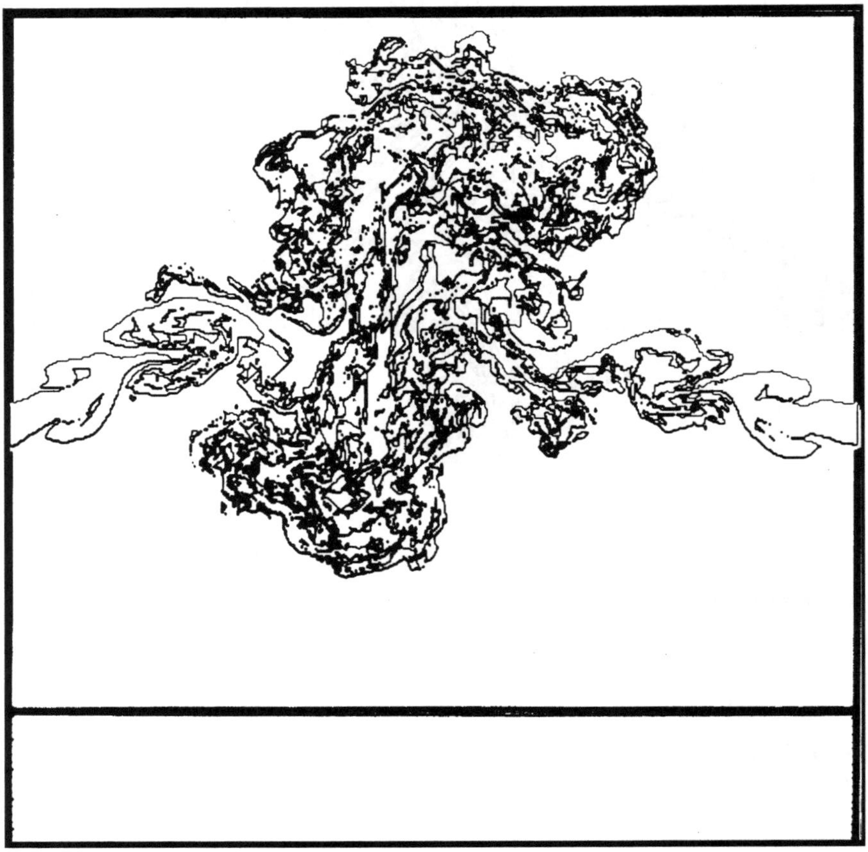

Fig. 6.18. Interfaces between the fluid of opposed jets and that in the channel

6.6.4 Turbulent combustion in rectangular channels

The tendency of the front to cave in, acquiring what is known in the literature as a tulip shape (Bazhenova et al 1968), is exhibited by Fig. 6.19 even upon asymmetric initiation of combustion. In a close-ended enclosure, without taking into account the baroclinic effects at the front, it is due to the zero velocity boundary condition imposed by the back wall.

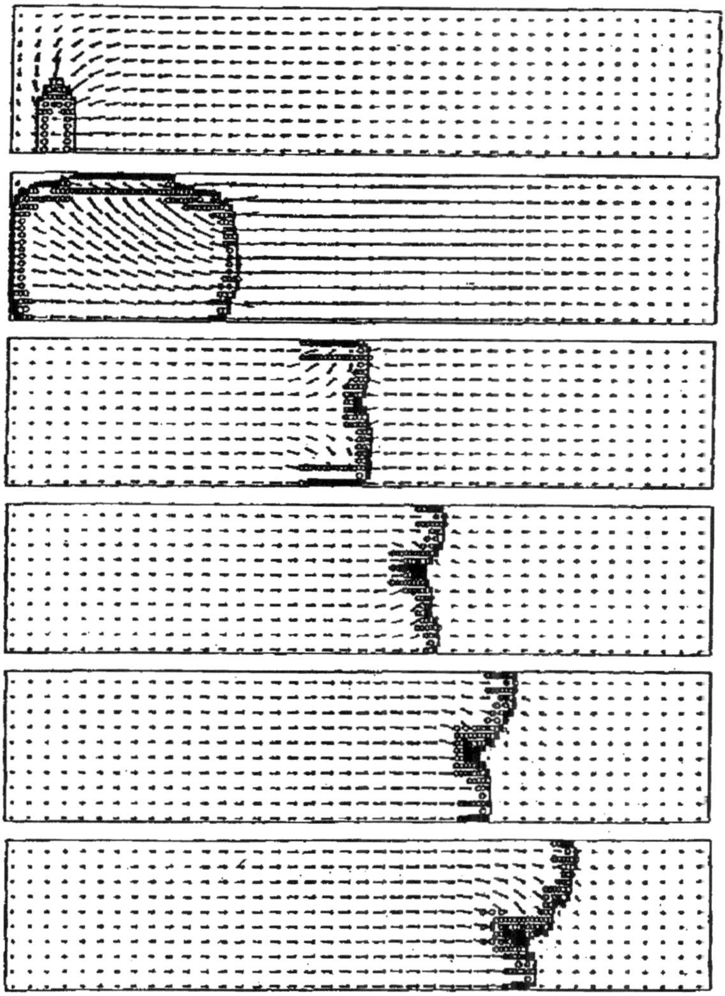

Fig. 6.19. Propagation of an exothermic front and the concomitant evolution of the velocity vector field without taking account of the baroclinic effect

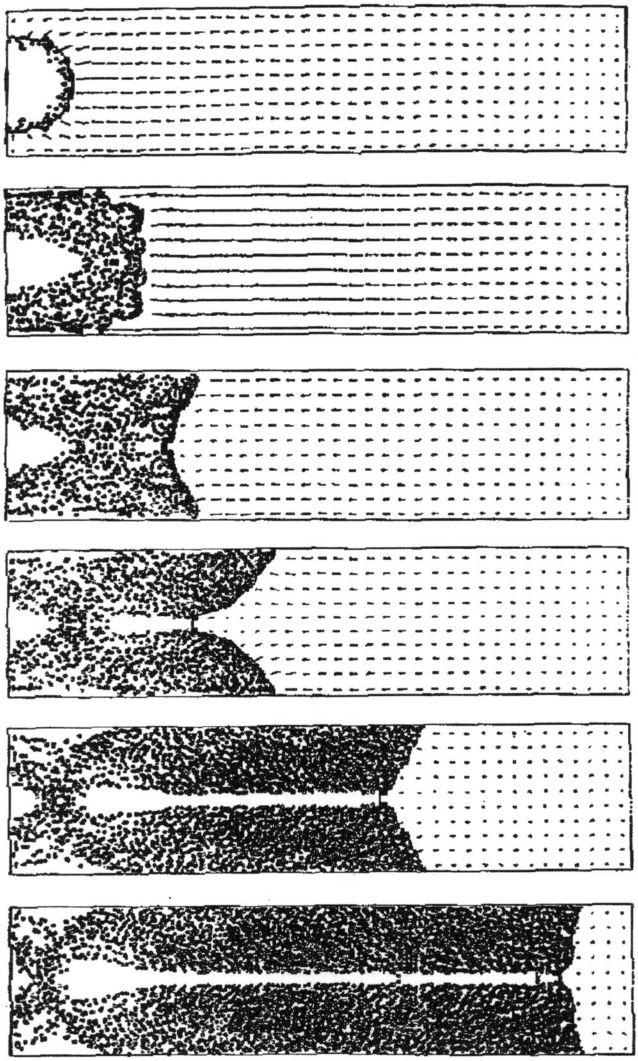

Fig. 6.20. Propagation of an exothermic front and the concomitant evolution of the velocity vector field with the baroclinic effect at the front taken into account (Rotman et al, 1988)

As evident above, due to the baroclinic effect, the combustion front acquires a tulip shape independently of the boundary condition imposed by the back wall.

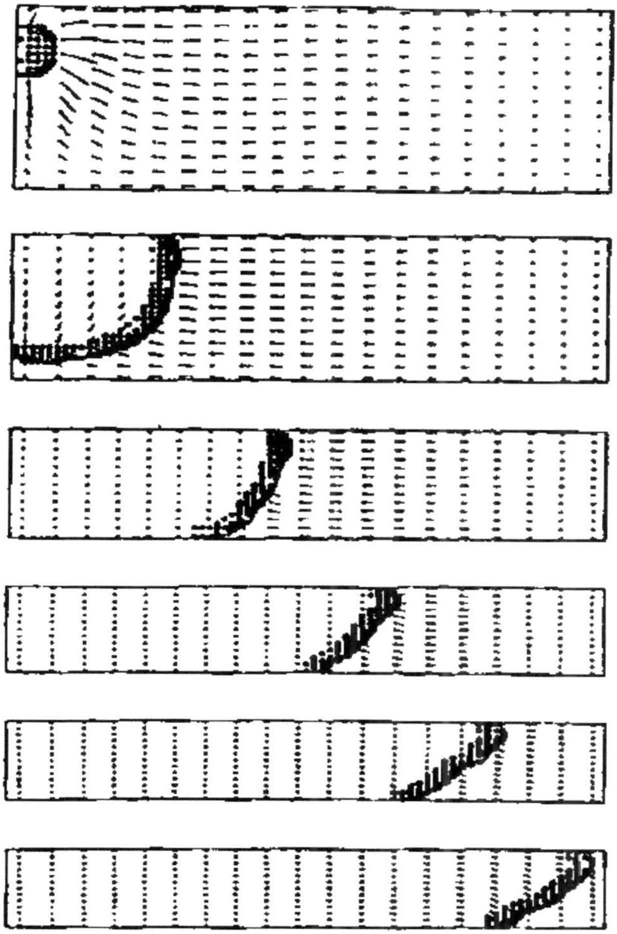

Fig. 6.21. Propagation of an exothermic front and the concomitant evolution of the velocity vector field in a piston compressed channel simulating conditions of a piston engine

The evolution of velocity field generated by a combustion front propagating a in a piston compressed channel is displayed by Fig. 6.21. Evidently in this case the tulip shape of the front is reduced to its one-sided caving-in as a consequence of asymmetry due to piston motion. Particularly pronounced here is the significant velocity produced by the advancing front immediately upon its inception.

7 Gasdynamic Aspects

7.1 Formulation

Gasdynamics deals with fast events where the effects of compressibility are of prime significance, while transport processes are of negligible influence because of their relatively short lifetimes. The flow field is considered then at its limit of the Reynolds number Re = ∞, the Peclet numbers for diffusion and thermal conduction Pe = ∞, and the Damköhler number Da = ∞.

Under such circumstances, the mass balance is expressed by the global continuity equation for a closed system:

$$\frac{\mathrm{D}\rho}{\mathrm{D}t} + \rho\nabla\cdot\mathbf{u} \equiv \frac{\partial\rho}{\partial t} + \nabla\cdot\rho\mathbf{u} = 0 \tag{7.1}$$

The force balance is expressed by the momentum equation:

$$\frac{\mathrm{D}\mathbf{u}}{\mathrm{D}t} + \frac{1}{\rho}\nabla p \equiv \frac{\partial\mathbf{u}}{\partial t} + \mathbf{u}\cdot\nabla\cdot\mathbf{u} + \frac{1}{\rho}\nabla p = F \tag{7.2}$$

where F is a force per unit mass exerted upon the system from an outside source.

The energy balance is expressed concisely as

$$\frac{\mathrm{D}e}{\mathrm{D}t} + w\nabla\cdot\mathbf{u} = \dot{q} \tag{7.3}$$

where w is the dynamic potential and $\dot{q}$ is the rate of irreversible energy acquired by the system per unit mass. The latter is expressed in terms of the Second Law, according to which

$$\dot{q} = T\frac{\mathrm{D}s}{\mathrm{D}t} \equiv \frac{\mathrm{D}e}{\mathrm{D}t} + p\frac{\mathrm{D}v}{\mathrm{D}t} = \frac{\mathrm{D}e}{\mathrm{D}t} - \frac{p}{\rho^2}\frac{\mathrm{D}\rho}{\mathrm{D}t} \tag{7.4}$$

– an expression deducible from (7.3) with the use of (7.1).

The formulation is closed by the relationship between the thermodynamic parameters – the density, ρ, involved in the continuity equation, the pressure, p, in the momentum equation, and the internal energy, e, in the energy equation –

$$e = \mathrm{e_o} + Cw \tag{7.5}$$

where $C \equiv (\frac{\partial e}{\partial w})_{\mathrm{p}}$,while the independent variable is the dynamic potential, $w \equiv p/\rho$. Practically, the variation of C is so small within the scope of interest, that, as expounded in Chapter 1 and demonstrated by Fig. 8.4, it is a constant, so that (7.5) is linear.

For most computational algorithms of the flow field, the momentum and energy equations are transformed, the former by adding (7.1) multiplied by **u** to (7.2), whence

$$\frac{\partial(\rho\mathbf{u})}{\partial t} + \nabla\cdot\rho\mathbf{u}\mathbf{u} + \nabla p = F \tag{7.6}$$

and the latter by adding (7.2) multiplied by **u** to (7.3), whence

$$T\frac{\mathrm{D}s}{\mathrm{D}t} = \rho\frac{\mathrm{D}}{\mathrm{D}t}(e + \frac{\mathbf{u}^2}{2}) + \nabla\cdot(p\mathbf{u}) = \dot{Q} \tag{7.7}$$

where $\dot{Q} \equiv \dot{q} + \mathbf{u}F$.

For gasdynamic analysis, (7.7) is modified by the use of the local velocity of sound, a, provided by (1.27), whence

$$\frac{1}{\rho(\gamma-1)}\frac{Dp}{Dt} - \frac{a^2}{\rho(\gamma-1)}\frac{D\rho}{Dt} = \dot{Q} \tag{7.8}$$

7.2 One-dimensional Flow Fields

Noting that, in a rectangular coordinate system

$$\nabla \cdot \mathbf{V} = \frac{\partial V_x}{\partial x} + \frac{\partial V_y}{\partial y} + \frac{\partial V_z}{\partial z}$$

in a cylindrical coordinate system

$$\nabla \cdot \mathbf{V} = \frac{1}{r}\frac{\partial (rV_r)}{\partial r} + \frac{1}{r}\frac{\partial V_\Theta}{\partial \Theta} + \frac{\partial V_z}{\partial z} = \frac{\partial V_r}{\partial r} + \frac{V_r}{r} + \frac{1}{r}\frac{\partial V_\Theta}{\partial \Theta} + \frac{\partial V_z}{\partial z}$$

and in a spherical coordinate system

$$\nabla \cdot \mathbf{V} = \frac{1}{r^2}\frac{\partial (r^2 V_r)}{\partial r} + \frac{1}{r\,sin\Theta}[\frac{\partial (V_\Theta sin\Theta)}{\partial \Theta} + \frac{\partial V_\phi}{\partial \phi}]$$

$$= \frac{\partial V_r}{\partial r} + 2\frac{V_r}{r} + \frac{1}{r\,sin\Theta}[\frac{\partial (V_\Theta sin\Theta)}{\partial \Theta} + \frac{\partial V_\phi}{\partial \phi}]$$

the conservation equations for one-dimensional flow fields are as follows. The continuity equation, according to (7.1), is

$$\frac{\partial \rho}{\partial t} + u\frac{\partial p}{\partial r} + \rho\frac{\partial u}{\partial r} = C \qquad (7.9)$$

where $C \equiv -\mathrm{j}\frac{\rho u}{r}$, while $\mathrm{j} \equiv \frac{\partial \ln A}{\partial lnr} = \begin{cases} 0 \text{ for plane symmetrical fields} \\ 1 \text{ for line symmetrical fields} \\ 2 \text{ for point symmetrical fields} \end{cases}$

The momentum equation, according to (7.2), is

$$\frac{\partial u}{\partial t} + u\frac{\partial u}{\partial r} + \frac{1}{\rho}\frac{\partial p}{\partial r} = F \qquad (7.10)$$

For example, in the case of flow though ducts, $F \equiv \frac{4\mathrm{f}}{\mathrm{D}}\frac{u^2}{2}\frac{u}{|u|}$ where, with τ_w denoting the wall shearing stress, $\mathrm{f} \equiv \frac{2\tau_w}{\rho u^2}$ is the conventional pipe friction coefficient.

The energy equation, according to (7.7), is

$$\frac{\partial}{\partial t}(e+\frac{u^2}{2})+u\frac{\partial}{\partial r}(e+\frac{u^2}{2})+u\frac{\partial p}{\partial r}+\frac{p}{\rho}\frac{\partial u}{\partial r}+j\frac{pu}{\rho r}=\dot{Q} \tag{7.11}$$

or, according to (7.8),

$$\frac{\partial p}{\partial t}+u\frac{\partial p}{\partial r}-a^2\frac{\partial \rho}{\partial t}-a^2u\frac{\partial \rho}{\partial r}=\Omega \tag{7.12}$$

where $\Omega \equiv \rho(\Gamma-1)\dot{Q}$ is usually ascribed to the energy derived from an outside source by heat transfer.

7.3 Method of Characteristics

In order to integrate the set of partial differential equations (PDE's) for given boundary conditions of the flow field, they are expressed in terms of algebraic relations. Such relations are provided by loci of singularities where the partial derivatives of the dependent variables, ρ, p, and u, are indeterminate, as they are along the characteristics that are real in the case of the hyperbolic set of PDE's specified above.

For this purpose, the partial derivatives, $\partial\psi/\partial\xi$, where $\psi = \rho, p, u$ and $\xi = t, r$, are considered as dependent variables. Three PDE's are then provided by the equations of continuity, momentum and energy, and three by the expressions for total derivates, $\frac{\partial\psi}{\partial t}\mathrm{d}t+\frac{\partial\psi}{\partial r}\mathrm{d}r=\mathrm{d}\psi$.

$$\frac{\partial \rho}{\partial t}+u\frac{\partial \rho}{\partial r}+\rho\frac{\partial u}{\partial r}=C \tag{7.13}$$

$$\frac{1}{\rho}\frac{\partial p}{\partial r}+\frac{\partial u}{\partial t}+u\frac{\partial u}{\partial r}=F \tag{7.14}$$

$$-a^2\frac{\partial\rho}{\partial t}-a^2u\frac{\partial\rho}{\partial r}+\frac{\partial p}{\partial t}+u\frac{\partial p}{\partial r} = \Omega \tag{7.15}$$

$$\mathrm{d}t\frac{\partial\rho}{\partial t}+\mathrm{d}r\frac{\partial\rho}{\partial r} = \mathrm{d}\rho \tag{7.16}$$

$$\mathrm{d}t\frac{\partial p}{\partial t}+\mathrm{d}r\frac{\partial p}{\partial r} = \mathrm{d}p \tag{7.17}$$

$$\mathrm{d}t\frac{\partial u}{\partial t}+\mathrm{d}r\frac{\partial u}{\partial r} = \mathrm{d}u \tag{7.18}$$

The loci of singularities for each of the dependent variables is thus given by

$$\frac{\partial\psi}{\partial x}=\frac{\mathrm{K}_\psi}{\mathrm{N}}=\frac{0}{0} \tag{7.19}$$

where, according to the set of (7.13) – (7.18),

$$N=\begin{vmatrix} 0 & u & 0 & 0 & 0 & \rho \\ 0 & 0 & 0 & 1/\rho & 1 & u \\ -a^2 & -a^2u & 1 & u & 0 & 0 \\ \mathrm{d}t & \mathrm{d}r & 0 & 0 & 0 & 0 \\ 0 & 0 & \mathrm{d}t & \mathrm{d}r & 0 & 0 \\ 0 & 0 & 0 & 0 & \mathrm{d}t & dr \end{vmatrix}=0 \tag{7.20}$$

while the representative numerator for $\Psi = \mathrm{u}$

$$\mathrm{K_u}=\begin{vmatrix} 0 & u & 0 & 0 & 0 & C \\ 0 & 0 & 0 & 1/\rho & 1 & F \\ -a^2 & -a^2u & 1 & u & 0 & \Omega \\ \mathrm{d}t & \mathrm{d}r & 0 & 0 & 0 & \mathrm{d}\rho \\ 0 & 0 & \mathrm{d}t & \mathrm{d}r & 0 & \mathrm{d}p \\ 0 & 0 & 0 & 0 & \mathrm{d}t & \mathrm{d}u \end{vmatrix}=0 \tag{7.21}$$

The roots of (7.19) are

$$\left(\frac{\mathrm{d}x}{\mathrm{d}t}\right)_{\mathrm{I,II}} = u \mp a \tag{7.22}$$

delineating Mach lines along which u - a = const for characteristics of positive slope, I, and u + a = const for characteristics of negative slope, I., Completing the set of characteristics at a point are the particle paths, III, for which

$$(\frac{dx}{dt})_{III} = u \tag{7.23}$$

With the use of (7.13), (7.14) and (7.15) obtained thereby are the compatibility conditions

$$\begin{aligned}(\frac{du}{dt})_{I,II} &= \mp\frac{1}{\rho a}(\frac{dp}{dt})_{I,II} \pm \frac{a}{\rho}C \pm \frac{1}{\rho a}\Omega + F \\ &= \mp\frac{2}{\gamma-1}(\frac{da}{dt})_{I,II} \pm \frac{a}{\rho}C \pm \frac{1}{\rho a}\Omega + F\end{aligned} \tag{7.24}$$

the second expression taking advantage of the isentropic relationship, $(\partial p/\partial a)_s = (2/(\gamma-1)\rho a$, provided by (1.27).

For a closed system of invariant mass $C = 0$, while, in the absence of irreversible effects, $\Omega = F = 0$. Under such circumstances, (8.21) can be integrated throughout the domain, yielding an a priori constraint

$$\frac{u}{a_o} \pm \frac{2}{\gamma-1}\frac{a}{a_o} = \text{const} \tag{7.25}$$

One has then a physical plane of x and t that is expressed usually in terms of normalized coordinates, $\xi \equiv x/x_o$ and $\tau \equiv (a_o/x_o)t$, where x_o is a reference dimension of space.

The method of characteristics provides a solution of the set of partial differential equations (7.9), (7.10) and (7.12), along the characteristic directions specified by (7.22) and (7.23). The physical space of ξ and τ is thereby expressed in terms of a set of characteristics at discrete intervals and mapped onto the thermodynamic state (or phase) diagram of u/a_o and a/a_o. It should be noted that the fundamental features of the method for unsteady one dimensional flow, apply to steady two dimensional fields by replacing the time coordinate, t, and the space radius, r, by the Cartesian coordinates x and y.

An implementation of the method of characteristics is provided by Fig. 7.1 – a copy of Fig. 60 in the classical book of Oswatitsch, 1956 – presenting a solution of a non-steady flow field generated in a closed tube, containing initially gas at high pressure, upon sudden opening of its end.

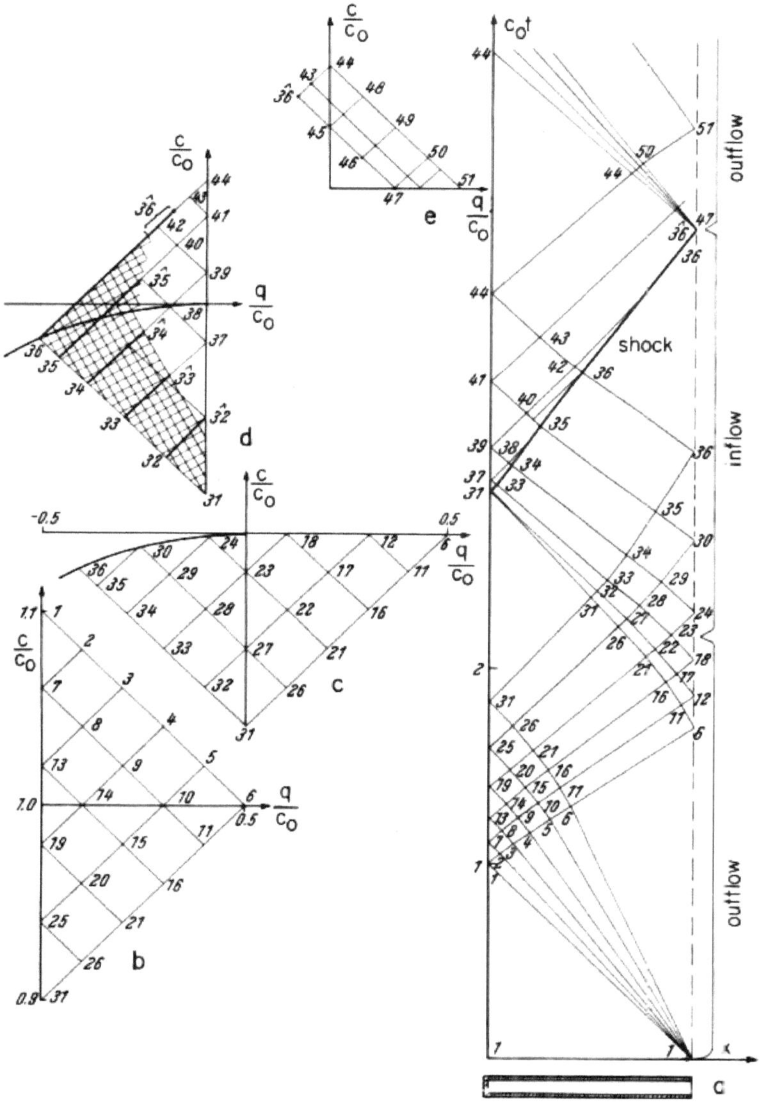

Fig. 7.1. Method of characteristics solution of a non-steady flow field in a tube, generated in a closed tube, containing initially gas at high pressure, upon sudden opening of its right end (Oswatitsch, 1956)

In Fig. 7.1, the velocity of sound, *a*, is denoted by *c*, the particle velocity, *u,* by *q*, while the normalized time coordinate is c_0t/x_o and the normalized space coordinate is x/x_o, subscript o referring to initial conditions of the undisturbed field, while x_o is the tube length. The physical space is presented by the x-c_0t diagram a, on the right, while the state (or phase) space is depicted by segments of c/c_o-q/c_o diagrams in different scales, b, c, d and e, on the left.

The moment the right end is opened, a rarefaction wave propagates to the left with its front moving at the initial velocity of sound, c_0. It is presented by a set of characteristics whose number is fixed by the assigned size of their mesh forming a rarefaction fan consisting of cells, each corresponding to a smaller velocity of sound and larger particle velocity according to the compatibility condition (7.24). Their locus of states is delineated by the descending diagonal on the upper edge of diagram. The rest of the solution is by the corresponding numbers of cells in diagram a and of the nodes in diagrams b, c, d and e. The rarefaction wave reflects from the open end as a compression wave presented by converging characteristics that, upon reflection from the closed end, are amplified to form a shock front delineated by a cluster of characteristics. Thereupon, the shock front reflects from the open end in the form of a rarefaction wave etc, establishing thereby subsequent periods of inflow and outflow marked on the right side of diagram *a.*

Of particular interest to the dynamics of combustion is the so-called pulse detonation engine made out of a tubular chamber where intermittent explosions take place at regular intervals. This type of a pulsating combustion system was developed towards the end of the Second World War to power the first unmanned flying bomb, the V1, referred to popularly as the buzz-bomb.

The intermittent combustion system was construed for it by an open-ended chamber fitted with a set of reed valves at the front. When pressure in the chamber was low, it was filled with ram air and when the pressure was elevated by combustion of air with fuel injected in the course of its intake, the intake valves were closed, while pulsating jets emanated from the open end providing thrust for propulsion.

A method of characteristics solution for a rudimentary example of such an operation is presented by Figs. 7.2 and 7.3 for a particular case of a pipe fitted with reed valves at the front where the Mach number of the flame front, $M_F \equiv \frac{S}{a} = 0.2$ (Oppenheim 1949).

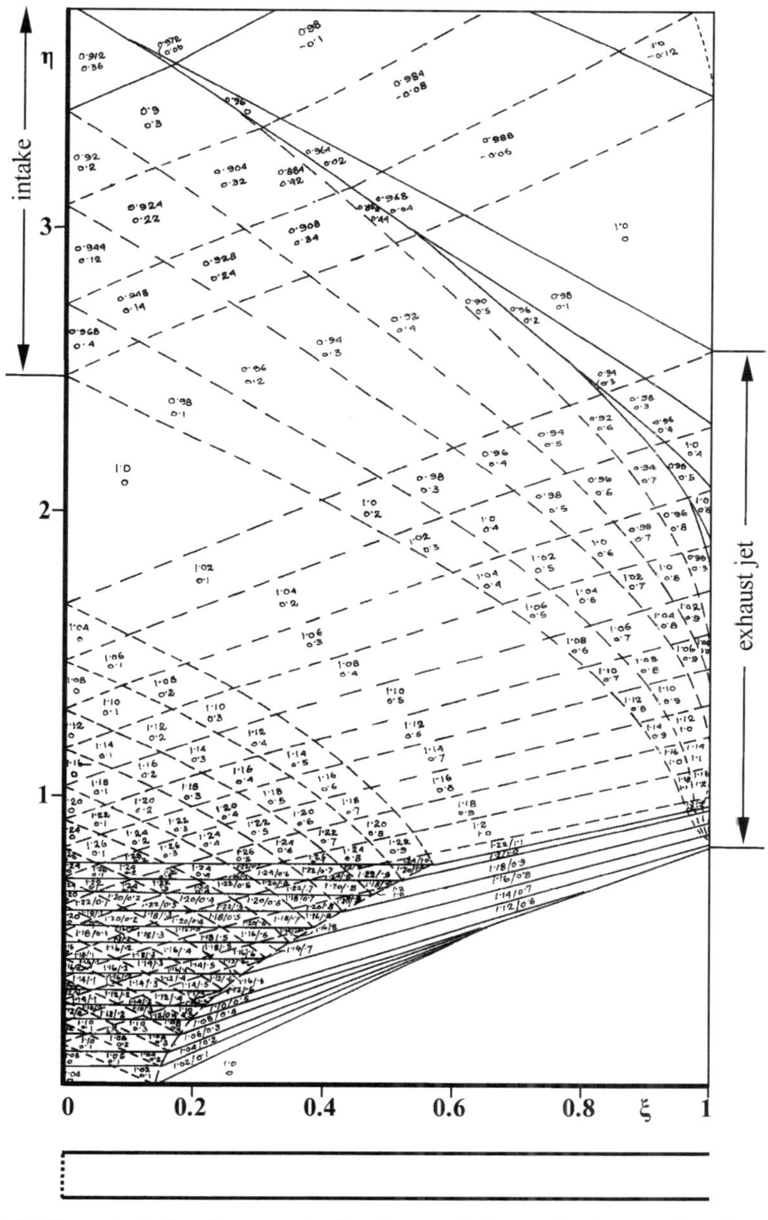

Fig. 7.2. Characteristics diagram for the first cycle in a pulsating combustion tube

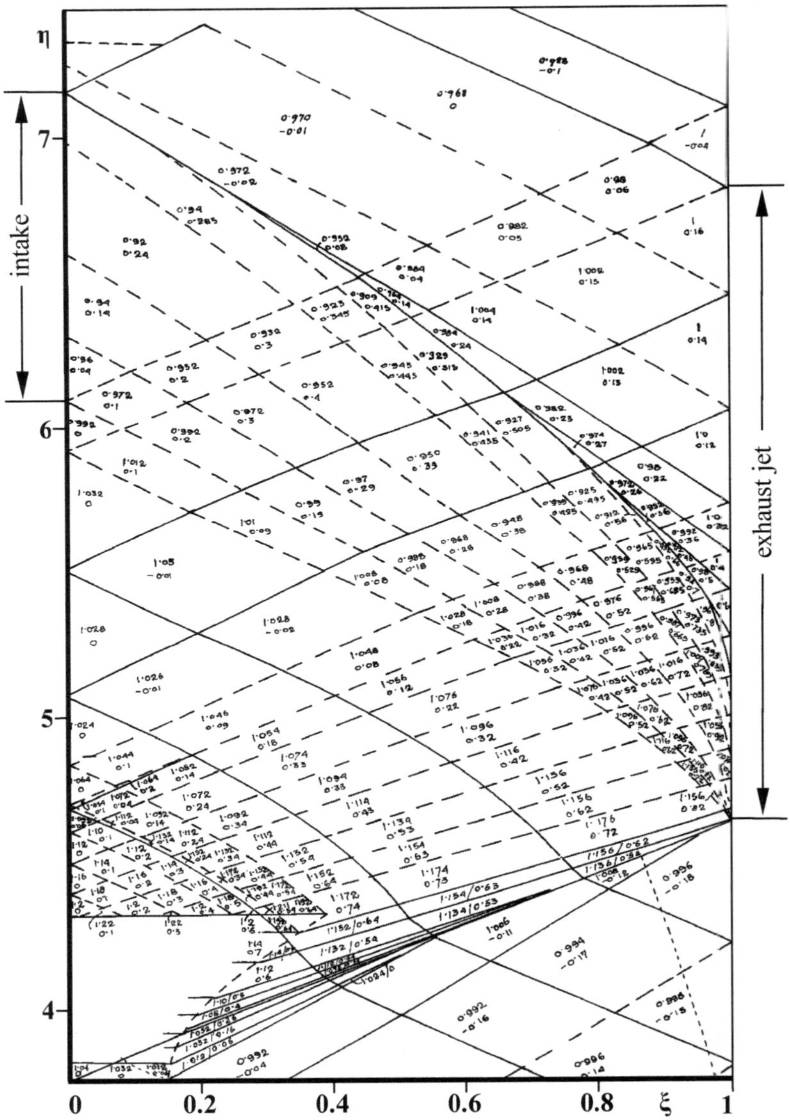

Fig. 7.3. Characteristic diagram for the second cCycle in a pulsating combustion tube

7.4 Exothermic Center

The gasdynamic properties of an exothermic center, introduced in section 1.3.2.2, are derived from constraints imposed by the compressible medium in which it resides. The latter are expressed by conservation equations in Lagrangian form for an inviscid gas. For mass, momentum, and energy, they are, respectively, as follows:

$$\frac{\partial v}{\partial t} = v[\frac{1}{r^j}\frac{\partial}{\partial r}(r^j u)] \tag{7.26}$$

$$\frac{\partial u}{\partial t} = -v\frac{\partial p}{\partial r} \tag{7.27}$$

$$\frac{\partial e}{\partial t} = -p\frac{\partial v}{\partial t} \tag{7.28}$$

where t and r are the time and space coordinates, v is the specific volume, u – the particle velocity, p – pressure, e – internal energy, while j = 0, 1, or 2 for, respectively, planar, cylindrical, or spherical geometry.

On this basis, computations were made for a stoichiometric hydrogen-oxygen mixture in a kernel of initial radius $r_i = 1$ mm, initial temperature, $T_i = 1336$ K, and initial pressure $p_i = 3.24$ atm, while the front of the center propagates into a gas of $\gamma_R = 1.14$ ($D_R = 8.1429$) and the local velocity of sound $a_o = 1$km/sec.

Front trajectories of exothermic centers are displayed by Fig. 7.4 in terms of non-dimensional parameters, $R \equiv r/r_i$ and $\tau \equiv (a_i / t)t$. Indicated by broken lines are their asymptotic limits corresponding to a volumetric expansion at constant pressure.

The process paths are presented on the pressure-specific volume diagram, Fig. 7.5. Time profiles of internal energy, $E \equiv e/w_i$, are shown by Fig. 7.6, while the power pulses of work done by the center in displacing the surrounding are depicted by Fig. 7.7. Displayed there, to convey a sense of magnitude, are pulses for processes at constant volume, V, and constant pressure, P, as well as along the Rayleigh line, R, for a Chapman-Jouguet deflagration.

As evident from Fig. 7.5, the process taking place in an exothermic center progresses along a path of a simultaneously increasing pressure and specific volume, bringing out its runaway nature.

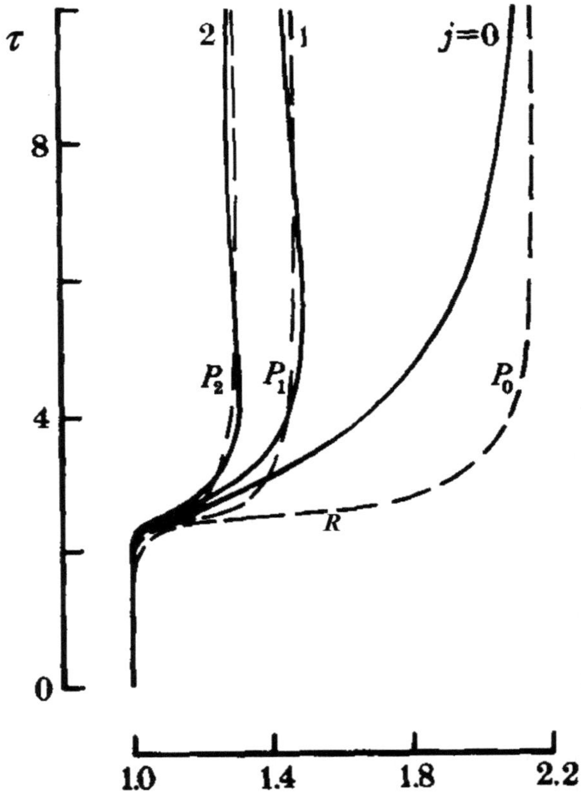

Fig. 7.4. Time-space trajectories of interfaces of exothermic centers in a stoichiometric hydrogen-oxygen mixture initially at T_i = 1336 K and P_i = 3.236 atm, in pPlanar (j = 0), cylindrical (j = 1), and spherical (j = 2) geometry (Cohen *et al.* 1975).
Broken lines delineate asymptotic limits of specific volume at constant pressure, P_0, its square root, P_1, and cube root, P_2.

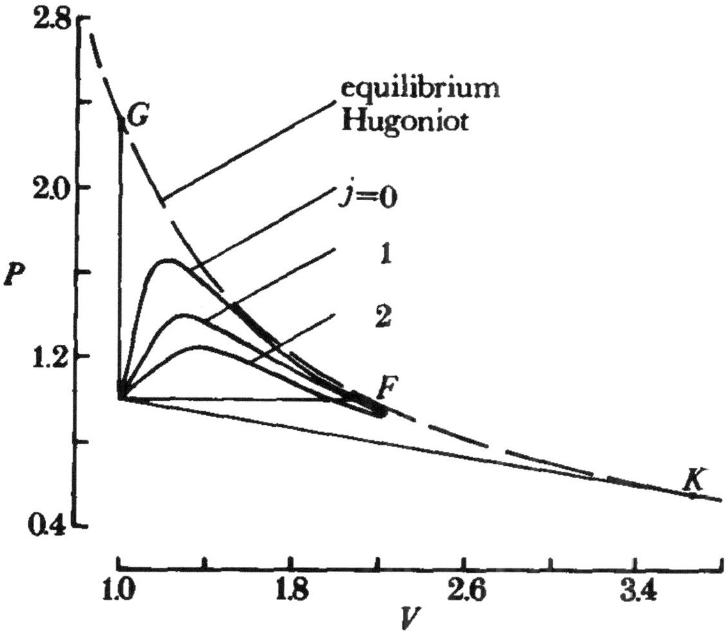

Fig. 7.5. Pressure-specific volume diagram of processes taking place in exothermic centers (Cohen *et at.* 1975)

When the energy generated by the exothermic reaction is deposited in the kernel of the center, its volume increases in conjunction with the temperature rise that, in turn, accelerates the reaction rate. The remarkable outcome of this process is the fact that the expenditure of energy for work in compressing the surroundings enhances the rate of chemical reaction.

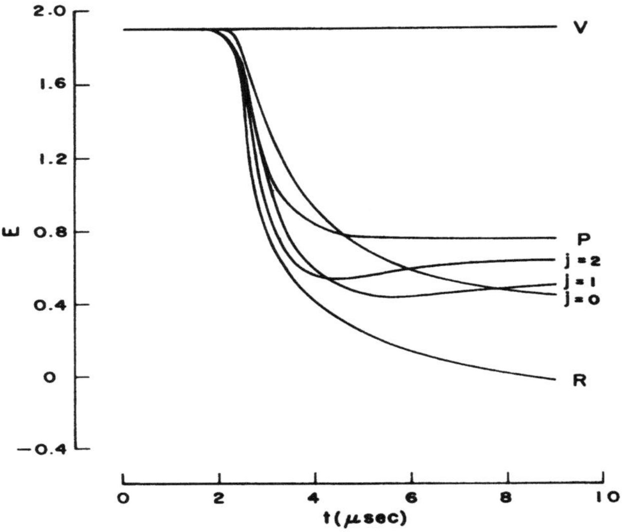

Fig. 7.6. Time profiles of internal energy in exothermic centers (Cohen et at. 1975)

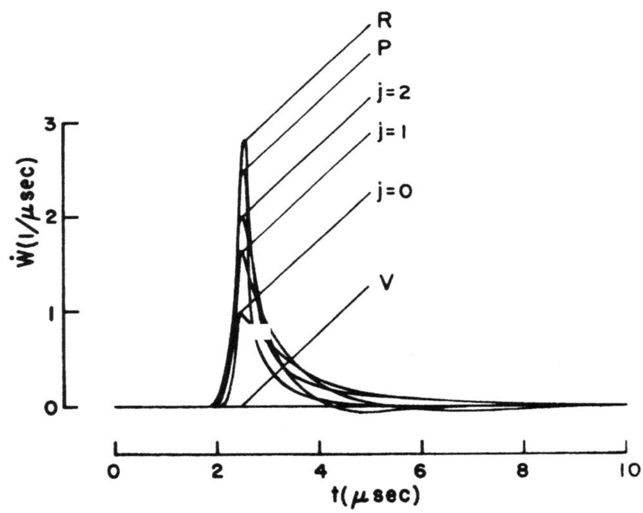

Fig. 7.7. Power pulses of work executed by exothermic centers (Cohen et at. 1975)

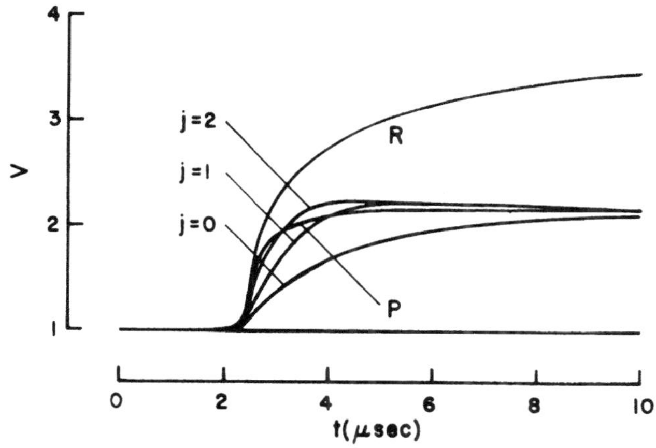

Fig. 7.8. Time profiles of specific volume in exothermic centers (Cohen et at. 1975)

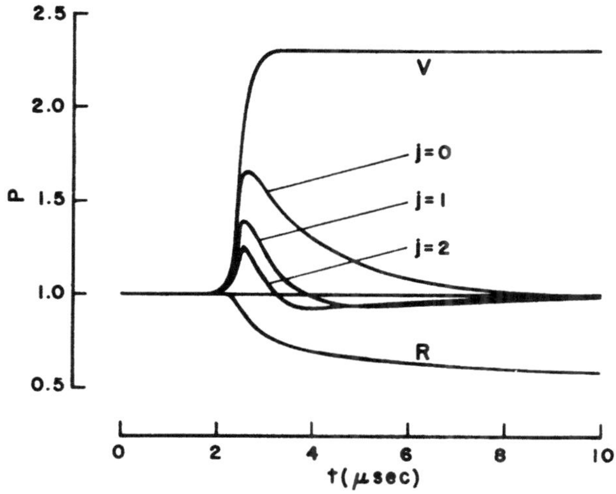

Fig. 7.9. Time profiles of pressure in exothermic centers (Cohen et at. 1975)

8 Fronts and Interfaces

8.1 Introduction

A gasdynamic front (referred to in the literature as a discontinuity) is a surface in the flow field across which a finite change of state takes place at constant mass flow rate and stream function per unit area; an interface is an impermeable front. Hence, according to these conditions, the mass rate per unit area

$$\dot{m} \equiv \rho_i \mathbf{v}_i \equiv \frac{\mathbf{v}_i}{v_i} = \rho_j \mathbf{v}_j \equiv \frac{\mathbf{v}_j}{v_j} \tag{8.1}$$

and constant stream force per unit area

$$f \equiv p_i + \dot{m}\mathbf{v}_i = p_j + \dot{m}\mathbf{v}_j \tag{8.2}$$

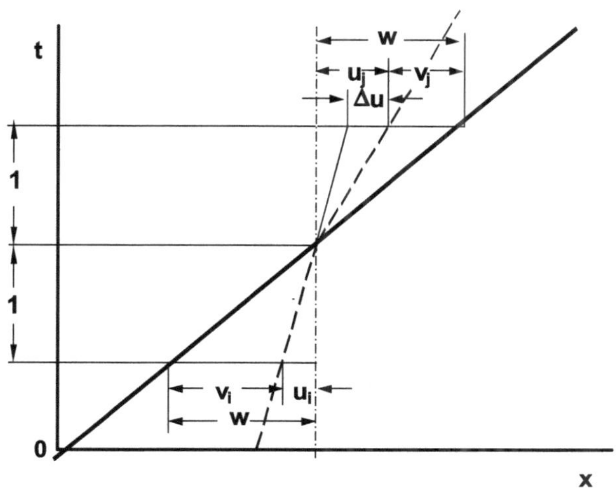

Fig. 8.1. Velocity change across a gasdynamic front in orthogonal plane to its front

As displayed in Fig. 8.1, the velocity at which the front propagates through the flow field

$$\mathbf{w} \equiv \mathbf{u}_i + \mathbf{v}_i = \mathbf{u}_j + \mathbf{v}_j \tag{8.3}$$

where $\mathbf{v}_k$ ($\mathbf{k} = \mathbf{i}, \mathbf{j}$) is the relative velocity with respect to the flow field immediately ahead of it and $\mathbf{u}_k$ is the particle velocity

Particular examples of gasdynamic fronts are shock fronts, rarefaction waves, detonation fronts and deflagration waves.

8.2 Change of State

According to (8.1) and (8.3), the change of particle velocity across a gasdynamic front,

$$\Delta\mathbf{u} \equiv \mathbf{u}_j - \mathbf{u}_i = \mathbf{v}_i - \mathbf{v}_j = \mathbf{v}_i\left(1 - \frac{\rho_i}{\rho_j}\right) = \mathbf{v}_i\left(1 - \frac{\mathbf{v}_j}{\mathbf{v}_i}\right) \tag{8.4}$$

whence the pressure jump across it, in view of (8.1) and, (8.2),

$$\Delta p \equiv p_j - p_i = \dot{m}(\mathbf{v}_i - \mathbf{v}_j) = \dot{m}\Delta\mathbf{u} = \dot{m}^2(v_j - v_i) \equiv -\dot{m}^2\Delta v \tag{8.5}$$

Thus,

$$\dot{m}^2 = -\frac{\Delta p}{\Delta v} \tag{8.6}$$

while

$$(\Delta\mathbf{u})^2 = \frac{(\Delta p)^2}{\dot{m}^2} = -\Delta p \Delta v \tag{8.7}$$

The energy conservation across a gasdynamic front is expressed by the condition

$$h_i + \frac{\mathbf{v}_i^2}{2} = h_j + \frac{\mathbf{v}_j^2}{2} \tag{8.8}$$

whence, the enthalpy change across a gasdynamic front, according to (8.1),

$$\Delta h \equiv h_j - h_i = \frac{1}{2}(\mathbf{v}_i^2 - \mathbf{v}_j^2) = \frac{\dot{m}^2}{2}(v_i^2 - v_j^2) = \frac{v_i + v_j}{2}\Delta p \tag{8.9}$$

and the internal energy change, by definition of enthalpy,

$$\Delta e \equiv \Delta h - \Delta pv = -\frac{p_i + p_j}{2}\Delta v \tag{8.10}$$

The relationships expressed by (8.9) and (8.10) are known as the Hugoniot equations.

8.3 Front Parameters

Front parameters are expressed in terms of variables normalized, as in Chapter 1, with respect to the coordinates of the initial state i, i.e. the pressure, $P \equiv p_j/p_i$, the specific volume, $\nu \equiv v_j/v_i$, the velocity of propagation, $\boldsymbol{V} \equiv \mathbf{v}_i/\sqrt{w_i} = \sqrt{\gamma_R}M_i$, the velocity change, $\boldsymbol{U} \equiv \Delta \boldsymbol{u}/\sqrt{w_i}$, and the velocity of sound, $\boldsymbol{A} \equiv \sqrt{a_j/\gamma_j w_i}$,

Thus, on the basis of (8.6),

$$\boldsymbol{V}^2 = \frac{P-1}{1-\nu} \tag{8.11}$$

according to (8.7),

$$\boldsymbol{U}^2 = (P-1)(1-\nu) \tag{8.12}$$

while

$$\boldsymbol{A}^2 = P\nu \tag{8.13}$$

For fixed velocity of propagation, $\mathbf{V} = \mathbf{V}_j/v = \sqrt{\gamma_R}\,M_i$ =const, (8.11) specifies a straight line on the pressure specific volume plane, known as the Rayleigh Line, inclined at a slope of

$$\Psi_R \equiv -(\frac{\partial P}{\partial v})_R = \frac{P-1}{1-v} = \frac{\mathbf{V}_j^2}{v^2} \tag{8.14}$$

that, at the limit of $j \Rightarrow i$, according to (1.47),

$$\Psi_s \equiv -(\frac{\partial P}{\partial v})_s = \gamma_j \frac{P}{v} = \frac{\mathbf{A}^2}{v^2} \tag{8.15}$$

Thus, a Rayleigh line stemming from state i is tangent to an isentrope passing through it and the local Mach number $M_j = \mathbf{V}_j / \mathbf{A} = 1$.

8.4 Hugoniot Curve

The Hugoniot Curve delineates the locus of states attained by a gas-dynamic front from a fixed initial state, i. Thus, in terms of $H \equiv h_j / w_i \equiv h_j / p_i v_i$, while $H_i = \text{const}$, its equation, (8.9), acquires a normalized form of

$$H - H_i = \frac{v+1}{2}(P-1) \tag{8.9'}$$

Its plots on the plane of pressure and specific volume are presented by Fig. 8.2, while Fig. 8.3 depicts it on the plane of temperature and entropy.

On curves denoted by H, state F is attained by an exothermic process at constant pressure, while state G by one at constant volume. The branch above point G represents the locus of states attained by detonation fronts – gasdynamic fronts propagating at Mach numbers $M_i > 1$, since the slopes of their Rayleigh Lines $\Psi_R > \Psi_S$. The branch below point F, corresponds to the states attained by deflagration fronts propagating at Mach numbers, $M_i < 1$. All the states between points G and F, are, according to (8.11), un-attainable by a single gasdynamic front.

In the particular case of $P_G = v_F = 1$, the curve portrays an adiabatic gas-dynamic front - a shock front - and it is referred to as the Rankine-Hugoniot curve denoted by RH. At the initial state, i, (8.9) is reduced to

(1.45), demonstrating that the RH-Curve is there tangent to the isentrope. The branch of the RH curve below point i has no physical meaning.

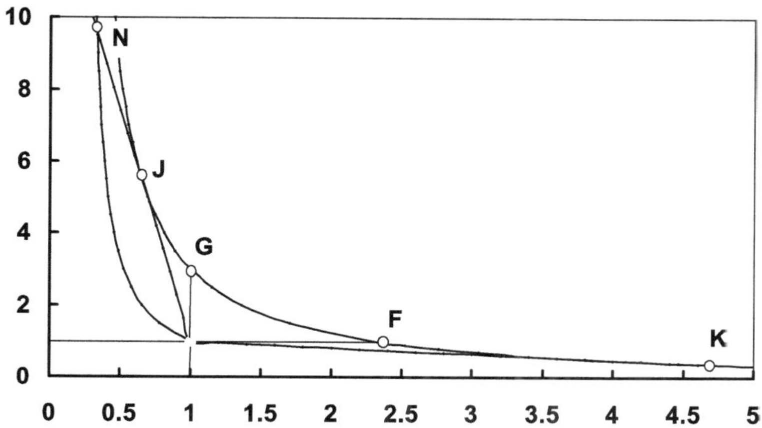

Fig. 8.2. Hugoniot curves on the plane of normalized pressure and specific volume

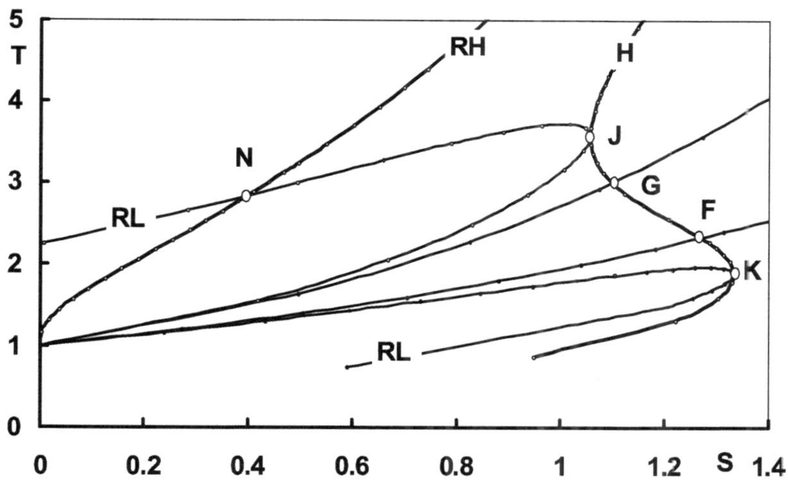

Fig. 8.3. Hugoniot curves on the plane of normalized temperature and entropy

Straight lines stemming from point i on the P-v plane of Fig. 8.2, are, according to (8.11) and (8.13), the Rayleigh Lines delineating the loci of

$\dot{m}$ = const. and f = const. Their slopes specify the Mach number at which a gasdynamic front propagates into the field immediately ahead of it. Point J represents the state attained by the Chapman-Jouguet detonation, whose Mach number $M_{CJ} > 1$. Point K depicts the state attained by the Chapman-Jouguet deflagration, whose Mach number $M_{CJ} < 1$. Point N is known as the von Neumann spike. It denotes the state attained by a shock font propagating at M_{CJ}. Point G represents the state attained by an exothermic process at constant volume. Point F represents the state attained by an exothermic process at constant pressure, referred to, therefore, as a constant pressure deflagration.

The local Mach numbers at states J and K are unity – a property referred to as the Chapman-Jouguet condition. Its proof, presented here on the basis of the classical arguments put forth by Jouguet (1917) and Becker (1922), is as follows.

By differentiating the Hugoniot equation, (8.9'), with respect to ν,

$$(\frac{\partial H}{\partial P})_{\mathrm{V}}(\frac{\partial P}{\partial \nu})_{\mathrm{H}} + (\frac{\partial H}{\partial \nu})_{\mathrm{P}} = \frac{\nu+1}{2}(\frac{\partial P}{\partial \nu})_{\mathrm{H}} + \frac{P-1}{2} \tag{8.16}$$

whence, the slope of the Hugoniot curve on the P-ν plane

$$\Psi_{\mathrm{H}} \equiv -(\frac{\partial P}{\partial \nu})_{\mathrm{H}} = \frac{(\frac{\partial H}{\partial \nu})_{\mathrm{P}} - \frac{P-1}{2}}{(\frac{\partial H}{\partial \nu})_{\mathrm{V}} - \frac{\nu+1}{2}} \tag{8.17}$$

The slope of the Rayleigh line, according to (8.11),

$$\Psi_{\mathrm{R}} \equiv -(\frac{\partial P}{\partial \nu})_{\mathrm{R}} = \frac{P-1}{1-\nu} \tag{8.18}$$

while for an isentrope, according to (1.45),

$$(\frac{\partial H}{\partial \nu})_{\mathrm{V}} \partial P + (\frac{\partial H}{\partial \nu})_{\mathrm{P}} \mathrm{d}\nu - \nu \mathrm{d}P = 0 \tag{8.19}$$

so that its slope

$$\Psi_s \equiv -(\frac{\partial P}{\partial v})_s = \frac{(\frac{\partial H}{\partial v})_P}{(\frac{\partial H}{\partial P})_v - v} \tag{8.20}$$

At the point where the Hugoniot curve is tangent to the Rayleigh line, $\Psi_H = \Psi_R$, according to (8.16) and (8.18),

$$[(\frac{\partial H}{\partial P})_v - \frac{v+1}{2}](P-1) = (\frac{\partial H}{\partial v})_P(1-v) - \frac{1-v}{2}(P-1) \tag{8.21}$$

whence

$$\frac{P-1}{1-v} = \frac{(\frac{\partial H}{\partial v})_P}{(\frac{\partial H}{\partial P})_v - v} \tag{8.22}$$

which, in view of (8.20), means that at this point, $\Psi_H = \Psi_R = \Psi_s$, as stipulated by the Chapman-Jouguet condition whose validity was thus demonstrated.

8.5 Linear State Trajectories

The significant states of the reactants, R and the products, P, for a stoichiometric hydrogen-oxygen mixture initially at NTP, are presented on the state diagram, $h(w)$, by Fig. 8.4

Their loci are expressed as straight state lines evaluated by linear regression. Their slopes

$$D_k \equiv \mathrm{d}h/\mathrm{d}w \equiv \frac{\gamma_k}{\gamma_k - 1} \qquad (\mathrm{k} = \mathrm{R, P}) \tag{8.23}$$

usher in the isentropic index, γ_K, that renders the algebraic relationships a familiar look of equations cited in the literature for the idealized case of a perfect gas with constant specific heats. However, unlike its consequent meaning as a ratio of specific heats, here it is a measure of the slope of a linear state trajectory.

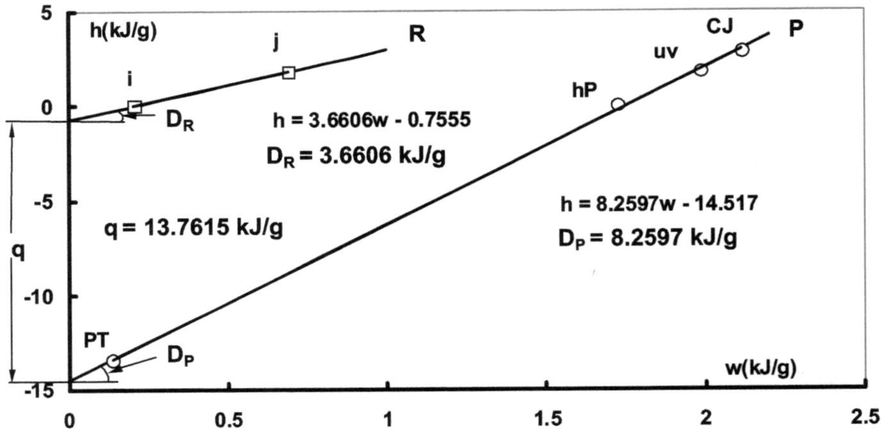

Fig. 8.4. State diagram for a stoichiometric hydrogen-oxygen mixture initially at NTP

As demonstrated by Fig. 8.4, straight state trajectories introduced in Chapter 1 for evaluation of thermodynamic parameters over their relatively short segments provides quite an accurate assessment magnitude of their magnitude over a full range of gasdynamic fronts from initial NTP conditions to the Chapman-Jouguet detonation. According to it, for a stoichiometric hydrogen oxygen mixture initially at NTP, $D_R = 3.660$ (k = i, j), whence $\gamma_R = 1.3759$, and $D_P = 8.2597$, whence $\gamma_P = 1.1377$, while their intercepts at w = 0 specifying the exothermic energy, q = 13.7615 kJ/g, that, with $w_i = 0.2063$ kJ/g, yields $Q \equiv q / w_i = 66.71$.

Since, at thermodynamic equilibrium, *H* is a unique function of *P* and *v*, the Hugoniot Curve, denoted by H, is, in effect, a locus of states at fixed exothermic energy, Q = const. Then, in terms of

$$\beta_k \equiv \frac{1}{2D_k - 1} = \frac{\gamma_k - 1}{\gamma_k + 1} \tag{8.24}$$

[whence $\gamma_k = (1+\beta_k)/(1-\beta_k)$], the Hugoniot equation (8.9') is expressed by the hyperbola

$$(P + \beta_P)(v - \beta_P) = C \tag{8.25}$$

where, with

$$\mathrm{P_G} = \frac{\mathrm{Q} + \mathrm{D_R} - 1}{\mathrm{D_P} - 1} = \frac{\beta_\mathrm{P}}{1 - \beta_\mathrm{P}}(\mathrm{Q} + \frac{1 + \beta_\mathrm{R}}{\beta_\mathrm{R}}) = (\gamma_\mathrm{P} - 1)(\mathrm{Q} + \frac{1}{\gamma_\mathrm{R} - 1}) \tag{8.26}$$

and

$$\nu_\mathrm{F} = \frac{\mathrm{Q} + \mathrm{D_R}}{\mathrm{D_P}} = \frac{2\beta_\mathrm{P}}{1 + \beta_\mathrm{P}}(\mathrm{Q} + \frac{1 + \beta_\mathrm{R}}{2\beta_\mathrm{R}}) = \frac{\gamma_\mathrm{P} - 1}{\gamma_\mathrm{P}}(\mathrm{Q} + \frac{\gamma_\mathrm{R}}{\gamma_\mathrm{R} - 1}) \tag{8.27}$$

$$\mathrm{C} = (\mathrm{P_G} + \beta_\mathrm{P})(1 - \beta_\mathrm{P}) = (1 + \beta_\mathrm{P})(1 - \nu_\mathrm{F}) \tag{8.28}$$

On this basis, according to (8.25) and (8.26),

$$\nu = \beta + (1 - \beta_\mathrm{P})\frac{\mathrm{P_G} + \beta_\mathrm{P}}{P + \beta_\mathrm{P}} = 1 - (1 - \beta_\mathrm{P})\frac{P - \mathrm{P_G}}{P + \beta_\mathrm{P}} \tag{8.29}$$

whence, by virtue of (8.11),

$$\boldsymbol{V}^2 = \frac{P + \beta_\mathrm{P}}{1 - \beta_\mathrm{P}}\frac{P - 1}{P - \mathrm{P_G}} \tag{8.30}$$

and, according to (8.12),

$$\boldsymbol{U}^2 = (1 - \beta_\mathrm{P})\frac{(P - \mathrm{P_G})}{P + \beta_\mathrm{P}}(P - 1) \tag{8.31}$$

while, in view of (8.13),

$$\boldsymbol{A}^2 = \frac{\mathrm{P_G} + \beta_\mathrm{P} + \beta_\mathrm{P}(P - \mathrm{P_G})}{(P + \beta_\mathrm{P})}P \tag{8.32}$$

Inverting (8.30) and taking into account the definition of $\boldsymbol{V}$,

$$P = \frac{(1 - \beta_\mathrm{P})(\gamma_\mathrm{R} M_\mathrm{i}^2 + 1)}{2} \pm \sqrt{\Delta} \tag{8.33}$$

where

$$\Delta = [\frac{(1-\beta_P)(\gamma_R M_i^2 + 1)}{2}]^2 - (1-\beta_P)P_G\gamma_R M_i^2 + \beta_P \tag{8.34}$$

Specified thus are the coordinates of two intersections between a straight Rayleigh Line stemming from the initial point, i, and the Hugoniot Curve specified by P_G. At the Chapman-Jouguet point, the two intersections coalesce into one, so that $\Delta = 0$, whence, according to (8.34),

$$\gamma_R M_{J,K}^2 = \frac{2P_G}{1-\beta_P} - 1 \pm \frac{2}{1-\beta_P}\sqrt{P_G^2 - (1-\beta_P)P_G - \beta_P} \tag{8.35}$$

Then, according to (8.25) and (8.26), the slope of the Hugoniot curve

$$\Psi_H \equiv -(\frac{\partial P}{\partial v})_H = \frac{P+\beta_P}{v-\beta_P} \tag{8.36}$$

At the Chapman-Jouguet state, J, where the Hugoniot curve is tangent to the Rayleigh line, (8.13) and (8.36) yield

$$(2P_J - 1 + \beta_P)(2v_J - 1 - \beta_P) = (1+\beta_P)(1-\beta_P) \tag{8.37}$$

The Rankine-Hugoniot curve, specifying the states created by an adiabatic shock front, corresponds to the locus of states of reactants, R, on Fig. 8.1 including the initial point, i, where the Mach number $M_i = 1$. It is expressed, therefore, by the Hugoniot relationships presented in the previous section with the stipulation that β_P is replaced by β_R, while $P_G = 1$ and $v_F = 1$. Its equation is thus

$$(P+\beta_R)(v-\beta_R) = (1+\beta_R)(1-\beta_R) \tag{8.38}$$

According to (8.37), in the particular case of $\beta_P = \beta_R$, the coordinates of the von Neumann spike, N, laying on the Rankine-Hugoniot curve, are related to those of the Chapman Jouguet state in terms of the following simple rule

$$P_N - 1 = 2(P_J - 1) \quad \text{and} \quad 1 - v_N = 2(1 - v_J) \tag{8.39}$$

This rule has been brought out by Langweiler 1938 upon the assumption that the products and the reactants behave as perfect gases with the same constant specific heats. Here it becomes clear appear that this is just a consequence of the assumption that their linear loci on the state diagram are parallel.

In view of (8.38), as a consequence of $P_G = 1$, eliminating γ_R by virtue of (8.24), (8.30) is reduced to

$$M_i^2 = \frac{P + \beta_R}{1 + \beta_R} \tag{8.40}$$

and, according to (8.31),

$$U_i^2 \equiv \frac{\Delta u}{a_i} = \frac{(1-\beta_R)^2 (P-1)^2}{(1+\beta_R)(P+\beta_R)} \tag{8.41}$$

while , (8.32) becomes

$$A_i^2 \equiv (\frac{a_j}{a_i})^2 = \frac{1+\beta_R P}{P+\beta_R} P \tag{8.42}$$

Inverting (8.40),

$$P = (1+\beta_R)M_i^2 - \beta_R \tag{8.43}$$

whence, by virtue of the above and (8.38),

$$\nu = \beta_R + \frac{1-\beta_R}{M_i^2} \tag{8.44}$$

while (8.41) becomes

$$U_i^2 = (1-\beta_R)(M_i - \frac{1}{M_i}) \tag{8.45}$$

and (8.42) yields

$$A_i^2 = (1+\beta_R)(1-\beta_R)(M_i^2 - \frac{\beta_R}{1+\beta_R})(\frac{1}{M_i^2} + \frac{\beta_R}{1-\beta_R}) \tag{8.46}$$

With (8.25) taken into account and omitting subscript R, the above expressions acquire the well-known forms of

$$P = \frac{2\gamma}{\gamma+1} M_i^2 - \frac{\gamma-1}{\gamma+1} \tag{8.43'}$$

$$v = \frac{\gamma-1}{\gamma+1} + \frac{2}{\gamma+1}\frac{1}{M_i^2} \tag{8.44'}$$

$$U_i^2 = \frac{2}{\gamma+1}(M_i - \frac{1}{M_i}) \tag{8.45'}$$

$$A_i^2 = (\frac{2}{\gamma+1})^2(\gamma M_i^2 - \frac{\gamma-1}{2})(\frac{1}{M_i^2} + \frac{\gamma-1}{2}) \tag{8.46'}$$

To elucidate the physical significance of these relationships, their graphical representation is provided in the following sections.

8.6 Propagating Fronts

Loci of states created by gasdynamic fronts are displayed by polar diagrams. For propagating fronts, their coordinates are made out of pressure ratio, P, or velocity of sound ratio, A, as ordinate, and flow velocity change, U, as abscissa. By expressing the ordinate in logarithmic scale, while the abscissa is in linear scale, the diagram is rendered a vector form, each polar acquiring the role of a hodograph. A font intersection event is then evaluated by vector addition.

Figure 8.5 presents shock polars on the P-U plane, evaluated by means of (8.43') and (8.45'), and a plot of Mach numbers at which the shock propagates, $M_i(P)$, computed by the use of (8.43') and (8.40), with respect to the isentropic index,γ, as a parameter.

Figure 8.6 displays shock polars on the A-U plane, evaluated by means of (8.46') and (8.40), and the Mach number of the front, M(P), with respect to isentropic index, γ, as a parameter. Figures 8.7 and 8.8 depict shock polars for a fixed isentropic index $\gamma = 1.4$, with respect to the velocity of sound in the field into which the front propagates, A_i, as a parameter

Figures 8.9 and to 8.10 present deflagration polars on, respectively, the P-U and A-U planes with an auxiliary plot of normalized velocities of propagation for a stoichiometric hydrogen-oxygen mixture, evaluated by the use of (8.30), (8.31) and (8.32) with data of Fig. 8.1.

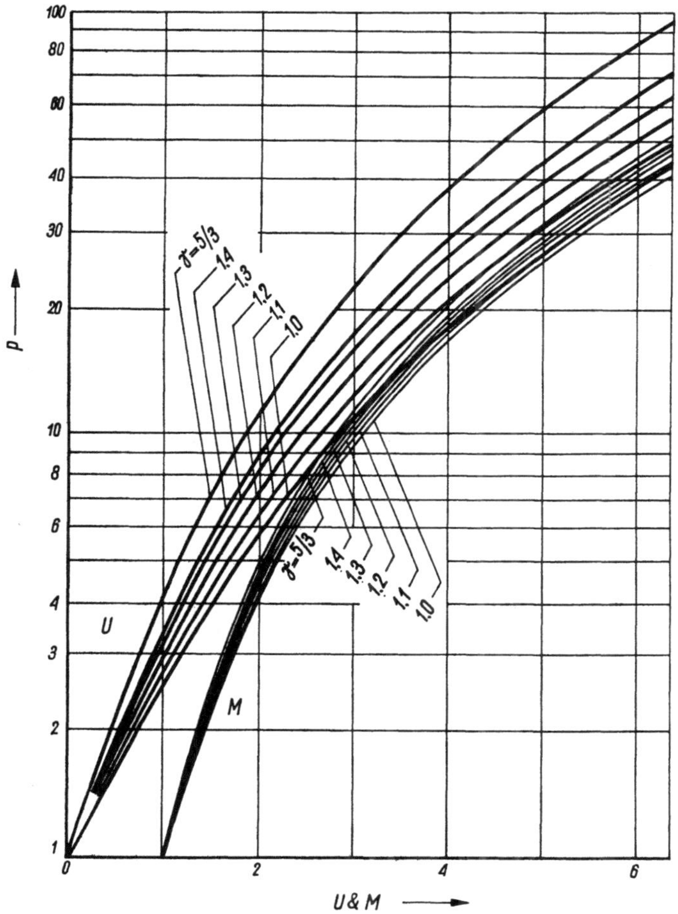

Fig. 8.5. Shock polars on the P-U plane and Mach numbers, M, with respect to the isentropic index, γ, as a parameter (Oppenheim 1970)

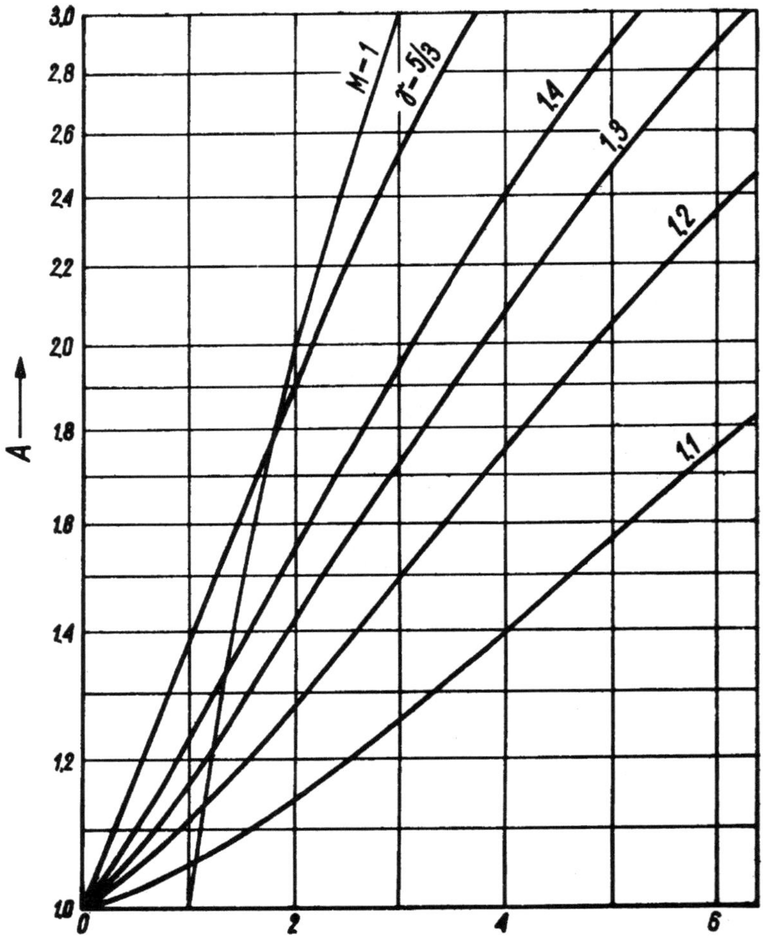

Fig. 8.6. Shock polars on the A-U plane with respect to the isentropic index, γ, as a parameter (Oppenheim 1970)

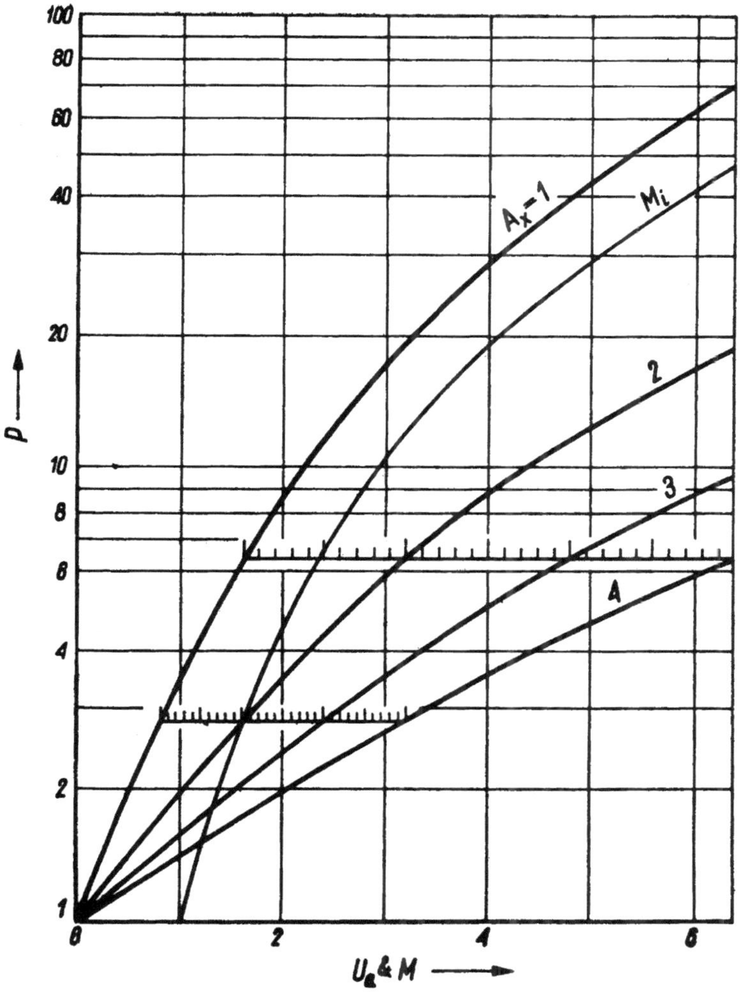

Fig. 8.7. Shock polars on the P-U plane and the Mach number, $M_i(P)$ for $\gamma = 1.4$ with respect to the velocity of sound, A_i as a parameter (Oppenheim 1970)

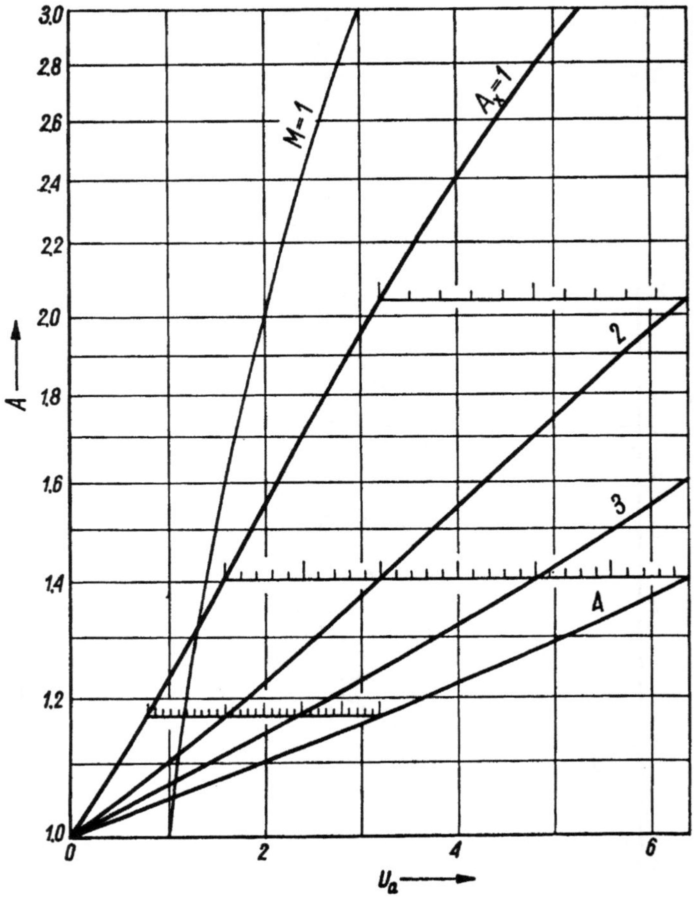

Fig. 8.8. Shock polars on the A-U plane for $\gamma = 1.4$ with respect to the velocity of sound, A_i, as a parameter (Oppenheim 1970)

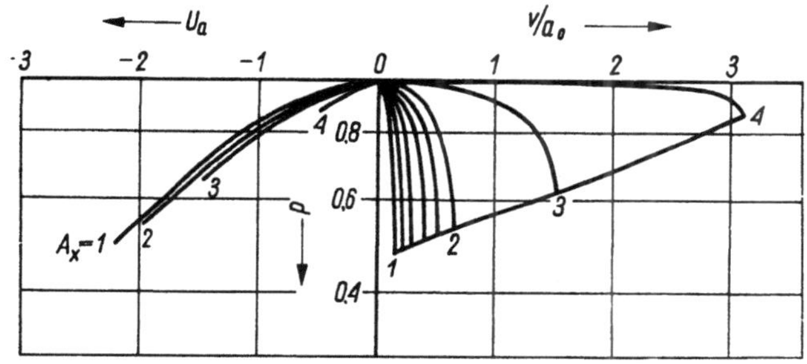

Fig. 8.9. Deflagration polars on the P-U plane for $\gamma = 1.4$ with respect to the velocity of sound, A_i, as a parameter (Oppenheim 1970)

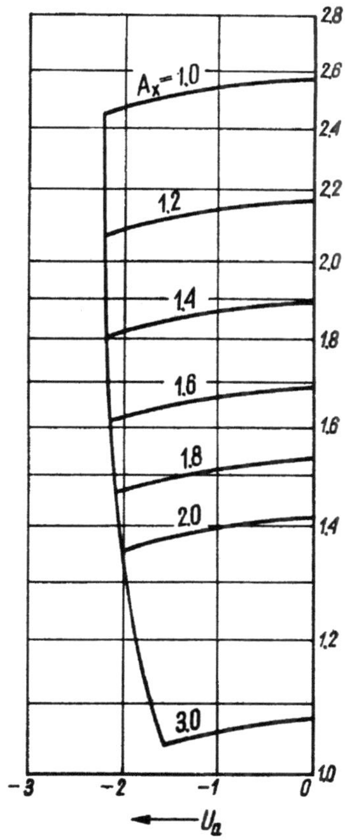

Fig. 8.10. Deflagration polars on the A-U plane with respect to the velocity of sound, A_i, as a parameter (Oppenheim 1970)

8.7 Simple Waves

A simple wave is made out of a set of sound waves, known as Mach waves, each propagating at a constant velocity of sound so that their world lines in time-space are straight. They model either the process of compression, where they coalesce into a shock front, or expansion, where they form a rarefaction fan within which an isentropic process takes place. In both cases a finite change of state is produced between two established regimes of flow. The effects of simple waves in a flow field are illustrated by Fig. 8.11, depicting an intersection between a compression wave and a rarefaction fan forming a sector of a continuous flow regime.

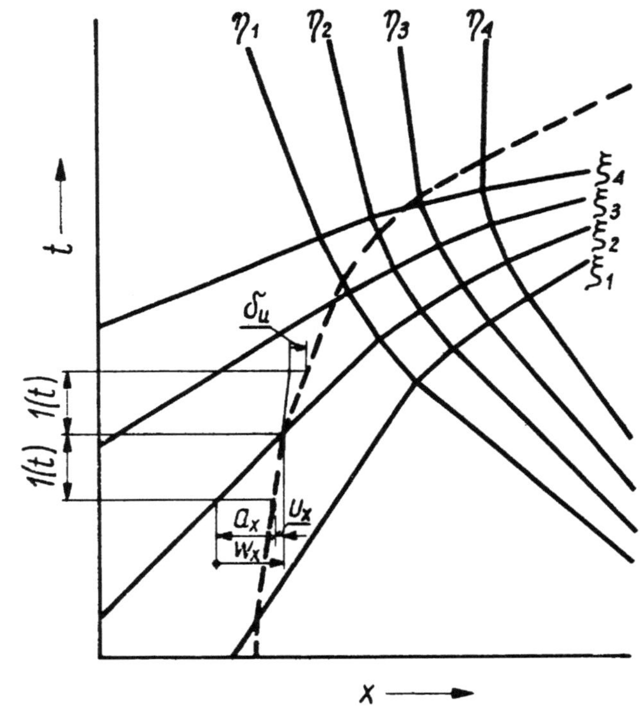

Fig. 8.11. Simple waves and their intersection in the physical plane

The change of state across a Mach wave is specified by (7.24) that, in its normalized form, provides a fundamental relationship between the local velocity of sound, $A \equiv a_j / a_i$, and the change in particle velocity, $U \equiv \Delta \boldsymbol{u}/a_i$,

$$A = \frac{\gamma - 1}{2} U + 1 \qquad (8.47)$$

while, since $A = P^{(\gamma-1)/2\gamma}$,

$$P = (\frac{\gamma - 1}{2} U + 1)^{2\gamma/(\gamma-1)} \qquad (8.48)$$

Rarefaction polars on the *U-P* coordinates, are displayed, according to (8.48), in Fig. 8.12 for a set of γ's, together with the corresponding plots of *A*(*P*) evaluated with the use of (8..47) specifying the *U-A* polars presented by Fig. 8.13.

Equivalent polar diagrams for a set of initial velocities of sound, with respect to which each rarefaction polar is normalized, are provided for γ = 1.4, by Figs. 8.14 and 8.15.

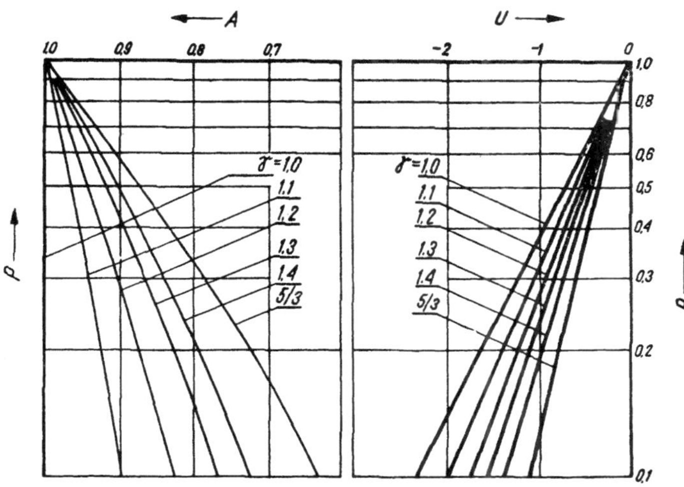

Fig. 8.12. Rarefaction polars on the P-U plane and the concomitant plot of *A*(*P*) with respect to the isentropic index, γ, as a parameter, (Oppenheim 1970)

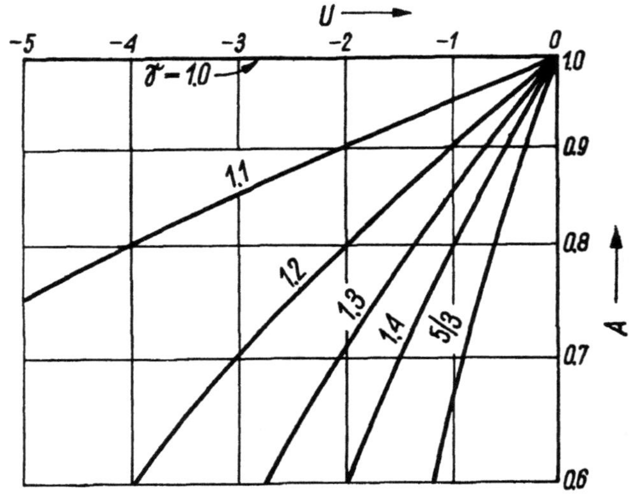

Fig. 8.13. Rarefaction polars on the A-U plane with respect to the isentropic index, γ, as a parameter (Oppenheim 1970)

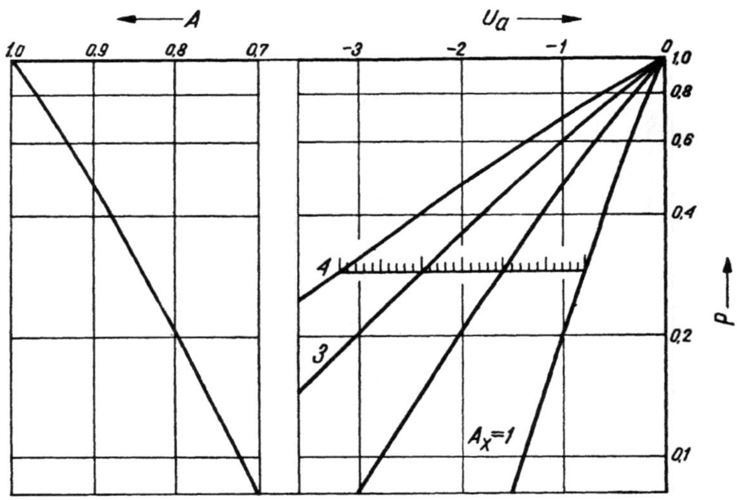

Fig. 8.14. Rarefaction polars on the P-U plane and the concomitant plot of $A(P)$ for the isentropic index of $\gamma = 1.4$, with respect to A_i as a parameter (Oppenheim 1970)

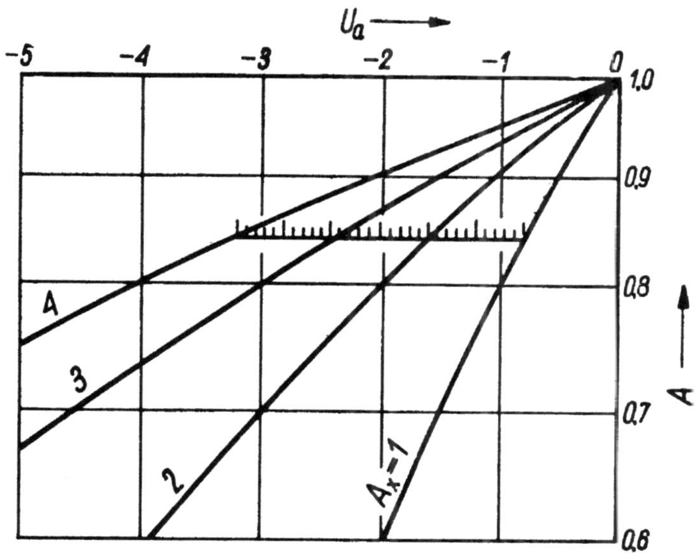

Fig. 8.15. Rarefaction polars on the A-U plane for the isentropic index of $\gamma = 1.4$, with respect of A_i as a Parameter, according to (8.47) (Oppenheim 1970)

8.8 Double Fronts

The concept of a double front system stems from classical contributions of von Neumann 1941, Döring 1943, and Zeldovich 1941, made during the Second World War to reveal the gasdynamic nature of detonation fronts that became known as the NDZ detonation structure. The model is made out of a shock front followed by a deflagration. Both of them move at the Chapman-Jouguet velocity, forming thus a steady state system. Subsequently, this model has been generalized by allowing the two fronts to propagate at different velocities in the form of a non-steady double front system (Oppenheim 1952, 1953), capable to provide a one-dimensional description of transition from deflagration to detonation – an elementary solution to what became known in the literature as the DTD problem.

Such a double front system is presented by Fig. 8.16. The process of transition is postulated as one involving first the products of deflagration to attain the Chapman-Jouguet state, K, of $M_K = 1$ and, thereupon, keep it at this level from up to the establishment of the Chapman-Jouguet detonation state, J.

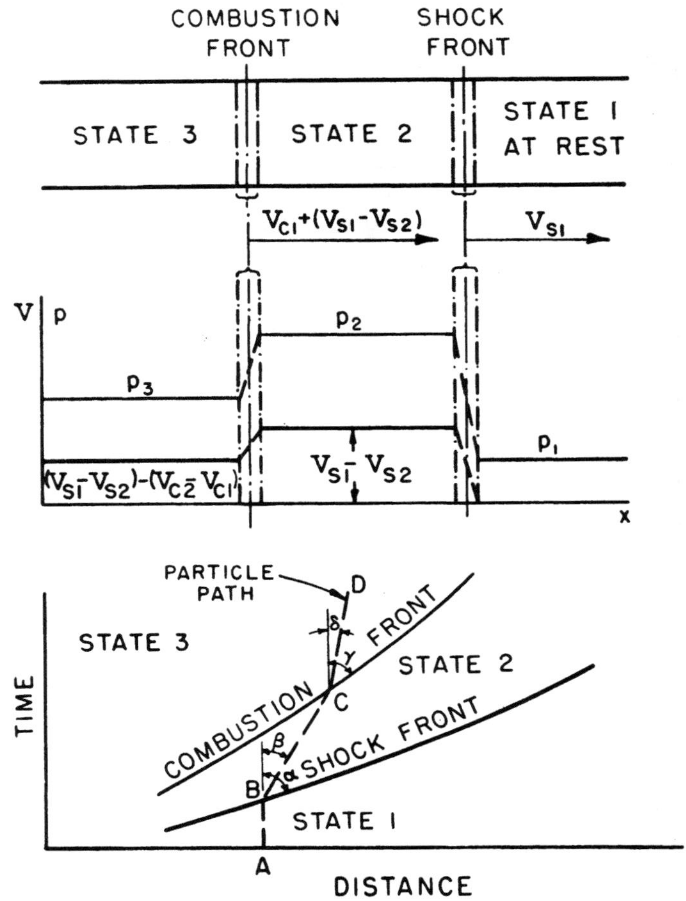

Fig. 8.16. A double front system (Oppenheim 1953)

The locus of states satisfying this condition is called the Q Curve. Its coordinates are evaluated by the use of the Rankine-Hugoniot equation, (8.38), whence

$$(P_2 + \beta_R)(\nu_2 - \beta_R) = (1 + \beta_R)(1 - \beta_R) \tag{8.49}$$

for transition across the shock front, the Hugoniot equation, (8.24), whence

$$(P_3 + \beta_P P_2)(v_3 - \beta_P v_2) = [(2Q + D_R P_2 v_2) - (1 + \beta_P)]\beta_P P_2 v_2 \qquad (8.50)$$

for transition across the deflagration front, and the Rayleigh Line equation, whence

$$\frac{P_3 - P_2}{v_2 - v_3} = \gamma_P \frac{P_3}{v_3} \qquad (8.51)$$

establishing the condition of $M_3 = 1$. Upon elimination of P_2 and v_2 from the above three equations, a unique function $P_3(v_3)$ is obtained.

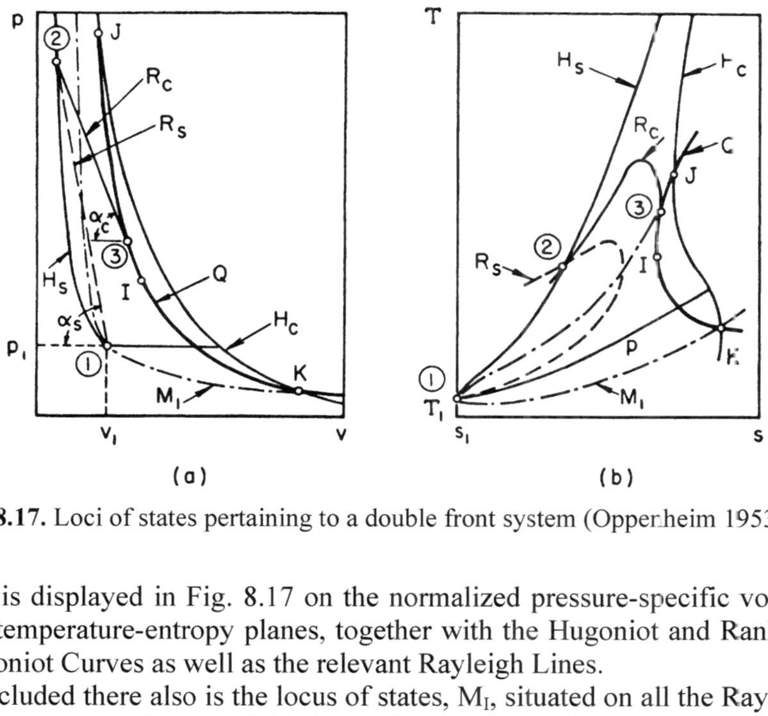

Fig. 8.17. Loci of states pertaining to a double front system (Oppenheim 1953)

It is displayed in Fig. 8.17 on the normalized pressure-specific volume and temperature-entropy planes, together with the Hugoniot and Rankine-Hugoniot Curves as well as the relevant Rayleigh Lines.

Included there also is the locus of states, M_I, situated on all the Rayleigh Lines stemming from the initial state, i. Its specification is provided by the condition of

$$\frac{P_2 - 1}{1 - v_2} = \gamma_R \frac{P_2}{v_2} \qquad (8.52)$$

whence

$$P_2 = [(1+\gamma_R(1-v_2^{-1})]^{-1} \tag{8.53}$$

The state of minimum entropy on the Q Curve is marked by symbol I. It corresponds to the case when the mass flow rates across the two fonts are the same, so that the velocity of state3 is equal to that of state 1(i.e. zero in the case illustrated by Fig. 8.16). Its coordinates are determined from the condition of $m_s^2 = m_c^2$, i.e., according to (8.6),

$$(v_3 - v_2)(p_2 - p_3) = (v_1 - v_2)(p_2 - p_1) \tag{8.54}$$

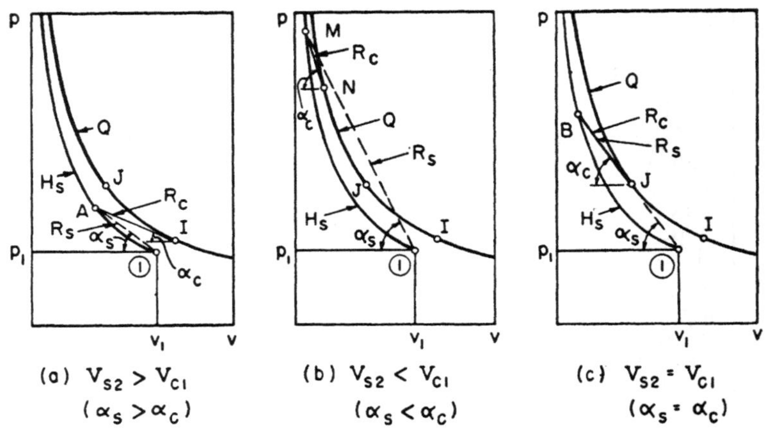

Fig. 8.18. Transition from deflagration to detonation of a double front system on the p-v diagram (Oppenheim 1953)

The Q Curve and the locus of M_I states pass through both the Chapman-Jouguet points, K and J, the former tracing the intermediate states attained in transition from K to J by a hypothetical, one-dimensional, double front system, and the latter delineating the states of maximum entropy on .the Rayleigh Lines stemming from the initial state **i**.

As displayed in Fig. 8.18a, as long as states 3 lies between states K and J, the slopes of the Rayleigh lines $\alpha_s > \alpha_c$. The velocity of state 2, created by the shock front, is therefore higher than that at which the deflagration propagates into its regime. Under such circumstances, the deflagration cannot catch up with shock front ahead of it. It is only when state 2 is above state J, as depicted in Fig. 8.18b, that $\alpha_c > \alpha_s$ and the deflagration can capture the shock front to establish the Chapman Jouguet detonation when $\alpha_s = \alpha_c$, as illustrated by Fig. 8.18c. For that reason, the attainment of overpressure with respect to the Chapman Jouguet pressure, P_J, appears to be a necessary prerequisite for a DTD transition to take place – a fact well documented by experimental evidence. A history of this event, in the case of such a double front system in a one-dimensional flow field is portrayed by Fig. 8.19.

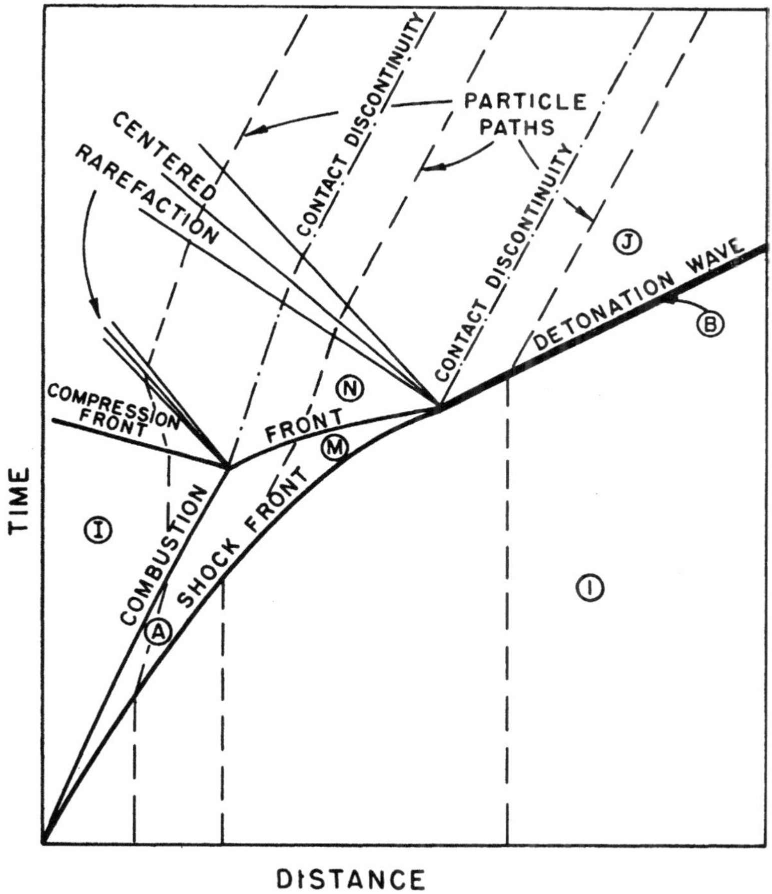

Fig. 8.19. Transition from deflagration to detonation by a double front system on the space-time diagram (Oppenheim 1953)

8.9 Oblique Fronts

The change of state across oblique fronts in a planar flow field is affected only by the geometry of the flow field. In an inviscid flow field gasdynamic fronts can be influenced only by the component of the velocity vector normal to their surfaces, while the tangential component remains unaltered. Kinematic consequences of this condition are portrayed by Fig. 9.20, where the normal velocity vector, $\mathbf{v}_i$ and the incident flow velocity vector, $\mathbf{w}_j$, have a common tangential component, $\mathbf{v}_t$.

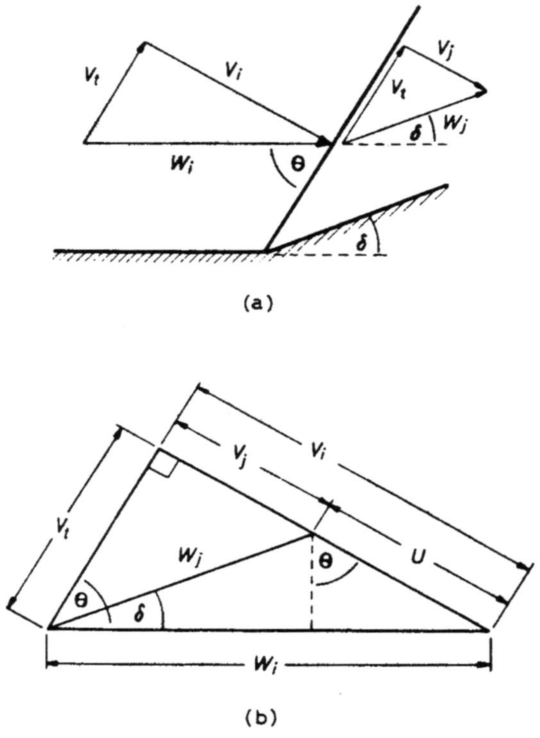

Fig. 8.20. Oblique front and its velocity vectors

As apparent from Fig. 8.20 , the angle of incidence

$$\Theta = \sin^{-1}\left(\frac{\mathbf{v}_i}{\mathbf{w}_i}\right) \tag{8.55}$$

and the flow deflection angle

$$\delta = \cot^{-1}[(\frac{\mathbf{w}_i}{\mathbf{u}\sin\Theta} - 1)\tan\Theta] \tag{8.56}$$

while the flow velocity immediately behind the oblique front

$$\mathbf{w}_j^2 = \mathbf{w}_i^2 + \mathbf{u}^2 - 2\mathbf{v}_i\mathbf{u} \tag{8.57}$$

Polar diagrams of oblique fronts are expressed similarly as those for propagating fronts except that, instead of the flow velocity change, *U*, the abscissa is expressed in terms of its deflection angle, *δ*.

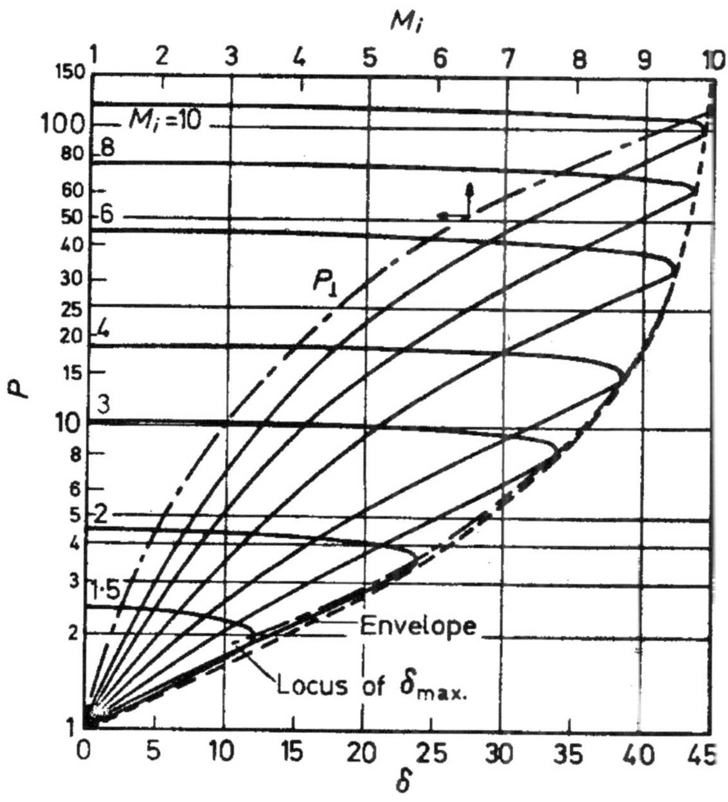

Fig. 8.21. Oblique shock polars on the P-δ plane for γ = 1.4, according to (8.59) with (8.43') (Oppenheim al 1968)

In terms of the Mach number of incident flow, $M_i \equiv \dfrac{\mathbf{w}_i}{a_i}$, (8.55), taking into account (8.11), becomes

$$\Theta = \sin^{-1}\sqrt{\frac{P-1}{\gamma_R M_i^2(1-\nu)}} \tag{8.58}$$

whence, in view of (8.12), (8.56) yields

$$\delta = \cot^{-1}[(\frac{\gamma_R M_i^2}{P-1}-1)\tan\Theta] \tag{8.59}$$

while, on the basis of (8.11) and (8.12),

$$M_j^2 = \frac{\gamma_R M_i^2-(P-1)(\nu+1)}{\gamma_j P\nu} \tag{8.60}$$

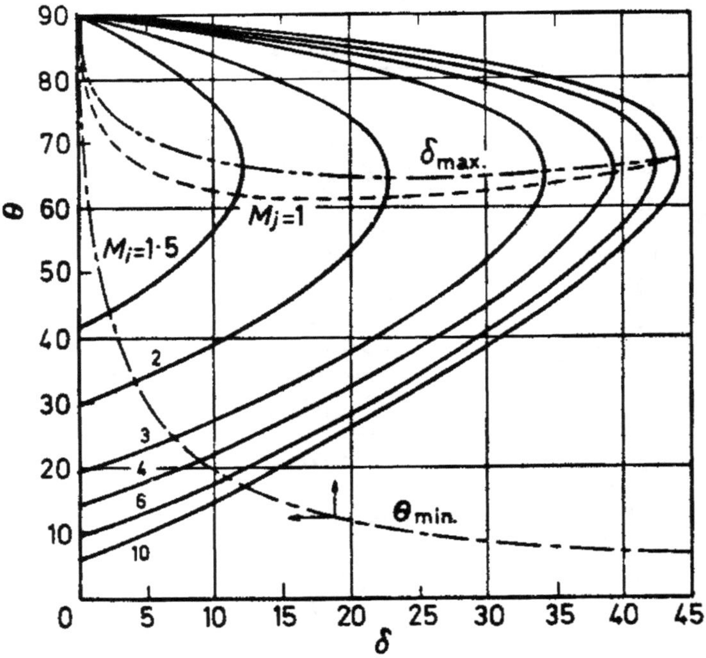

Fig. 8.22. Flow deflection angle across oblique shock fronts for γ = 1.4, according to (8.59) with (8.43') (Oppenheim al 1968)

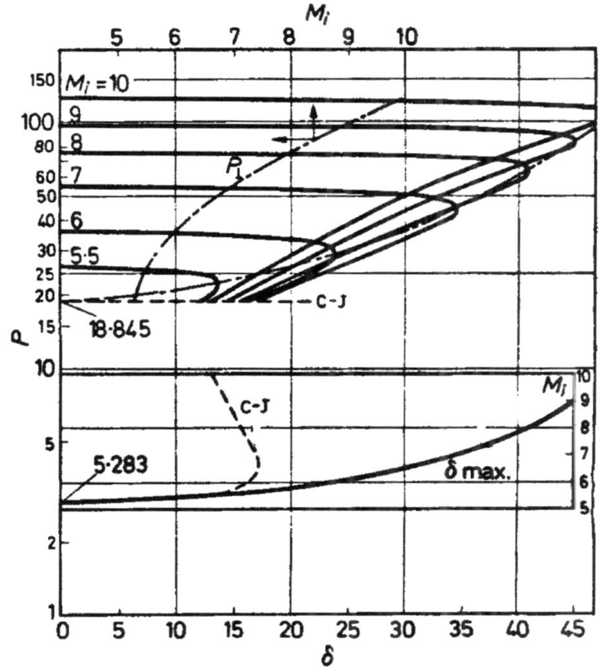

Fig. 8.23. Oblique detonation polars on the P-d plane for a stoichiometric hydrogen-oxygen mixture, according to (8.59), (8.58) (8.30) and (8.43') (Oppenheim et al 1968)

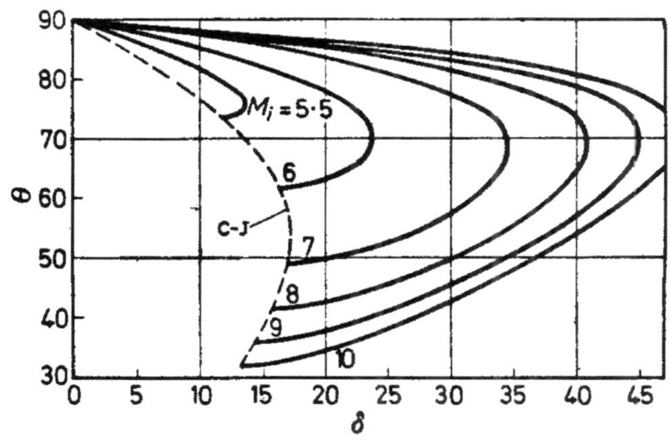

Fig. 8.24. Flow deflection angles across oblique detonation fronts for a stoichiometric hydrogen-oxygen mixture, according to (8.59), (8.58). (8.30) and (8.43) (Oppenheim et al 1968)

8.10 Prandtl-Meyer Expansion

8.10.1 Kinematic Relations

Prandtl-Meyer expansion is a rarefaction fan in a steady two-dimensional flow field whose velocity diagram is presented by Fig. 8.23, together with its hodograph and expansion fan

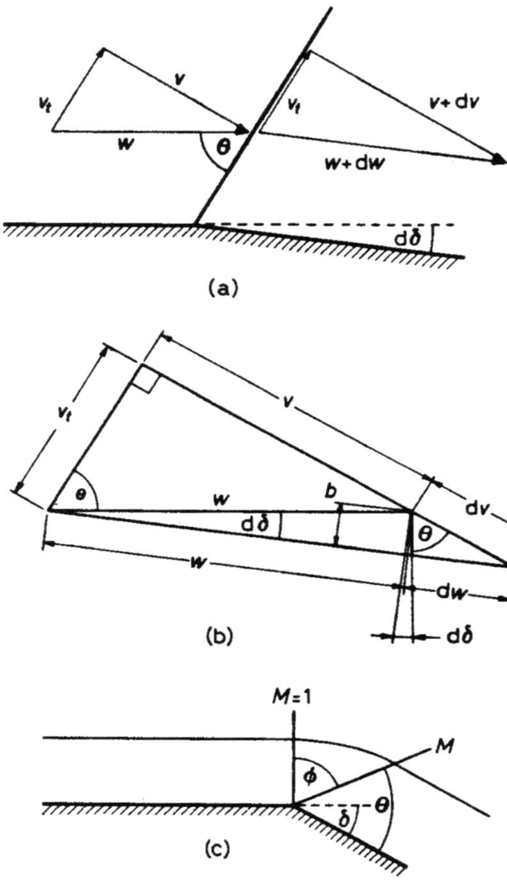

Fig. 8.25. The Prandtl-Meyer Expansion
(a) *Velocity diagram*); (b) *Velocity hodograph*; (c) *Expansion fan*

The height of the lower triangle in Fig. 8.25b

$$b = (2\boldsymbol{w}\sin^2\frac{\mathrm{d}\delta}{2} + \mathrm{d}\boldsymbol{w})\cot(\theta + \mathrm{d}\theta) \tag{8.61}$$

or, in the limit of vanishing second order terms,

$$b = \cot\theta\,\mathrm{d}\boldsymbol{w} \tag{8.62}$$

whence

$$\mathrm{d}\delta = \frac{b}{\boldsymbol{w}} = \cot\theta\frac{\mathrm{d}\boldsymbol{w}}{\boldsymbol{w}} \tag{8.63}$$

Since, as evident from Fig. 8.5b, $\frac{\mathrm{d}\boldsymbol{w}}{\boldsymbol{w}} = \frac{\mathrm{d}\boldsymbol{v}}{\boldsymbol{v}}$,while $\boldsymbol{v} = a$ and, hence, $\sin\theta = \mathrm{M}^{-1}$, the above becomes

$$\mathrm{d}\delta = \sqrt{M^2 - 1}\frac{\mathrm{d}\boldsymbol{w}}{\boldsymbol{w}} \tag{8.64}$$

and, noting that $\mathrm{d}\theta \equiv \frac{\mathrm{d}\sin\theta}{\sqrt{1-\sin^2\theta}}$,

$$\mathrm{d}\theta = -\frac{\mathrm{d}M}{M\sqrt{M^2 - 1}} \tag{8.65}$$

8.10.2 Thermodynamic Relations

The energy balance across a Mach wave in the expansion fan is

$$\mathrm{d}h + \boldsymbol{w}\mathrm{d}\boldsymbol{w} = 0 \tag{8.66}$$

whence

$$\frac{\mathrm{d}\boldsymbol{w}}{\boldsymbol{w}} = -\frac{\mathrm{d}h}{\mathrm{d}w}\frac{\mathrm{d}w}{\boldsymbol{w}^2} \tag{8.67}$$

Since, according to (8.23), $\frac{dh}{dw} \equiv D_R \equiv \frac{\gamma}{\gamma - 1}$, while $dw = \frac{2a}{\gamma} da$ and $\frac{da}{a} \equiv \frac{d\mathbf{w}}{\mathbf{w}} - \frac{dM}{M}$,

$$\frac{d\mathbf{w}}{\mathbf{w}} = -\frac{\alpha}{M^2}\frac{da}{a} = -\frac{\alpha}{M^2}\left(\frac{d\mathbf{w}}{\mathbf{w}} - \frac{dM}{M}\right) \tag{8.68}$$

where $\alpha \equiv \frac{2}{\gamma - 1}$. Thus,

$$\frac{d\mathbf{w}}{\mathbf{w}} = -\frac{\alpha}{M^2 + \alpha}\frac{dM}{M} \tag{8.69}$$

and (8.64) becomes

$$d\delta = \frac{\alpha\sqrt{M^2 - 1}}{M^2 + \alpha}\frac{dM}{M} \tag{8.70}$$

8.10.3 Expansion Fan

The width of an expansion fan, from its leading edge to trailing edge, is, according to Fig. 8.25c,

$$\phi = \frac{\pi}{2} + \delta - \theta \tag{8.71}$$

whence, with (8.70) and (8.65),

$$d\phi = d\delta - d\theta = \frac{\alpha + 1}{\mu^2 + \alpha + 1} d\mu \tag{8.72}$$

where $\mu \equiv \sqrt{M^2 - 1}$. Integrating the above, subject to initial condition of $\phi = 0$ at $M = 1$,

$$\phi = \sqrt{\alpha + 1} \tan^{-1} \frac{\mu}{\sqrt{\alpha + 1}} \tag{8.73}$$

while, according to Fig. 8.25b,

$$\theta = \frac{\pi}{2} - \cos^{-1} M^{-1} \tag{8.74}$$

whence, according to (8.71), in terms of $M = \sqrt{\mu^2 + 1}$ and, $\gamma = \dfrac{2 + \alpha}{\alpha}$

$$\delta = \sqrt{\frac{\gamma + 1}{\gamma - 1}} \tan^{-1} \sqrt{\frac{\gamma - 1}{\gamma + 1}(M^2 - 1)} - \cos^{-1} M^{-1} \tag{8.75}$$

Finally, a relationship between the local ach number and pressure is obtained from (8.68), according to which

$$\frac{dM}{M} = -\frac{M^2 + \alpha}{M^2} \frac{da}{a} \tag{8.76}$$

and, since the flow field is isentropic, so that, according to (1.49), $\dfrac{da}{a} = \dfrac{\gamma - 1}{2\gamma} \dfrac{dP}{P}$,

$$\frac{MdM}{M^2 + \alpha} = -\frac{\gamma - 1}{2\gamma} \frac{dP}{P} \tag{8.77}$$

whose integral, subject to initial conditions of $M=1$ at $P=1$, is

$$M^2 = \frac{\gamma + 1}{\gamma - 1} P^{-\frac{\gamma - 1}{\gamma}} - \frac{2}{\gamma - 1} \tag{8.78}$$

Shock Polar on the P-δ Plane for the Prandtl-Meyer expansion and its auxiliary functions $M(P)$ and $\Theta(P)$, according to (8.74), (8.75) and (8.78) with $\gamma = 1.4$, are displayed by Fig. 8.26.

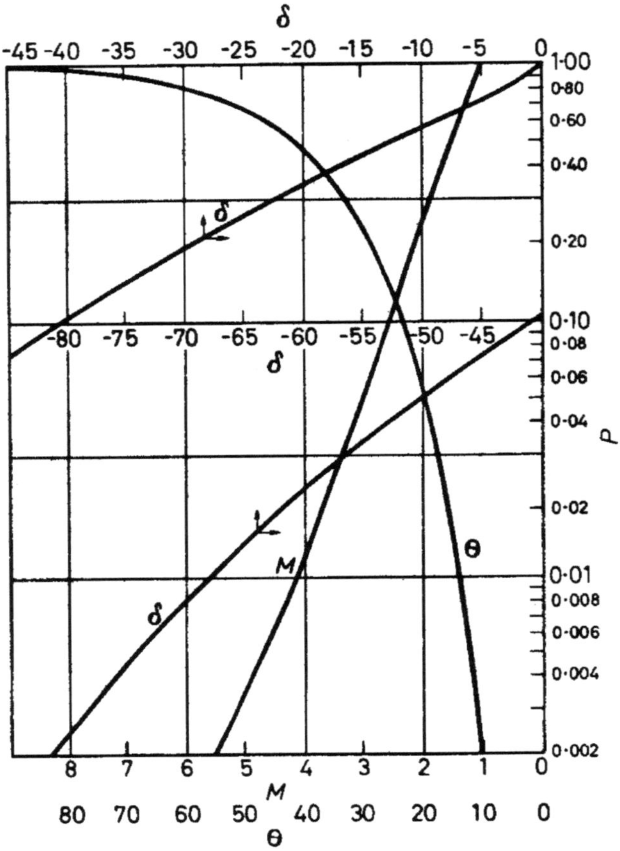

Fig. 8.26. Front polar on the P-δ plane for Prandtl-Meyer expansion and its auxiliary functions *M*(*P*) and *Θ*(*P*), according to (8.74), (8.75) and (8.79), for y = 1.4

8.11 Front Interactions

Interactions between gasdynamic fronts take place in one-dimensional unsteady flow fields as either head-on collisions or rear-end merging (vid. von Neumann 1943, Courant and Friedrichs 1948, Glass and Hall 1958). Presented here are four cases: (1) interactions in a shock tube; (2) head-on collision of a shock front with a deflagration; (3) shock merging; (4) merging of a shock front with a rarefaction wave.

Each of them is illustrated by space time wave diagrams on the physical plane of *T*(*X*), where $T \equiv \dfrac{t}{a_0 L}$ and $X \equiv \dfrac{x}{L}$, and by state planes of *P*(*U*) and $A(U)$, where $P \equiv \frac{p}{p_o}$, $U \equiv \frac{u}{a_o}$. and $A \equiv \frac{a}{a_o}$. Shock fronts are delineated by continuous lines, gasdynamic interfaces by broken lines, fronts of rarefaction fans by chain-dotted lines, and their rear borders by chain-double-dotted lines. Solutions were obtained by the vector polar method presented by Oppenheim et al (1964).

8.11.1 Shock Tube

A shock tube, in its simplest form, is a constant cross-section area duct filled with a gas, separated by a diaphragm into two sections, one at a higher pressure than the other. Upon a sudden removal of the diaphragm, a shock front is formed, propagating into the low pressure section while a rarefaction fan expands into the high pressure section.

The initial sequence of front interactions taking place in the shock tube immediately after the diaphragm is removed is illustrated in Fig. 8.27 for the particular case of the initial pressure ratio across the diaphragm, P_4 = 2.62. The gases on both sides of the diaphragm are initially at the same temperature and their isentropic index, $\gamma = 1.4$. The shock propagating into state 0 is modeled by shock polars of Figs.8.9 and 8.10 for $A_x = 1$, starting from point U = 0, P = 1 and A = 1. The rarefaction fan expanding into state 4 is modeled by rarefaction polars of Figs. 8.14 ands 8.15 for A_x = 1, starting from point U = 0, P = 2.62 and A = 1. The intersection between the two polars on the P-U determines state 1, while on the A-U plane it specifies states 1 and 1a, corresponding to two different temperatures on the two sides of the gasdynamic interface that stems from the diaphragm section. Presented, moreover, in Fig. 8.27 is the reflection of the shock front from the closed front end, establishing state 2, and the reflection of the rarefaction fan from the back end, producing state 3. These two states are determined by the intersection with the U = 0 axis of the shock polar starting from state 1 for $A_1 = 1.067$, and of a rarefaction wave starting from state 1a for $A_{1a} = .933$. The parameters of Fig. 8.27 are listed in Table 8.1.

The solution of front interactions in a simple shock tube is, in fact, so straightforward that it can be expressed in algebraic form. Thus, according to (8.31) with $P_G = 1$, the particle velocity behind the shock is

$$U_1^2 = \frac{2}{\gamma(\gamma-1)} \frac{(P_1-1)^2}{(P_1-1)-2\gamma/(\gamma+1)} \tag{8.79}$$

while behind the rarefaction fan, according to (8. 48),

$$\frac{U_{1a}}{A_4} = \frac{2}{\gamma - 1}[1 - (\frac{P_{1a}}{P_4})^{\frac{\gamma-1}{2\gamma}}] \tag{8.80}$$

Since $u_1 = u_{1a}$ and $p_1 = p_{1a}$, it follows from the above that

$$\frac{2A_4}{\gamma - 1}[1 - (\frac{P_1}{P_4})^{\frac{\gamma-1}{2\gamma}}] = (P_1 - 1)_o \sqrt{\frac{2\gamma(\gamma+1)}{(P_1 - 1) + 2\gamma/(\gamma+1)}} \tag{8.81}$$

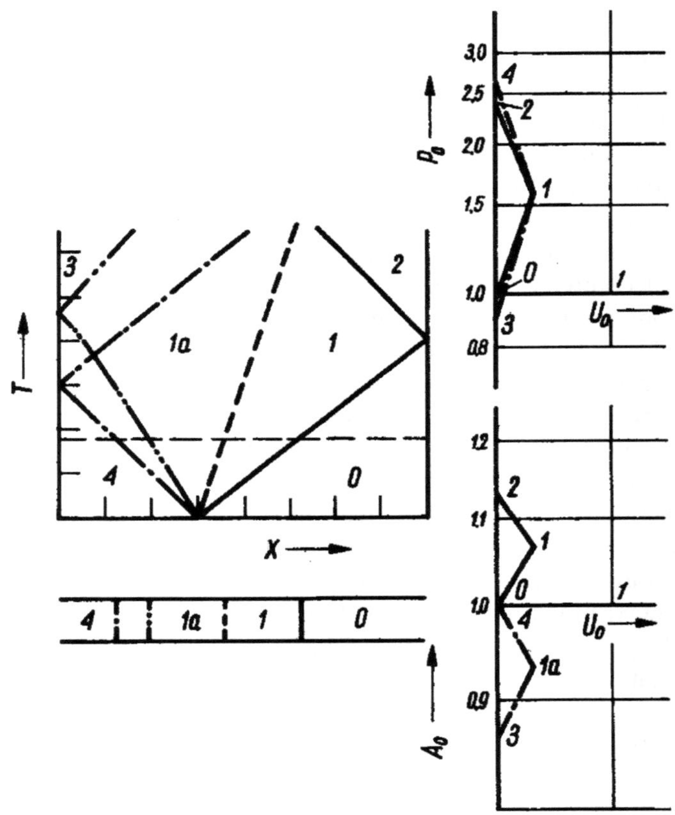

Fig. 8.27. Front interactions in a shock tube

Table 8.1. Parameters of states in Fig. 8.2

State	M	P	A	U	V	W
4	-	2.62	1	-	-	-
1a	-1.00	1.584	0.933	0.33	-1	-1
	-1.00	-	-	-	-0.93	-0.60
1	1.25	1.584	1.067	0.33	1.25	1.25
2	-1.20	2.41	1.133	0	-1.28	-0.95
3	1	0.884	0.86	0	0.93	1.263
	1	-	-	-	0.86	0.86

The initial pressure ratio across the diagram, P_4, that is required to produce a desired shock pressure ratio, P_1, or, by means of (8.43), a desired Mach number, M_1, of the incident shock, is

$$P_4 = P_1\{1 - \frac{(\gamma-1)(P_1-1)}{A_4\sqrt{2\gamma(\gamma+1)[(P_1-1)+2\gamma/(\gamma+1)]}}]^{-\frac{2\gamma}{\gamma-1}} \tag{8.82}$$

Upon reflection from the back wall, the particle velocity is brought to zero and, as a consequence of this requirement, the pressure attained there is

$$P_2 = P_1\frac{(3\gamma-1)P_1-(\gamma-1)}{(\gamma-1)P_1+(\gamma+1)} \tag{8.83}$$

8.11.2 Head-on Collision of a Shock Front with a Deflagration

Consider the case of head-on collision between a shock moving to the right at $M_S = 1.6$ with a deflagration moving to the left at $M = 0.14$, illustrated by Fig. 8.28. As it appears there, the transmitted deflagration polar reaches the Chapman-Jouguet state, K, expressed by point 2, before intersecting the polar 10-20 of the transmitted shock. The only compatible wave system of this interaction has to include a rarefaction wave 2-20, propagating immediately behind the Chapman-Jouguet state 2.

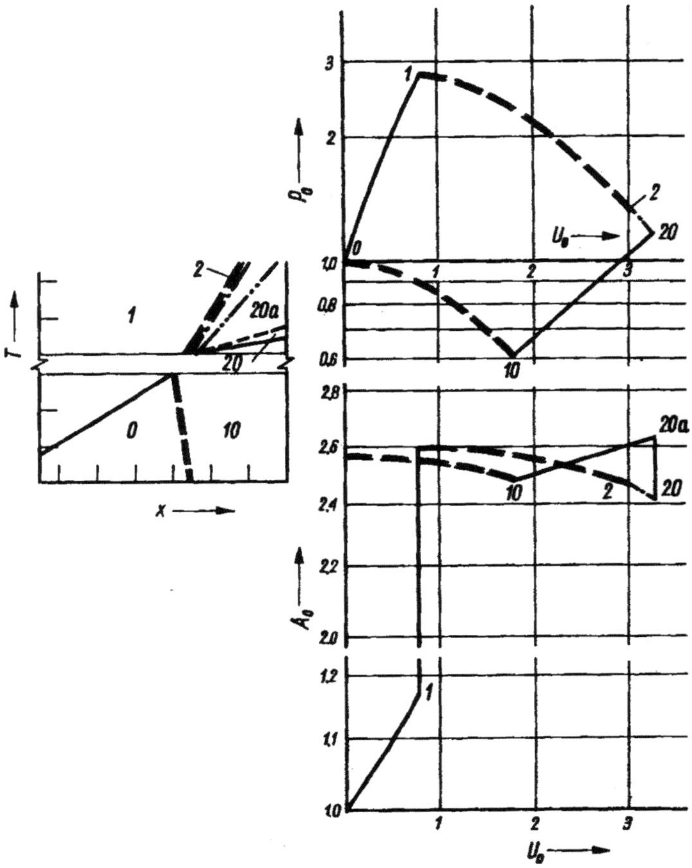

Fig. 8.28. Collision of a shock with a deflagration front

Table 8.2. Parameters of states in Fig. 8.28

State	M	P	A	U	V	W
1	1.6	2.75	1.173	0.8	1.6	1.6
10	-0.14	0.61	2.47	1.8	-0.14	-1.4
2	-0.08	1.336	2.44	3.02	-0.21	0.59
20	-1	1.15	2.41	3.28	-0.244	0.58
	-1				-0.241	0.87
20a	1.37	1.15	2.62	3.28	3.38	5.18

The parameters of Fig. 8.28 are listed in Table 8.2. As evident there, the shock raises the pressure by a factor of 2.75, while the deflagration brings about a pressure drop of ~40%.

8.11.3 Shock Merging

Two shocks propagating in the same direction must merge because the second shock front propagates at supersonic velocity into a subsonic field left behind the first shock. Therefore, as demonstrated by von Neumann 1942, such an interaction results in a transmitted shock and a reflected rarefaction fan. This is illustrated by Fig. 8.29.

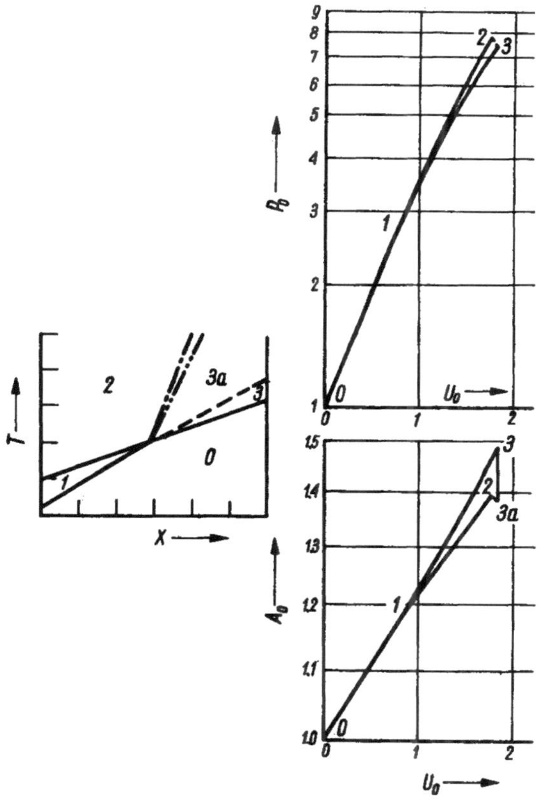

Fig. 8.29. Shock Merging

This outcome is brought about as a consequence of the fact that on the state plane of P(U), at state 1 produced by the first shock, the second shock polar that creates state 2 is steeper than the first polar between states 0 and 3, specifying the change of states formed by the transmitted shock front. Under such circumstances, a dynamically compatible closure between states 2 and 3 can be provided only by a rarefaction fan, as demonstrated in

Fig. 8.29. Since the opposite holds true for the corresponding shock polars on the A-U plane, the velocity of sound in state 2 is lower than that in state 3, as appropriate for a rarefaction. The coordinates of Fig. 8.29 are presented by Table 8.3.

Table 8.3. Parameters of states in Fig. 8.29

State	M	P	A	U	V	W
1	1.60	2.77	1.176	0.80	1.60	1.60
2	1.60	7.70	1.390	1.77	1.88	2.68
3a	-1	7.24	1.377	1.83	-1.39	0.38
	-1		-		-1.38	0.45
3	7.24	7.24	1.49	1.83	2.54	2.54

8.11.4 Merging of a Shock Front with a Rarefaction Fan

Merging of a rarefaction moving to the right with a shock front propagating ahead of it is displayed by Fig. 8.30.

The front ray of a rarefaction fan propagating at the local velocity of sound is bound to catch up with a presiding shock front that leaves behind a locally subsonic flow field

At the point of intersection, 1, reached by the polar of the incident shock, 0-1, at the top of the state diagram of Fig. 30, the front ray of the rarefaction polar is steeper and flatter than the shock polar.

Consequently, there are three dynamically compatible sets of wave fronts depending on the extent of the rarefaction fan

(a) a transmitted shock, 0-4, with a reflected shock front, 2-3, and an inteface, 34, between them

(b) a transmitted shock front, 0-4’, with a gasdynamic interface, 2’4’

(c) a transmitted shock front,0-4”, with a reflected rarefaction fan, 2”-3” , and a gasdynamic interface, 3”4”, between them.

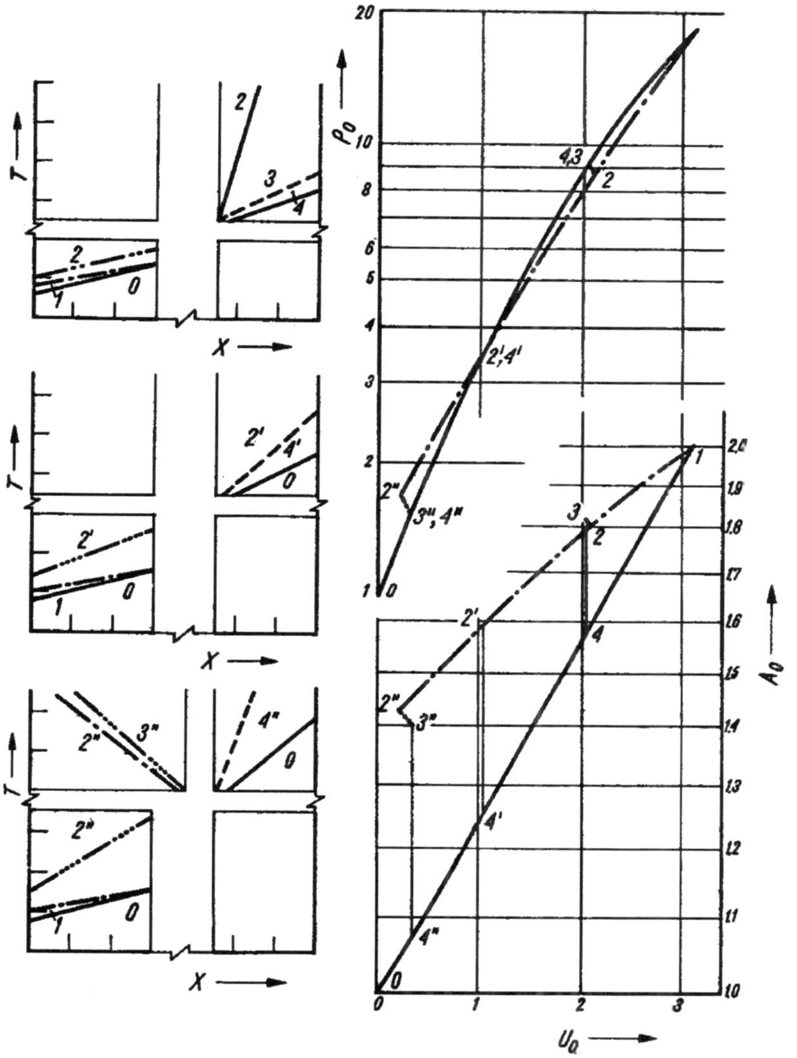

Fig. 8.30. Merging of a Shock Front with a Rarefaction Fan ($\gamma = 1.4$)

Table 8.4. Parameters of states in Fig. 8.30

State	M	P	A	U	V	W
1	3.92	18	2	3.1	3.92	3.92
2	+1	8.6	1.8	2.1	2.0	5.1
	+1				1.8	3.9
3	1.02	0.05	1.81	2.04	-1.834	0.27
4	2.77	8.9	1.57	2.04	2.77	2.77
2'	+1	3.56	1.585	1.04	2.0	5.1
	+1				1.585	2.625
4'	1.78	3.56	1.246	1.04	1.78	1.78
2"	+1	1.65	1.425	.0.2	2.0	5.1
	+1				1.425	1.625
3"	-1	1.506	1.405	0.32	-1.425	-1.225
	-1				-.1.405	-1.085
4"	1.2	1.506	1.067	0.32	1.2	1.2

8.12 Front Intersections

Intersections of gasdynamic fronts are displayed in planar flow fields oriented in orthogonal direction to their surfaces. Typical intersections are illustrated on Fig. 8.31, and their polar diagrams of P(δ) are presented by Fig. 8.32.

They consist of the following (a) regular quadruple intersection; (b) inverse Mach intersection; (c) normal Mach intersection;(d) conventional Mach intersection; (e) limit of conventional Mach intersections;(f) conventional arrowhead intersection; (g) chocked arrowhead intersection. Solutions were obtained by the vector polar method presented by Oppenheim et al (1970).

The dynamic compatibility condition of invariant pressure and flow direction in the field generated by the intersection give rise to a slip interface between flow sectors of different particle velocities, temperatures, hence velocities of sound and densities, on its sides, generating shear that, in a viscous fluid, creates turbulence engendering vorticity.

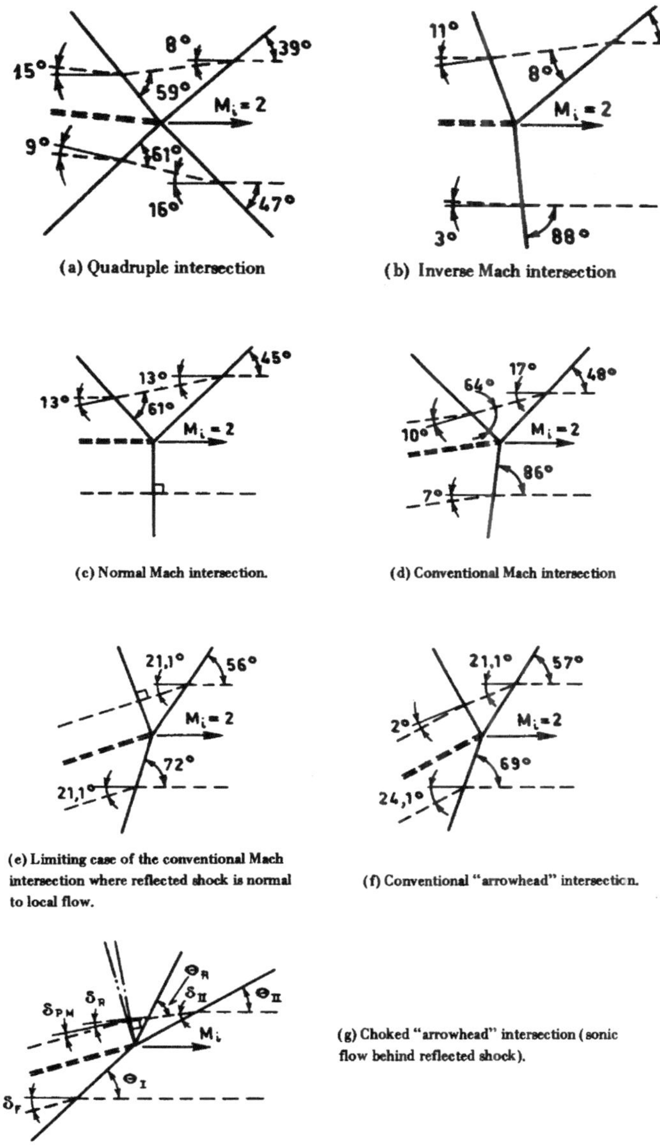

Fig. 8.31. Shock intersections

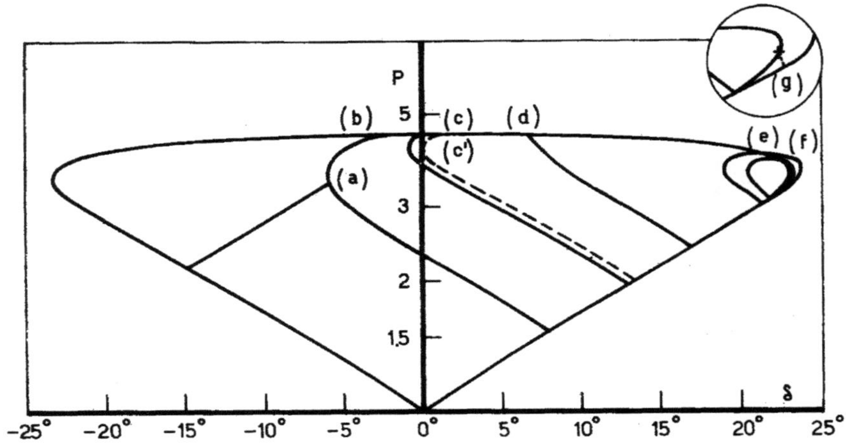

Fig. 8.32. Polar diagrams of front intersections presented by Fig. 8.31

To provide details of solutions, four typical cases of intersections are presented, namely:

1. Arrowhead intersection or shock merging
2. Mach intersection of shock fronts
3. Mach intersection of a shock front with a detonation wave
4. Collision of a Mach intersection with a wall.

8.12.1 Arrowhead Intersection

An arrowhead intersection is displayed by Fig. 8.33 for a representative case of M_1 = 8-0, δ_{12} = 10° and δ_{23} = 15° in a gas of γ = 1-4 (β = 1/6). In general, according to Courant and Friedrichs 1948, Chapter IV, Fig. 68, such an intersection is created by two intersecting shock fronts. The problem is specified by the Mach number of the incident front and the deflection angles, δ_{12} and δ_{23}, imposed by the walls.

With respect to the configuration of gasdynamic fronts on the physical plane of Fig. 8.33(a) and their state diagram provided by Fig. 8.33(b), the Mach number of the incident shock, M_1, identifies the polar on the P-δ plane, and angle δ_{12} fixes state (2), while the auxiliary function $M_j(P)$, specifies the Mach number M_2 and, hence, the polar for the second shock front. State (3) is then located on this polar by the second deflection angle δ_{23}. Since the transmitted shock must propagate into state (1), its state must be located on the same polar as state (2). A dynamically compatible configuration is obtained by a rarefaction fan producing state (4a), concomitantly with a shock front producing state (4b). In the particular case of the second shock front producing state (3'), lying on the shock polar stemming from state (1), there is only a transmitted shock front. Parameters of Fig. 8.33 are provided by Table 8.5.

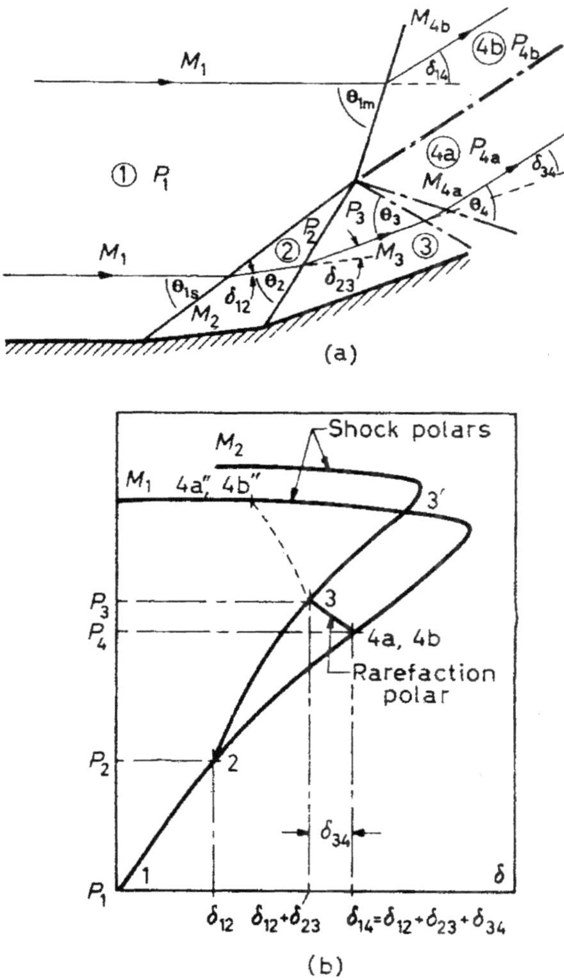

Fig. 8.33. Arrowhead intersection

Table 8.5. Parameters of Fig. 8.33

State	1	2	3	4a	4b
Mach number, M	8·0	5·77	3·75	3·9	2·93
Pressure ratio, P	1·0	5·25	29·3	24·5	24·5
Wave angle, θ	θ_{1m}	θ_{1s}	θ_2	θ_3	θ_4
Degrees	35·9	16·3	23·4	15·2	14·6
Deflection angle, δ	δ_{12}	δ_{23}		δ_{34}	δ_{14}
Degrees	10·0	15·0		2·2	27·2

8.12.2 Mach intersection of Shock Fronts

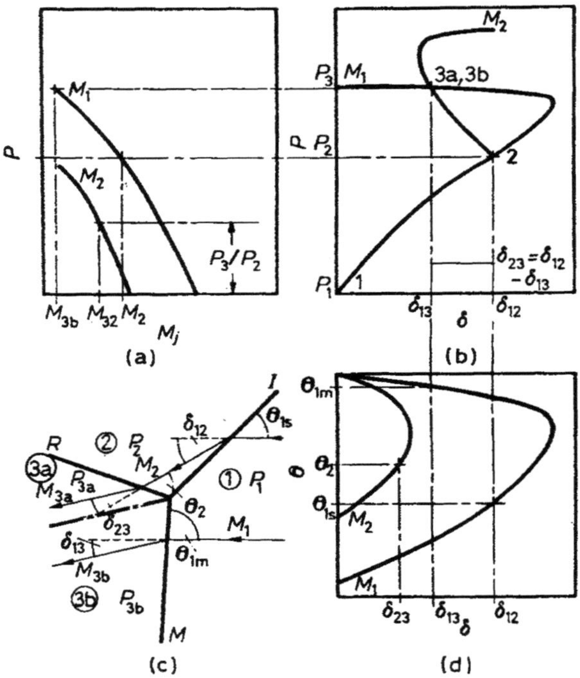

Fig. 8.34. Mach intersection of shock fronts

A Mach intersection is fully defined in terms of two parameters: Mach number of the incident shock, M_1, and the angle by which its front is inclined with respect to the incident flow, Θ_1. A representative case of M_1= 6 and Θ_1 = 45°.in a gas of γ = 1-4 (β = 1/6) is illustrated by Fig.

8.34, where the configuration of the shock fronts in the physical plane is displayed in (c), the solution on the state diagram $P(\delta)$ is presented in (b), and the auxiliary functions $\Theta(\delta)$ in (d) and $P(M_j)$ in (a). The deflection angle, δ_{12}, brought about by the incident shock front, is determined by the auxiliary function $\Theta(\delta)$, while the Mach number of the flow field behind it, M_2, is obtained from the function $P(M_j)$. Specifies thereby is the polar of the reflected wave, R, between states 2 and 3a. Concomitantly, at the same point on the polar of the incident shock front, is state 3b, reached by a third shock front that must be brought about in order to satisfy the dynamic compatibility condition of uniform pressure and deflection angle in the field behind the intersection. This front, as evident from the state diagram, (b), is stronger than the other two. It is referred to the Mach stem.

The parameters of Fig. 8.34 is provided by Table 8.6.

Table 8.6. Parameters of Fig. 8.34

State	1	2		3a	3b
Mach number, M	6·0	2·15		1·81	0·91
Pressure ratio, P	1·0	19·5		31·0	31·0
Wave angle, θ	θ_{1m}		θ_{1s}		θ_2
Degrees	73·75		43·0		35·25
Deflection angle, δ	δ_{12}		δ_{23}		δ_{13}
Degrees	31·5		−8·3		23·2

8.12.3 Mach intersection of shock and detonation fronts

In an explosive gas, the Mach stem, engendered by a triple intersection, may be a detonation front, as illustrated in Fig. 8.35. The solution is, in principle, the same as that of Fig. 8.34, except for the use of a detonation polar instead of a shock polar propagating into state where the incident Mach number is M_1. Since, for the same Mach number, the pressure reached by detonation is lower than that attained by the shock front, as depicted on the state diagram in Fig. 8.35, there is an upper limit in the strength of the incident shock front, beyond which a detonation cannot be established. For a stronger incident shock, the only dynamically compatible solution is a triple intersection of shock fronts described in the previous section.

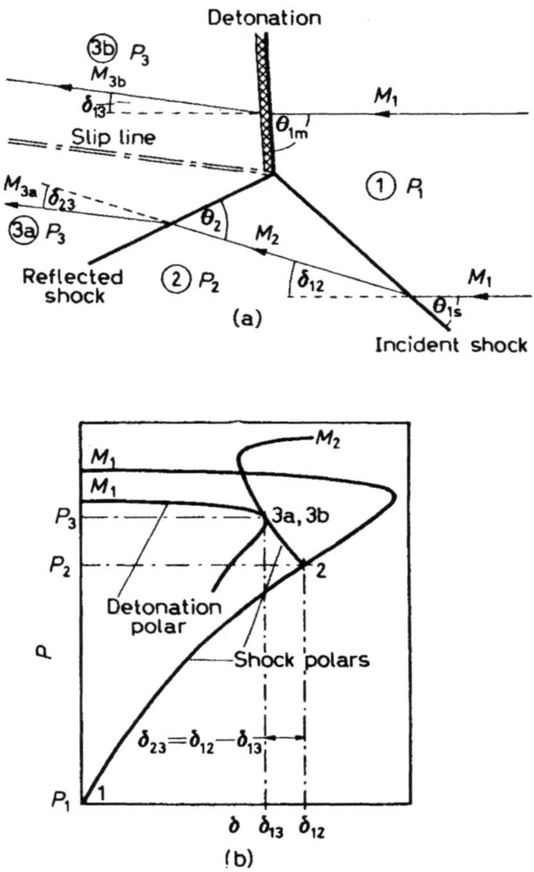

Fig. 8.35. Mach intersection between shock and a detonation fronts

Table 8.7. Parameters of Fig. 8.35

State	1	2	3a	3b
Mach number, M	6·0	2·15	1·81	0·91
Pressure ratio, P	1·0	19·5	31·0	31·0
Wave angle, θ	θ_{1m}	θ_{1s}		θ_2
Degrees	73·75	43·0		35·25
Deflection angle, δ	δ_{12}	δ_{23}		δ_{13}
Degrees	31·5	−8·3		23·2

Illustrated by Fig. 8.35 is the limiting case of the strongest incident shock that can produce a Mach intersection with a detonation front. As evident there on the P-δ state diagram, for stronger shock fronts the deflection angle, δ_{12}, is too large for the polar of the reflected shock to intersect with the detonation polar.

Specifically, Fig. 8.35 displays the case of $M_1 = 6$ and $\Theta_1 = 43^{\circ}$.in a gas of $\gamma = 1\text{-}4$ and the Hugoniot curve of $\beta = 0.06$ and $C = 9$. Its parameters are provided by Table 9.7.

8.12.4 Collision of a Mach intersection with a wall

In a medium at rest, the triple point of a Mach intersection moves in the direction of the incident velocity, corresponding to M_1. If the incident shock front is vertical to a plane wall, it travels toward it, resulting in a collision, as illustrated on the physical plane, (c), in Fig. 8.36. Upon this event, the triple pint is reflected from the plane, traveling along the Mach stem that acquires thereby the role of an incident shock, while a new Mach stem, vertical to the plane, is formed.

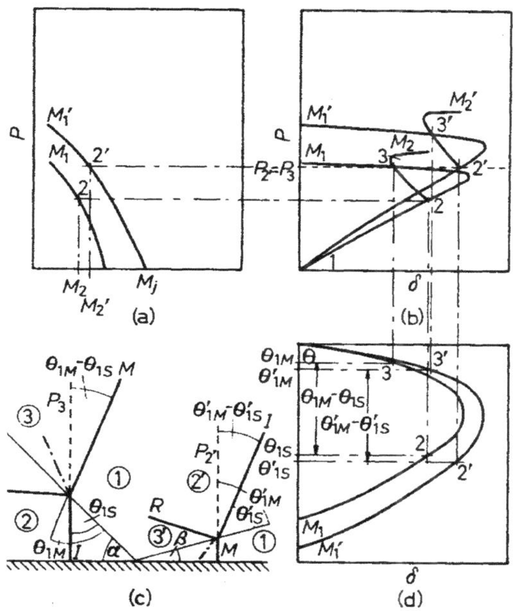

Fig. 8.36. Collision of Mach intersection with a wall

The solution of this intersection is based on the premise that the new incident shock front that was formerly a Mach stem is not aware of the reflection until it is reached by the reflected triple intersection point.

Consequently the pressure ratio across the incident shock of the reflected intersection is equal to that across the Mach stem of the incident intersection, and the angle between the incident shock and the Mach stem remains unchanged. The dynamic compatibility conditions of a triple point reflection are expressed, therefore, by the invariance of pressure ratio across this front and its geometric orientation, so that, as demonstrated on the P-δ state diagram, Fig. 8.36(b), $P_{2'} = P_3$, while, as displayed on the physical plane, Fig. 8.36(c), as well as on the Θ-δ diagram, Fig. 8.36(d), $\Theta'_{1M} - \Theta'_{1S} = \Theta_{1M} - \Theta_{1S}$.

The problem is formulated, as before, in terms of the Mach number of the incident shock front, M_1, and its angle of incidence, Θ_{1S}. Since the incident shock is perpendicular to the plane wall, it follows that the inclination of the triple point trajectory to the plane wall is $\alpha = 90^o - \Theta_{1S}$.

The condition of pressure invariance $P_{2'} = P_3$ fixes point 2' in Fig. 8.36(c) of the incident shock front in the reflected triple point system. The condition of its invariant direction, $\Theta'_{1M} - \Theta'_{1S} = \Theta_{1M'} - \Theta_{1S}$, imposes further restriction on state 2' that, with reference to Fig. 8.36(d), identifies its shock polar corresponding to, M_1'. Thus, all the states of the reflected intersection, 1, 2' and 3' are determined, including, in particular, the angle of reflection, $\beta = 90^o - \Theta'_{1S}$.

One should note that angle α of the incident triple point trajectory has a lower bound and angle β of the reflected point trajectory has an upper bound. This is due to the fact that, in a Mach intersection, the incident shock polar is limited by δ_{max}.

Parameters of an intersection displayed by Fig. 8.36 for the case of M_1 = 6 and $\Theta_{1S} = 45°$, in a gas of $\gamma = 1.4$, is provided by Table 8.8.

Table 8.8. Parameters of Fig. 8.36

State	1	2	3	4a	4b
Mach number, M	8·0	5·77	3·75	3·9	2·93
Pressure ratio, P	1·0	5·25	29·3	24·5	24·5
Wave angle, θ	θ_{1m}	θ_{1s}	θ_2	θ_3	θ_4
Degrees	35·9	16·3	23·4	15·2	14·6
Deflection angle, δ	δ_{12}	δ_{23}		δ_{34}	δ_{14}
Degrees	10·0	15·0		2·2	27·2

8.12.5 Intersections of Fronts of Explosions

Consider double explosions forming blast waves headed by shock fronts that intersect, upon either head-on or rear-end collision. The ensuing front configurations are displayed by Fig. 8.37, where the trajectories of propagating intersection points are delineated by broken lines.

The head-on collision is displayed as case (a), where, at first, the fronts of two blast waves collide on the line of their centers, forming, at first, regular shock intersections propagating on both sides along the center line of collision. As they progress, the angle of intersection increases until it acquires a critical value, beyond which the regular shock intersection can no longer satisfy the dynamic compatibility conditions, and, as a consequence, they are transformed into two Mach intersections spreading apart from each other along their Mach stems.

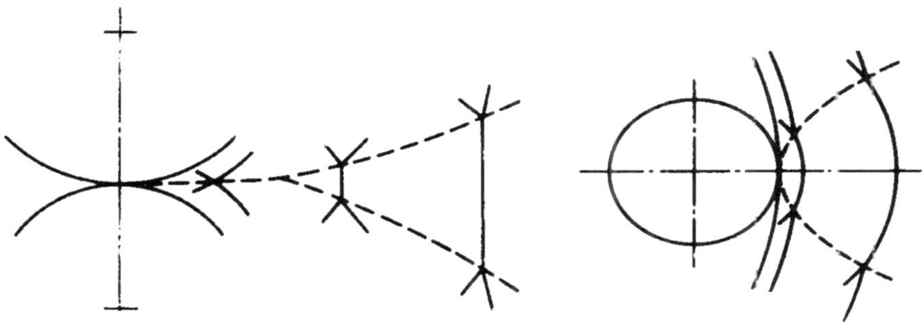

(a) (b)

Fig. 8.37. Intersections between shock fronts of colliding and merging blast waves

The merging of blast waves is illustrated by Fig. 8. 37b. It takes place when the second explosion is initiated inside the first, or, what amounts to the same, when an explosion is created behind a plane shock. In this case, at first the two shock fronts merge along the line of their centers, producing, as in the previous case, two turbulence generating Mach intersections.

PART 3

EXPLOSION

9 Blast Waves

9.1 Background

Blast waves are non-steady flow fields of compressible media bounded by gasdynamic fronts. In general, they are formed by explosions – flow fields generated by the deposition of energy at high power density. They can be generated also by accelerating pistons forming shock fronts at the heads of the fields they create. In this respect, they are of direct relevance to prominent dynamic combustion instability phenomena, exemplified prominently by the onset of knock in internal combustion engines.

Although explosions were known for centuries, the comprehensive scientific study of blast waves has been developed in the course of nascent research on atom bomb. The primary incentive for formation of their theory was furnished by the cinematographic record of the Trinity explosion test that took place at Alamogordo, New Mexico, in 1945, reported by Mack 1947 and published by G.I. Taylor 1950. Its copy is presented by Fig. 9.1.

The only qualitative information provided by this record is the trajectory of the shock front propagating at the head of the blast wave generated by the explosion. This has been, nonetheless, sufficient to deduce the value of its energy, as brilliantly demonstrate by G.I. Taylor (vid. Chapter 10).

A more revealing insight into the evolution of a blast wave is provided by the cinematographic schlieren record of the blast wave created in air by a beam of a sharp neodymium laser pulse focused upon a thin steel wire, illustrated by Fig. 9.2. The experimental apparatus consisted of a cubical chamber fitted on both sides with optical quartz window plates of 8.25 mm in diameter, providing a field of view restricted by a 1.3 cm wide slit aperture for cinematographic framing, a schlieren set up with a pulsating ruby laser as the light source, a neodymium pulse laser as a source of ignition, and a rotating mirror camera operating with open shutter*.

---

* design and construction of the apparatus is provided by Oppenheim and Kamel 1971

Since the energy deposited thus in air is quite small, provided thereby is an experimental demonstration that it is not the critical energy that matters for the generation of a blast wave, but the critical power density at which it is deposited in the field.

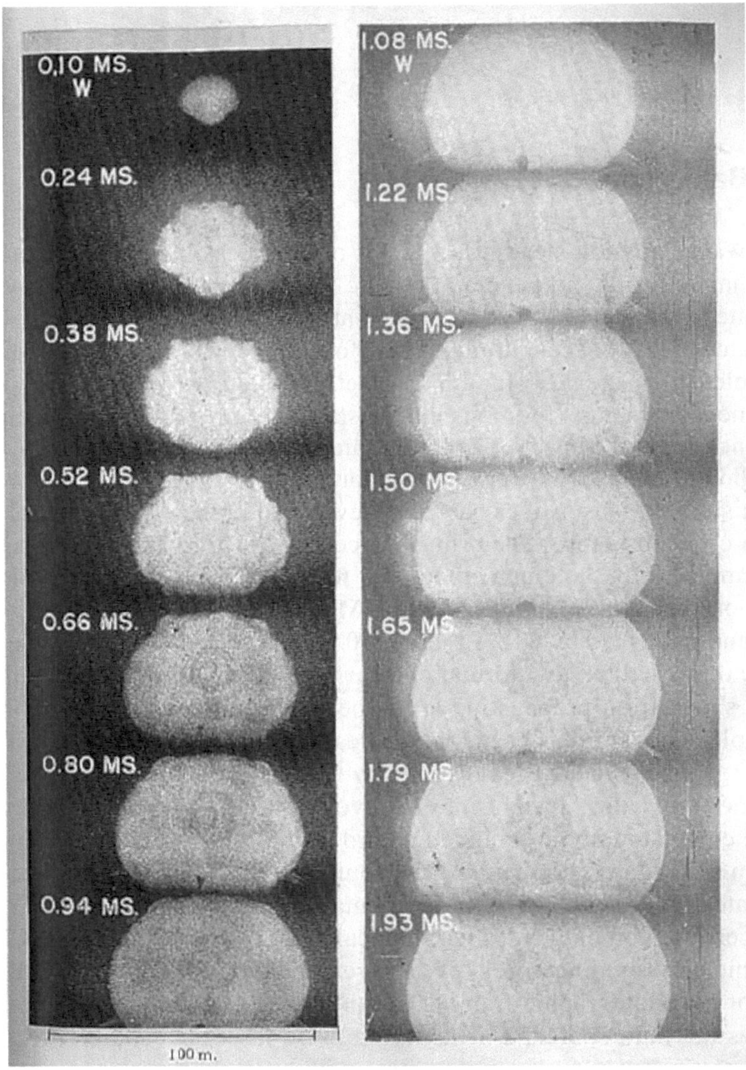

Fig. 9.1. Cinematographic record of the Trinity atom bomb explosion (Taylor 1950)

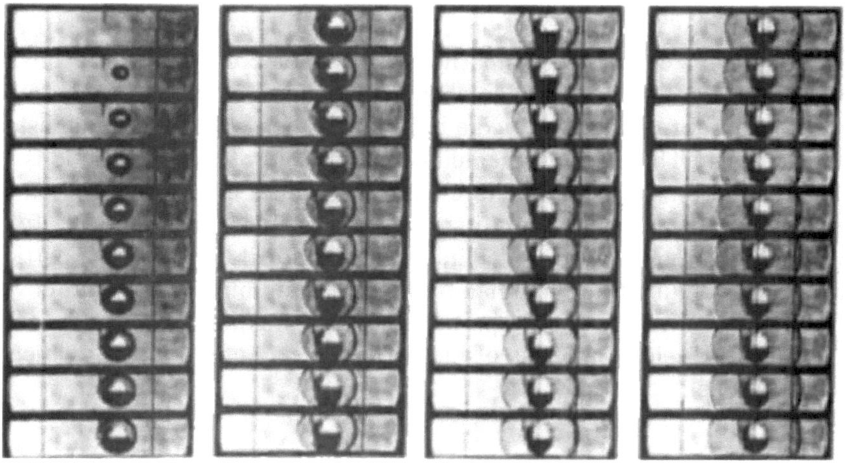

Fig. 9.2. Cinematographic schlieren record of a blast wave created in air by a beam of sharp neodymium laser pulse focused upon a thin steel wire (Oppenheim and Kamel 1972)
Time interval between frames: 1 μsec

The theory of blast waves has been founded in the 1940's by notable contributions of J. von Neumann 1941, G.I. Taylor 1946, and L.I Sedov 1946, 1959a.

Among comprehensive expositions of blast wave theory, are the authoritative chapters in the books of Courant arid Friedrichs 1948 and of Zel'dovich and Raizer 1963, as well as in the monographs of Sedov 1959b, Sakurai 1965, and Korobenikov 1985. All of them are concerned with point explosion – the far flow fields that exist a sufficiently large distant from their source to become independent of its shape. The fields are then either spherical, cylindrical or plane symmetric and their formulation is one-dimensional. The exposition of theory presented here is based on the book of Oppenheim 1970 and the paper of Oppenheim et al. 1971.

Blast waves are geometrically symmetrical. Their analysis is formulated, therefore, in terms of spatially one-dimensional equations expressing the conservations of mass, momentum and energy, subject to imposed boundary conditions.

The theory of is expressed conventionally in the Eulerian frame of reference. Being concerned primarily with the outcome of explosions, it deals primarily with decaying adiabatic blast waves, while the constitutive equations are formulated for media behaving as perfect gases with constant specific heats. The theory presented here is devoid of these restrictions. It is formulated for the field structure expressed either in space at a fixed instant of time, corresponding to Eulerian space profiles, or in time at a fixed point in space, corresponding to Eulerian time profiles, or along a particle path, corresponding to Lagrangian profiles. Taken into account, moreover, are the actual thermodynamic properties of the medium, rather than their idealized expressions in terms of the perfect gas assumption with which it is conventionally treated.

9.2 Coordinates

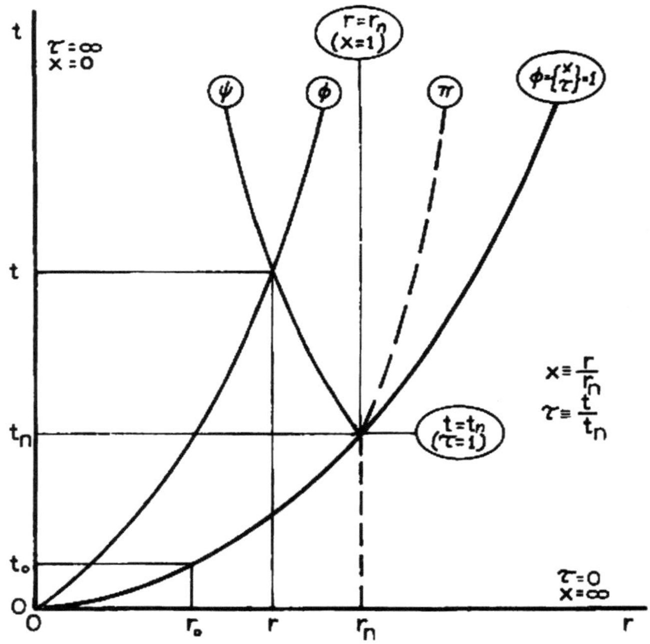

Fig. 9.3. Blast wave coordinates

The coordinates of a blast wave are displayed in Fig.9.3. Their roles are specified in the next two sections.

9.2.1 Front

The front coordinates r_n and t_n, are provided by experimental observation. In non-dimensional form, normalized with respect to a fixed point of reference at the front denoted by subscript o, they are expressed by

$$\xi \equiv \frac{r_n}{\mathrm{r_o}} \quad \text{and} \quad \eta \equiv \frac{t_n}{\mathrm{t_o}} \tag{9.1}$$

The trajectory of the front is described by $\xi(\eta)$ and its propagation velocity

$$w_n \equiv \frac{\mathrm{d}r_n}{\mathrm{d}t_n} = \frac{\mathrm{r_o d}\xi}{\mathrm{t_o d}\eta} \tag{9.2}$$

is expressed in terms of the velocity modulus

$$\mu \equiv \frac{d \ln \xi}{d \ln \eta} \equiv \frac{d \ln r_n}{d \ln t_n} \equiv \frac{w_n t_n}{r_n} \tag{9.3}$$

as well as in terms of the decay coefficient

$$\lambda \equiv \frac{\mathrm{d} \ln y}{\mathrm{d} \ln \xi} = -2\frac{\mathrm{d} \ln w_n}{\mathrm{d} \ln \mathrm{r_n}} = -2\frac{r_n \dot{r}_n}{\ddot{r}_n^2} \tag{9.4}$$

where, $y \equiv a_a^2 / w_n^2$ - the reciprocal of Mach number squared – where a_a is the velocity of sound in the medium into which the blast wave propagates, a constant in combustion fields.

The velocity modulus, μ, as well as the decay coefficient, λ, are functions of y. According to their definitions specified by (9.3) and (9.4),

$$\frac{d \ln \mu}{d \ln \xi} = \frac{1}{\mu} - \frac{\lambda + 2}{2} \tag{9.5}$$

If μ = constant, a commonly encountered case of a power-law, the front trajectory expressed by

$$\xi = \eta^{\mu} \tag{9.6}$$

and (9.5) is reduced to

$$\mu = \frac{2}{\lambda + 2} \tag{9.7}$$

whence $\lambda = 2/\mu - 2$ = constant.

9.2.2 Field

The field coordinates are the independent variables of blast waves. In non-dimensional form, normalized with respect to a point, n, at the front trajectory and expressed as

$$x \equiv \frac{r}{r_n} \qquad \text{and} \qquad \tau \equiv \frac{t}{t_n} \tag{9.8}$$

while the dependent variables in normalized form are

$$h \equiv \frac{\rho}{\rho_a} \qquad g \equiv \frac{p}{\rho_a w_n^2} \qquad \varepsilon \equiv \frac{e}{w_n^2} \qquad f \equiv \frac{u}{w_n} \tag{9.9}$$

9.3 Formulation

Blast wave theory for point explosions is formulated on the basis of conservation equations for one dimensional flow fields, expressed in terms of radius and time as their time-space coordinates. By transforming them into generalized form expressed in terms of a field coordinate and a front coordinate, they are rendered a comprehensive character, applicable to any of three frames of reference: Eulerian space, Eulerian time, and Lagrangian time. Thereupon, by transformation into normalized blast wave coordinates, the physical coordinates of time and space are eliminated and they are cast into an autonomous form. The exposition of blast wave theory is culminated by specification of boundary conditions, imposed by gasdynamic fronts of a shock or a detonation, and by formulation of integral functions.

A schematic diagram of plane, line or point symmetrical blast waves is displayed in Fig. 9.3. Their conservation equations, formulated by (7.9), (7.10) and (7.11), are expressed in form of a divergence that, comprehensively in a vector form, is as follows

$$\frac{\partial}{\partial t}a_k + \frac{\partial}{\partial r}ub_k = \dot{c}_k \tag{9.10}$$

where, with

$$k = \begin{vmatrix} \text{1 for conseration of mass} \\ \text{2 for conservation of momentum} \\ \text{3 for conservation of energy} \end{vmatrix}$$

a_k, b_k, $\dot{c}_k$ are components of a conservation vector specified in Table 9.1.
where

$$\mathrm{j} = \begin{vmatrix} \text{0 for plane symmetrical blast waves} \\ \text{1 for line symmetrical blast waves} \\ \text{2 for point symmetrical blast waves} \end{vmatrix}$$

The line of $\pi(r,t) =$ constant in Fig. 9.1 delineates the trajectory of a particle. Its velocity is, therefore, $u \equiv (\frac{\partial r}{\partial t})_\pi$. Any particle trajectory can be identified with the motion of an accelerating piston face generating the blast wave.

Then, in terms of

$$A_k \equiv \frac{a_k}{\rho_a \mathrm{r}_\mathrm{n}^\mathrm{j} \mathrm{w}_\mathrm{n}^{\mathrm{k}-1}}, \quad B_k \equiv \frac{b_k}{\rho_a \mathrm{r}_\mathrm{n}^\mathrm{j} \mathrm{w}_\mathrm{n}^{\mathrm{k}-1}}, \quad \dot{C}_k \equiv \frac{\dot{c}_k}{\rho_a \mathrm{r}_\mathrm{n}^{\mathrm{j}-1} \mathrm{w}_\mathrm{n}^{\mathrm{k}}},$$

(9.10) is cast in a normalized form of

$$\frac{\partial}{\partial \tau}A_k + \frac{\partial}{\partial x}fB_k = \dot{C}_k \tag{9.11}$$

for which the components of vector $(A_k, B_k, \dot{C}_k)$ are provided by Table 9.2.

Table 9.1. Components of the conservation vector

k	$\dfrac{a_k}{\rho r^j}$	$\dfrac{b_k}{\rho r^j}$	$\dfrac{\dot{c}_k}{\rho r^j}$
1	1	1	0
2	u	$u + \dfrac{p}{\rho u}$	$\dfrac{jp}{\rho r}$
3	$e + \dfrac{u^2}{2}$	$e + \dfrac{u^2}{2} + \dfrac{p}{\rho}$	0

Table 9.2. Components of the normalized conservation vector

k	$\dfrac{A_k}{hx^j}$	$\dfrac{B_k}{hx^j}$	$\dfrac{C_k}{hx^j}$
1	1	1	0
2	f	$f + \dfrac{g}{fh}$	$\dfrac{jg}{xh}$
3	$\sigma + \dfrac{f^2}{2}$	$\sigma + \dfrac{f^2}{2} + \dfrac{g}{h}$	0

9.4 Blast Wave Coordinates

The independent variables, x and τ, are transformed into a front coordinate, $\psi(x,\tau)$ and a field coordinate, $\phi(x,\tau)$. A line of $\psi(x,\tau)$ = constant can be identified either with $\tau = 1$, or with $x = 1$, or with $\pi =$ constant. In the first case $\psi = 1$ specifies the Eulerian space profiles expressed as functions of $\phi = x$. In the second case $\psi = 1$ specifies the Eulerian time profiles expressed as functions of $\phi = \tau$. In the third case $\psi = \pi =$ constant, along which $(\frac{\partial x}{\partial \tau})_\psi = f$, specifies the Lagrangian time profiles expressed as functions of $\phi = \tau$.

Since, for a continuous flow field of a blast wave, the functions $\psi(x,\tau)$ and $\phi(x,\tau)$ are smooth, $\frac{\partial^2 x}{\partial\psi\partial\phi} = \frac{\partial^2 x}{\partial\phi\partial\psi}$ and $\frac{\partial^2 \tau}{\partial\psi\partial\phi} = \frac{\partial^2 \tau}{\partial\phi\partial\psi}$ and the transformation of any variable, $\mathscr{F}(x,\tau)$ is prescribed by

$$\left.\begin{aligned} \frac{\partial\mathscr{F}}{\partial\tau} &= -\frac{\frac{\partial}{\partial\psi}(\mathscr{F}\frac{\partial x}{\partial\phi}) - \frac{\partial}{\partial\phi}(\mathscr{F}\frac{\partial x}{\partial\psi})}{\mathbf{J}(x,\tau;\psi,\phi)} \\ \frac{\partial\mathscr{F}}{\partial x} &= -\frac{\frac{\partial}{\partial\psi}(\mathscr{F}\frac{\partial \tau}{\partial\phi}) - \frac{\partial}{\partial\phi}(\mathscr{F}\frac{\partial \tau}{\partial\psi})}{\mathbf{J}(x,\tau;\psi,\phi)} \end{aligned}\right| \tag{9.12}$$

where the Jacobian $\mathbf{J}(x,\tau;\psi,\phi) \equiv \frac{\partial(x,\tau)}{\partial(\psi,\phi)} \equiv \frac{\partial x}{\partial\psi}\frac{\partial\tau}{\partial\phi} - \frac{\partial\tau}{\partial\psi}\frac{\partial x}{\partial\phi}$.

Thus, the depended variables of (9.10) are transformed into

$$\left.\begin{aligned} A_k &= -a_{\mathrm{k}}\frac{\partial r}{\partial \phi} + ub_{\mathrm{k}}\frac{\partial t}{\partial \phi} \\ B_k &= a_{\mathrm{k}}\frac{\partial r}{\partial \psi} + ub_{\mathrm{k}}\frac{\partial t}{\partial \psi} \\ \dot{C}_k &= \dot{c}_{\mathrm{k}}\left(\frac{\partial r}{\partial \psi}\frac{\partial t}{\partial \phi} - \frac{\partial t}{\partial \psi}\frac{\partial r}{\partial \phi}\right) \end{aligned}\right| \tag{9.13}$$

and (9.10) is expressed by

$$\frac{\partial}{\partial \psi}\mathcal{A}_k + \frac{\partial}{\partial \phi}f\mathcal{B}_k = \dot{\mathcal{C}}_k \tag{9.14}$$

where

$$\mathcal{A}_k = \rho_a \mathrm{r}_\mathrm{n}^\mathrm{j}\mathrm{w}_\mathrm{n}^{\mathrm{k}-1}\frac{M_k}{\phi}, \mathcal{B}_k = \rho_a \mathrm{r}_\mathrm{n}^\mathrm{j}\mathrm{w}_\mathrm{n}^{\mathrm{k}-1}\frac{N_k}{\psi}, \dot{\mathcal{C}}_k = \rho_a \mathrm{r}_\mathrm{n}^\mathrm{j}\mathrm{w}_\mathrm{n}^{\mathrm{k}-1}\frac{\dot{K}_k}{\phi\psi} \tag{9.15}$$

while

$$\left.\begin{aligned} M_k &= A_k x\frac{\partial \ln x}{\partial \ln \phi} - fB_k\tau\mu\frac{\partial \ln \tau}{\partial \ln \phi}] \\ N_k &= A_k x\left(\frac{\partial \ln x}{\partial \ln \xi}+1\right) - fB_k\tau\left(\mu\frac{\partial \ln \tau}{\partial \ln \xi}+1\right) \\ \dot{K}_k &= \dot{C}_k x\tau\left[\frac{\partial \ln x}{\partial \ln \phi}\left(\mu\frac{\partial \ln \tau}{\partial \ln \xi}+1\right) - \mu\frac{\partial \ln \tau}{\partial \ln \phi}\left(\frac{\partial \ln x}{\partial \ln \xi}+1\right)\right] \end{aligned}\right| \tag{9.16}$$

Hence, upon incorporating (9.16) into (9.14), via (9.15), and invoking (9.4), (9.5) and (9.6), yields

$$[\mathrm{j}+1-(\mathrm{k}-1)\frac{\lambda}{2}]M_k + \frac{\partial M_k}{\partial \ln \xi} - \frac{\partial N_k}{\partial \ln \phi} = \dot{K}_k \tag{9.17}$$

9.5 Eulerian Space Profiles

The Eulerian space profiles, as pointed out in previous section, are delineated along the line of $\tau = 1$, so that a line of constant ϕ corresponds to $x =$ constant, whence

$$\frac{\partial \ln x}{\partial \ln \phi} = 1 \quad \text{while} \quad \frac{\partial \ln \tau}{\partial \ln \phi} = \frac{\partial \ln \tau}{\partial \ln \xi} = \frac{\partial \ln x}{\partial \ln \xi} = 0 \tag{9.18}$$

The components of the transformed conservation vector are reduced then to

$$\left.\begin{aligned} M_k &= A_k x \\ N_k &= A_k x - fB_k \\ \dot{K}_k &= \dot{C}_k x \end{aligned}\right| \tag{9.19}$$

for which, on the basis of Table 9.2, their explicit expressions, are provided by Table 9.3.

Table 9.3. Components of conservation vector for Eulerian space profiles

k	$\frac{M_k}{hx^{j+1}}$	$\frac{N_k}{hx^{j+1}}$	$\frac{K_k}{hx^{j+1}}$
1	1	$1 - \frac{f}{x}$	0
2	f	$f\left(1 - \frac{f}{x}\right) - \frac{g}{xh}$	$\frac{jg}{xh}$
3	$\sigma + \frac{f^2}{2}$	$\left(\sigma + \frac{f^2}{2}\right)\left(1 - \frac{f}{x}\right) - \frac{gf}{xh}$	0

9.6 Eulerian Time Profiles

The Eulerian time profiles are delineated along the line of $x = 1$, so that a line of constant ϕ corresponds to τ= constant, whence

$$\frac{\partial \ln \tau}{\partial \ln \phi} = 1 \quad \text{while} \quad \frac{\partial \ln x}{\partial \ln \phi} = \frac{\partial \ln x}{\partial \ln \xi} = \frac{\partial \ln \tau}{\partial \ln \xi} = 0 \tag{9.20}$$

The components of the transformed conservation vector are reduced then to

$$\begin{aligned} M_k &= fB_k \tau \mu \\ N_k &= A_k - fB_k \tau \\ \dot{K}_k &= \dot{C}_k x \tau \mu \end{aligned} \tag{9.21}$$

Their explicit expressions, derived from Table (9.2), are provided by Table (9.4)

Table 9.4. Components of conservation vector for Eulerian time profiles

k	$\frac{M_k}{\mu h \tau}$	$\frac{N_k}{\mu h \tau}$	$\frac{K_k}{\mu h \tau}$
1	f	$\frac{1 - f\tau}{\tau}$	0
2	$f^2 + \frac{g}{h}$	$f\left(\frac{1 - f\tau}{\tau}\right) - \frac{g}{h}$	$j\frac{g}{h}$
3	$f\left(\sigma + \frac{f^2}{2}\right) + \frac{fg}{h}$	$\left(\sigma + \frac{f^2}{2}\right)\left(\frac{1 - f\tau}{\tau}\right) - \frac{fg}{h}$	0

9.7 Lagrangian Time Profiles

The Lagrangian time profiles are delineated along a line of ϕ corresponding π= constant, along which $(\frac{\partial r}{\partial t})_\pi = u$, so that

$$(\frac{\partial x}{\partial \tau})_\pi = \frac{w_n t_n}{r_n} f = \mu f \tag{9.22}$$

while

$$\frac{\partial \ln \tau}{\partial \ln \phi} = 1; \quad \frac{\partial \ln \tau}{\partial \ln \xi} = 0; \quad \frac{\partial \ln x}{\partial \ln \phi} = \frac{\partial \ln x}{\partial \ln \xi} = \mu \frac{\tau}{x} f \tag{9.23}$$

The components of the transformed conservation vector are reduced then to

$$\left. \begin{aligned} M_k &= \mu \tau f (B_k - A_k) \\ N_k &= A_k x \zeta + \tau f (A_k - B_k) \\ \dot{K}_k &= \mu \dot{C}_k x \tau \zeta \end{aligned} \right| \tag{9.24}$$

where

$$\zeta \equiv 1 - \frac{\tau}{x} f \tag{9.25}$$

and their expressions, derived from Table (9.2) are provided by Table 9.5.

Table 9.5. Components of conservation vector for Lagrangian profiles

k	$\frac{M_k}{\mu h x^{j+1} \zeta \tau}$	$\frac{N_k}{h x^{j+1} \zeta \tau}$	$\frac{K_k}{\mu h x^{j+1} \zeta \tau}$
1	0	$\frac{1}{\tau}$	0
2	$\frac{g}{x\zeta h}$	$\frac{f}{\tau} - \frac{g}{x\zeta h}$	$\frac{jg}{xh}$
3	$\frac{fg}{x\zeta h}$	$\frac{\sigma + \frac{f^2}{2}}{\tau} - \frac{fg}{x\zeta h}$	0

9.8 Expanded Form

By expressing the logarithmic gradients of σ by logarithmic gradients of g and h by virtue of (1.22) and (1.23), so that

$$\frac{\partial \sigma}{\partial \ln \phi} = \frac{1}{\Gamma - 1} \frac{g}{h} \left(\frac{\partial \ln g}{\partial \ln \phi} - \frac{\partial \ln h}{\partial \ln \phi} \right) \tag{9.26}$$

$$\frac{\partial \sigma}{\partial \ln \xi} - \lambda \sigma = \frac{1}{\Gamma - 1} \frac{g}{h} \left(\frac{\partial \ln g}{\partial \ln \xi} - \frac{\partial \ln h}{\partial \ln \xi} - \lambda \right) \tag{9.27}$$

and carrying out the differentiation of each term in (9.16), one obtains the expanded form of the conservation equations

$$\alpha_k + f_k^{\phi} \frac{\partial \ln f}{\partial \ln \phi} + f_k^{\xi} \frac{\partial \ln f}{\partial \ln \xi} + h_k^{\phi} \phi \frac{\partial \ln h}{\partial \ln \phi} + h_k^{\psi} \psi \frac{\partial \ln h}{\partial \ln \xi} + g_k^{\phi} \phi \frac{\partial \ln h}{\partial \ln \phi} + g_k^{\psi} \psi \frac{\partial \ln h}{\partial \ln \xi} = 0 \tag{9.28}$$

whose coefficients are specified by Tables 9.6, 9.7 and 9.8, for, respectively, the Eulerian space, Eulerian time and Lagrangian time profiles.

Table 9.6. Parameters of expanded equations for the Eulerian space profiles

k	α_k	f_k^{ϕ}	f_k^{ξ}	h_k^{ϕ}	h_k^{ξ}	g_k^{ϕ}	g_k^{ξ}
1	$j\frac{f}{x}$	$\frac{f}{x}$	0	$-\left(1-\frac{f}{x}\right)$	1	0	0
2	$-\frac{\lambda}{2}$	$-\left(1-\frac{f}{x}\right)$	1	0	0	$\frac{g}{hfx}$	0
3	$-\lambda$	0	0	$\gamma\left(1-\frac{f}{x}\right)$	$-\gamma$	$-\left(1-\frac{f}{x}\right)$	1

Table 9.7. Parameters of expanded equations for Eulerian time profiles

k	α_k	$f_k{}^{\phi}$	$f_k{}^{\xi}$	$h_k{}^{\phi}$	$h_k{}^{\xi}$	$g_k{}^{\phi}$	$g_k{}^{\xi}$
1	$j - \frac{\lambda}{2}$	$-\frac{1}{\mu}$	1	$\frac{1-\tau f}{\mu\tau f}$	1	0	0
2	$(\frac{g}{hf^2} - \frac{1}{2})\lambda$	$\frac{1-\tau f}{\mu\tau f}$	1	0	0	$-\frac{g}{\mu hf^2}$	$\frac{g}{hf^2}$
3	$-\lambda$	0	0	$-\gamma\frac{1-\tau f}{\mu\tau f}$	$-\gamma$	$\frac{1-\tau f}{\mu\tau f}$	1

Table 9.8. Parameters of expanded equations for the Lagrangian time profiles

k	α_k	$f_k{}^{\phi}$	$f_k{}^{\xi}$	$h_k{}^{\phi}$	$h_k{}^{\xi}$	$g_k{}^{\phi}$	$g_k{}^{\xi}$
0	$-\frac{\lambda}{2} - \zeta$	$-\frac{1}{\mu}$	1	0	0	0	0
1	$-\frac{\lambda}{2} + j\zeta$	$-\frac{1}{\mu}$	1	$\frac{x\zeta}{\mu\tau f}$	0	0	0
2	$-\lambda$	$\frac{\zeta hf}{\mu g\tau}$	0	0	0	$-\frac{1}{\mu}$	1
3	0	0	0	$-\frac{\gamma}{\mu}$	0	$\frac{1}{\mu}$	0

9.9 Autonomous Form

Thereupon, the conservation equations for blast waves are compressed by combining the depended and independent variables into two phase coordinates

$$F \equiv \frac{\tau}{x} f = \frac{t}{\mu r} u \qquad \text{and} \qquad Z \equiv (\frac{\tau}{x})^2 \frac{g}{h} = (\frac{t}{\mu r})^2 \frac{p}{\rho} \qquad (9.29)$$

which, for the space profiles where $\tau = 1$, become $F \equiv \dfrac{f}{x}$ and , $Z \equiv \dfrac{1}{x^2}\dfrac{g}{h}$,

while, for the time profiles where $x = 1$, become $F \equiv \tau f$ and $Z \equiv \tau^2 \dfrac{g}{h}$.

Obtained thus are autonomous equations in the form of

$$\beta \mathbf{D} \frac{\partial \ln \psi}{\partial \ln \phi} = \mathbf{\Psi}_{\varphi} + \mathbf{\Psi}_{\xi} \tag{9.30}$$

where, for the Eulerian system, $\boldsymbol{D} = (1 - F)^2 - Z$, while, for the Lagrangian system $\boldsymbol{D} = [1 - (1-\mu)F]^2 - Z$, with $\beta = 1$ for space profiles and $\beta = \mu^{-1}$ for time profiles, whereas ψ expresses the normalized variables, F, h, g, Z, while $\mathbf{\Psi}_{\mathrm{n}}$ (n = ϕ, ξ) denotes three types of function, **F, H, G, Z.** listed in Tables 9.9, 9.10 and 9.11, for the Eulerian space, Eulerian time and the Lagrangian time profiles, respectively.

Table 9.9. Functions in autonomous equations for Eulerian space profiles

$\mathbf{\Psi}$	Subscript	
	$\phi = x$	ξ
F	$(1-F)\left(F - \dfrac{\lambda+2}{2}\right) + \dfrac{Z}{F}[(j+1)\Gamma F - \lambda]$	$(1-F)\dfrac{\partial \ln F}{\partial \ln \xi} + \dfrac{Z}{F}\dfrac{\partial \ln g}{\partial \ln \xi}$
H	$\dfrac{F}{1-F}\left[\mathbf{F}_x + (j+1)\mathbf{D}\right]$	$\dfrac{F}{1-F}\left[\mathbf{F}_\xi + \dfrac{\mathbf{D}}{F}\dfrac{\partial \ln h}{\partial \ln \xi}\right]$
G	$\gamma \mathbf{H}_x + \lambda \dfrac{\mathbf{D}}{1-F}$	$\gamma \mathbf{H}_\xi + \dfrac{\mathbf{D}}{1-F}\left[\dfrac{\partial \ln g}{\partial \ln \xi} - \gamma \dfrac{\partial \ln h}{\partial \ln \xi}\right]$
Z	$(\gamma - 1)\mathbf{H}_x + [\lambda - 2(1-F)]\dfrac{\mathbf{D}}{1-F}$	$(\gamma - 1)\mathbf{H}_\xi + \dfrac{\mathbf{D}}{1-F}\left[\dfrac{\partial \ln g}{\partial \ln \xi} - \gamma \dfrac{\partial \ln h}{\partial \ln \xi}\right]$

Table 9.10. Functions in autonomous equations for the Eulerian time profiles

Ψ	Subscript	
	$\phi = \tau$	ξ
F	$(1-F)\left(\frac{\lambda+2}{2}-F\right) - \frac{Z}{F}[(j+1)\Gamma F - \lambda]$	$\mu'\mathbf{D} - [F(1-F)+\Gamma Z]\frac{\partial \ln F}{\partial \ln \xi} - \frac{Z}{F}\frac{\partial \ln g}{\partial \ln \xi}$
H	$\frac{F}{1-F}[\mathbf{F}_\tau - (j+1)\mathbf{D}]$	$\frac{F}{1-F}\left[\mathbf{F}_\xi - \mu'\mathbf{D} - \mathbf{D}\frac{\partial \ln F}{\partial \ln \xi} - \mathbf{D}\frac{\partial \ln h}{\partial \ln \xi}\right]$
G	$\gamma \mathbf{H}_\tau - \lambda\frac{F}{1-F}\mathbf{D}$	$\gamma \mathbf{H}_\xi - \frac{F}{1-F}\mathbf{D}\left[\frac{\partial \ln g}{\partial \ln \xi} - \gamma\frac{\partial \ln h}{\partial \ln \xi}\right]$
Z	$(\gamma-1)\mathbf{H}_\tau + \lambda\frac{F}{1-F}\mathbf{D} + (\lambda+2)\mathbf{D}$	$(\gamma-1)\mathbf{H}_\xi - \frac{F}{1-F}\mathbf{D}\left[\frac{\partial \ln g}{\partial \ln \xi} - \gamma\frac{\partial \ln h}{\partial \ln \xi}\right] + 2\mu'\mathbf{D}$

Table 9.11. Functions in autonomous equations for the Lagrangian time profiles

Ψ	Subscript	
	$\phi = \tau$	ξ
ζ	$-\frac{F}{\zeta}\mathbf{F}_\tau$	$\left(1-\frac{F}{\zeta}\right)\mathbf{D}\frac{\partial \ln F}{\partial \ln \xi} - \mathbf{H}_\xi$
F	$-\zeta\left[\zeta\left\{F - \frac{\lambda+2}{2}\right\} + \frac{Z}{F}\{(j+1)\Gamma F + \rho_a' - \lambda\}\right]$	$\mu'\mathbf{D} - Z\left[\Gamma\frac{\partial \ln F}{\partial \ln \xi} + \frac{\zeta}{F}\frac{\partial \ln g}{\partial \ln \xi}\right]$
H	$\frac{F}{\zeta}[\mathbf{F}_\tau - (j+1)\zeta\mathbf{D}]$	$\frac{F}{\zeta}\left[\mathbf{F}_\xi - \mu'\mathbf{D} - \mathbf{D}\frac{\partial \ln F}{\partial \ln \xi}\right]$
G	$\gamma \mathbf{H}_\tau$	$\gamma \mathbf{H}_\xi$
Z	$(\gamma-1)\mathbf{H}_\tau - 2\mathbf{D}\left(F - \frac{\lambda+2}{2}\right)$	$(\gamma-1)\mathbf{H}_\xi + 2\mu'\mathbf{D}$

9.10 Boundary Conditions

The change of state taking place at the front denoted by subscript n - the gasdynamic front that, by definition, forms the outer boundary of a blast wave - is prescribed by the Hugoniot equation expressed, according to (8.25) and (8.28), by

$$(P_n + \beta_P)(v_n - \beta_P) = (1 - \beta_P)(P_G + \beta_P) \tag{9.31}$$

whence

$$h_n = v_n^{-1} = \frac{P_n + \beta_P}{(1 - \beta_P)P_G + \beta_P(P_n + 1)} \tag{9.32}$$

$$g_n = \frac{1 - v_n}{P_n - 1} P_n = (1 - \beta_P)\frac{P_n - P_G}{P_n + \beta_P}\frac{P_n}{P_n - 1} \tag{9.33}$$

$$F_n = f_n = 1 - v_n = (1 - \beta_P)\frac{P_n - P_G}{P_n + \beta_P} \tag{9.34}$$

$$\begin{aligned} Z_n &= \frac{g_n}{h_n} = \frac{1 - v_n}{P_n - 1} P_n v_n \\ &= (1 - \beta_P)[(P_G + \beta_P) + \beta_P(P_n - P_G)]\frac{P_n - P_G}{P_n + \beta_P}\frac{P_n}{P_n - 1} \end{aligned} \tag{9.35}$$

The pressure ratio, P_n, is expressed in terms of the Mach number, M_n, by virtue of (8.33) and (8.34), whence

$$P_n = \mathbf{M} \pm \sqrt{\mathbf{M}^2 - (1 + \beta_P)P_G M_n^2 + \beta_P} \tag{9.36}$$

where $\mathbf{M} \equiv \dfrac{(1 + \beta_P)M_n^2 + 1 - \beta_P}{2}$.

Then, in terms of $y \equiv \dfrac{1}{M_n^2}$, (9.34) and (9.35) yield

$$F_n = \frac{2(P_n - P_G)}{(\gamma+1)P_n + \gamma - 1} \tag{9.37}$$

and

$$Z_n = \frac{2\gamma}{\gamma+1}(\gamma P_n + P_G + \gamma - 1)\frac{(P_n - P_G)P}{(\gamma+1)P + \gamma - 1} \tag{9.38}$$

where β is expressed, according to (8.24), in terms of γ, upon the understanding that for a shock $\gamma = \gamma_R$, while for detonation $\gamma = \gamma_P$.

The equation for the Hugoniot curve on the phase plane is obtained by eliminating P_n from (9.37) and (9.38) yielding

$$Z_n = \frac{\gamma}{2}\frac{(\gamma-1)F_n + 2P_G}{\gamma F_n + P_G - 1}(1 - F_n)F_n \tag{9.39}$$

that, for $P_G = 1$, is reduced to the Rankine–Hugoniot equation

$$Z_n = \frac{1}{2}[(\gamma-1)F_n + 2](1 - F_n) \tag{9.40}$$

while, for $P_G = \infty$ - the limit of zero counter pressure - it becomes

$$Z_n = \gamma F_n(1 - F_n) \tag{9.41}$$

The trajectory of the Rayleigh line is based on (8.4), whence $F_n = 1 - \nu_n$, according to which, in view of (8.11),

$$y = \frac{1 - \nu_n}{\gamma(P_n - 1)} \tag{9.42}$$

whence, with $Z_n = P_n \nu_n y$, it follows from its definition that

$$Z_n = (\gamma F_n + y)(1 - F_n) \tag{9.43}$$

which, as it should, is identical to (9.41) for $y = 0$.

The Hugoniot curves for a set of P_G and the Rayleigh lines for as set of y, evaluated by the use of (9.39) (9.40) and (9.43) are presented, for the case of $\gamma = 1.4$, by Fig.9.4.

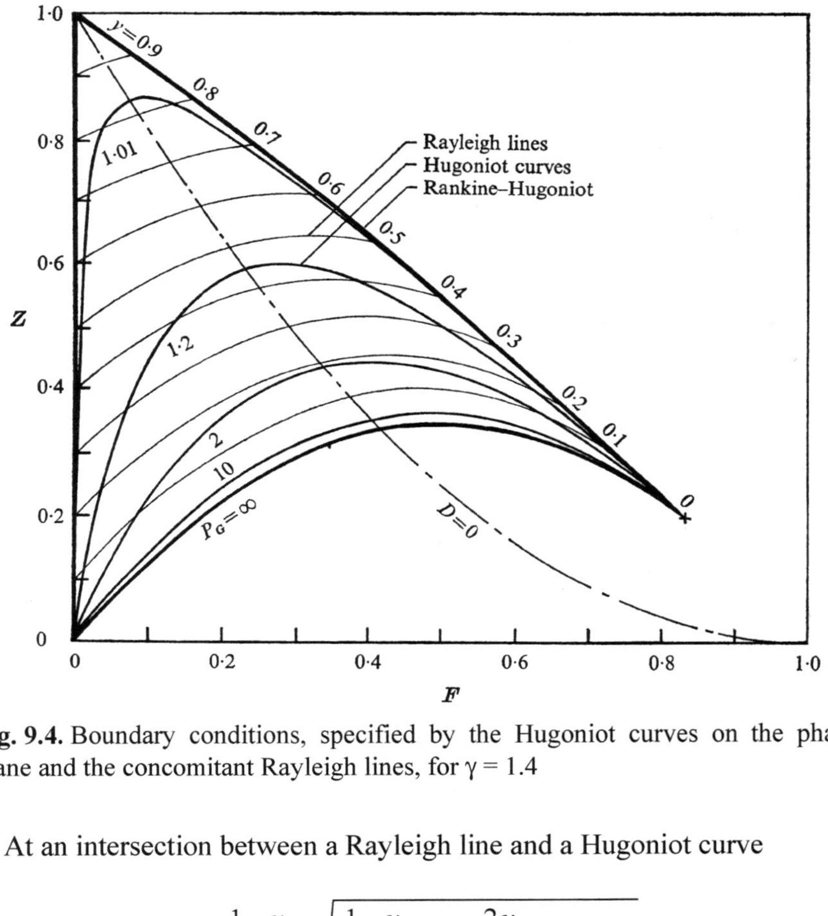

Fig. 9.4. Boundary conditions, specified by the Hugoniot curves on the phase plane and the concomitant Rayleigh lines, for $\gamma = 1.4$

At an intersection between a Rayleigh line and a Hugoniot curve

$$F_n = \frac{1-y}{1+\gamma} \pm \sqrt{\left(\frac{1-y}{1+\gamma}\right)^2 - \frac{2y}{(1+\gamma)\gamma}(P_G - 1)} \tag{9.44}$$

-a relation obtained by eliminating Z_n from (9.39) and (9.43). For its intersection with a Rankine-Hugoniot curve, for which $P_G = 1$, it follows from the above that

$$F_n = 2\frac{1-y}{1+\gamma} \tag{9.45}$$

In Fig. 9.4, the boundary conditions at the front of blast waves propagating into an atmosphere of uniform density are restricted to states within a region bounded by the Rankine-Hugoniot curve of $P_G = 1$ on one side and the Hugoniot curve of $P_G = \infty$ on the other. It is of interest to note how rapidly an increase in the value of P_G causes the Hugoniot curves to approach the limiting case of $P_G = \infty$, the curve corresponding to $P_G = 10$ being practically coincident with it.

The $D = 0$ parabola, marked by a chain-broken line, represents the locus of the Chapman-Jouguet conditions that separates the region of strong detonations on the right from that of weak detonations on the left. Since the latter cannot be sustained by detonations, the physically relevant boundary conditions are confined for them to the relatively narrow region between the $D = 0$ parabola and the Rankine-Hugoniot curve.

9.11 Integral Functions

Integral functions express the global mass and global energy conservation principles. Their conventional expressions, presented here, refer to point explosions.

The conservation of mass stipulates that the total mass of a blast wave is equal to the mass of the atmosphere engulfed by its front. Thus, at a given instant, in an atmosphere of uniform density, with $\sigma_j = 1, 2\pi, 4\pi$ for, respectively, $j = 0$, 1 and 2.

$$M = \sigma_j \int_0^{r_n} \rho r^j dr = \sigma_j \rho_a \frac{r_n^{j+1}}{j+1} \tag{9.46}$$

or, in normalized form, the mass integral

$$I_M \equiv \frac{M}{\sigma_j \rho_a r_n^{j+1}} = \frac{1}{j+1} \tag{9.47}$$

The conservation of energy stipulates that the total energy of the blast wave derives from the internal energy deposited initially for its generation, E_d, and the internal energy of the atmosphere engulfed by its front, i.e.

$$E = \sigma_j \int_0^{r_n} (e + \frac{u^2}{2}) \rho r^j dr = E_d + \sigma_j \int_0^{r_n} e_a \rho_a r^j dr \tag{9.48}$$

or, in terms of normalized variables defined by (9.9),

$$E = \sigma_j \rho_a w_n^2 r_n^{j+1} \int_0^{r_n} (\varepsilon + \frac{f^2}{2}) h x^j dx = E_d + \sigma_j \int_0^{r_n} \varepsilon_a \rho_a r^j dr \tag{9.49}$$

In non-dimensional form, the energy integral is

$$I_E \equiv \frac{E}{\sigma_j \rho_a w_n^2 r_n^{j+1}} = \int_0^{1_n} (\varepsilon + \frac{f^2}{2}) h x^j dx \tag{9.50}$$

that, in terms of the phase coordinates defined by (9.29), is expressed as

$$I_E \equiv \int_0^{1_n} (\frac{Z}{\gamma - 1} + \frac{F^2}{2}) h x^{j+2} dx \tag{9.51}$$

The energy deposited in the medium to generate the blast wave of a point explosion is, according to (9.49),

$$\begin{aligned} E_d &= \sigma_j \rho_a w_n^2 r_n^{j+1} I_E - \sigma_j r_n^{j+1} \frac{p_a}{(j+1)(\gamma_a - 1)} \\ &= \sigma_j r_n^{j+1} [\frac{\gamma_a}{p_a y} I_E - \frac{1}{(j+1)(\gamma_a - 1)}] \end{aligned} \tag{9.52}$$

Its value is expressed in terms of the reference radius

$$r_o \equiv (\frac{E_d}{\sigma_j p_a})^{\frac{1}{j+1}} \tag{9.53}$$

a constant for a constant energy blast wave introduced, on the basis of dimensional analysis, by Taylor 1946, 1950, 1950a, as well as by Sedov 1946, 1959a,b.

In general, the energy of a blast wave is variable. Its integral, I_E, as well as the Mach number, and hence y, are therefore variable. Thus, in view of (9.1), according to (9.52),

$$\xi^{-(j+1)} = \gamma \frac{I_E}{y} - \frac{1}{(j+1)(\gamma-1)} \tag{9.54}$$

whence, from its definition specified by (9.5),

$$\lambda = \frac{(j+1)I_E - y/(\gamma-1)\gamma}{(j+1)dI_E/d\ln y} \tag{9.55}$$

for which, on the basis of (9.51),

$$\frac{dI_E}{d\ln y} = \gamma \int_0^1 \Big[\frac{Z}{\gamma(\gamma-1)}\Big(\frac{\partial \ln Z}{\partial \ln y} + \frac{\partial \ln h}{\partial \ln y}\Big) + F^2\Big(\frac{\partial \ln F}{\partial \ln y} + \frac{1}{2}\frac{\partial \ln h}{\partial \ln y}\Big)\Big] h x^{j-2} dx \tag{9.56}$$

10 Self-Similar Blast Wave

10.1 Introduction

The concept of self-similarity played a key role in the development of blast wave theory. It has been formulated to explore the effects of atom bomb explosion. For that purpose, the blast wave is considered as a gasdynamic flow fields generated by the deposition of a finite amount of energy in zero time at a geometric point in an unbounded compressible medium – an event referred to as a point explosion: Ipso facto, the flow field is geometrically symmetric, while its extent is bounded by a gasdynamic front, whereas its structure is governed by conservation principles expressed in terms of time dependent, spatially one-dimensional partial differential equations.

In the pioneering solutions of von Neumann 1941; Taylor 1941; Sedov 1945, the conservation equations were transformed into ordinary differential equations by reducing the independent variables, on the basis of dimensional analysis, into a single self-similarity coordinate - an obvious first step in treating a novel class of physical problems. Thereupon, the concept of self-similarity has been prominently featured in most of the early studies (Taylor 1946; Sedov 1946; Stanyukovich 1946) and texts on this subject (Courant & Friedrichs 1948; Sedov 1957; Stanyukovich 1955; Korobeinikov *et al.* 1961; Zel'dovich & Raizer 1963, Korobeinikov 1985), achieving eventually a significant role in fluid mechanics (Barenblatt 1994, 1996, 2003). In all of these publications, the thermodynamic properties of the medium were expressed in terms of inviscid perfect gases with constant specific heats, while the flow field was considered as adiabatic. Here, the idealization of a perfect gas is replaced pragmatically by the straight state line formulation introduced in Chapter 1 (based on Oppenheim et al 1972), while, as appropriate for combustion, the state of the medium into which the blast wave propagates is considered to be uniform.

10.2 Formulation

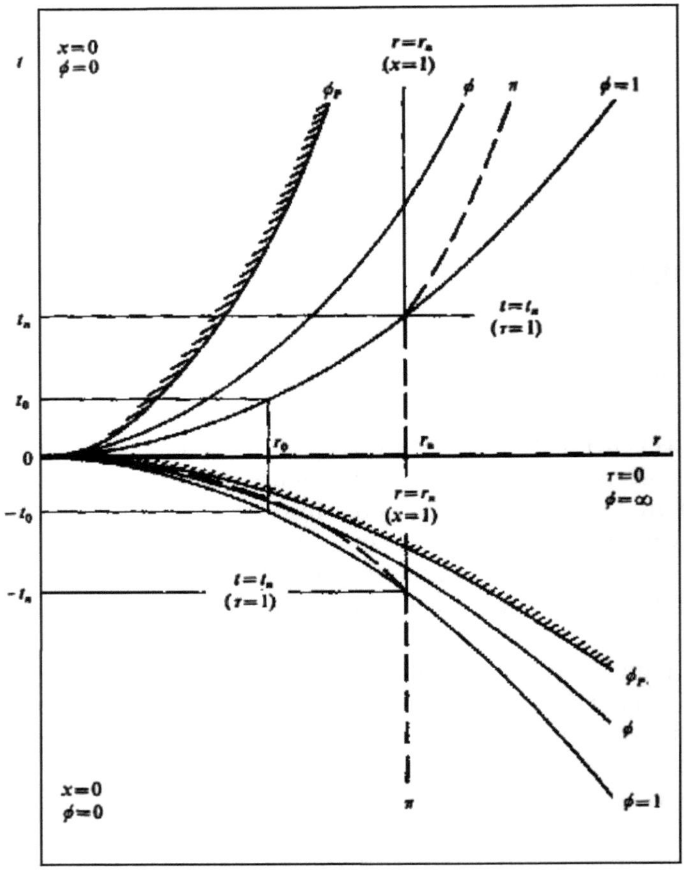

Fig. 10.1. Trajectories of a self-similar blast wave generated by explosion and implosion

The trajectories of a self-similar blast wave are displayed in Fig. 10.1. The conservation equations, expressed in an autonomous form by (9.30) with Tables 9.9, 9.10 and 9.11, are reduced to ordinary differential equations by postulating that the front derivatives, $\partial/\partial\xi$ expressed by ψ_ξ, are negligible, so that they are restricted to just the first term, Ψ_ϕ.

Moreover, since a self-similar flow field in one coordinate system is self-similar in any other system, all the equations are expressed in terms of a single independent variable

$$\phi = x / \tau^{\mu} \tag{10.1}$$

where

$$\mu = \frac{2}{\lambda + 2} \tag{10.2}$$

For the Eulerian space profiles, $\tau = 1$, so that $\phi = x$; for the Eulerian time profiles $x = 1$, so that $\phi = \tau^{-\mu}$, while for the Lagrangian time profiles the slope of the particle path at a point is equal to its local velocity. The transformation of the independent variable into these three coordinate systems, respectively, is expressed by

$$\frac{\mathrm{d}}{\mathrm{d}\ln\phi} = \frac{\mathrm{d}}{\mathrm{d}\ln x} = -\mathrm{a}\frac{\mathrm{d}}{\mathrm{d}\ln\tau} = -\frac{\mathrm{a}}{(1-F)}\frac{\mathrm{d}}{\mathrm{d}\ln\tau} \tag{10.3}$$

where

$$\mathrm{a} = \frac{1}{\mu} = \frac{\lambda + 2}{2} \tag{10.4}$$

Thus, the autonomous blast wave equations are reduced to the following ordinary differential equations

$$\frac{\mathrm{d}F}{\mathrm{d}\ln\phi} = -\frac{\mathbf{Q}(F,Z)}{\mathbf{D}(F,Z)} \tag{10.5}$$

$$\frac{\mathrm{d}Z}{\mathrm{d}\ln\phi} = -\frac{Z}{1-F}\frac{\mathbf{P}(F,Z)}{\mathbf{D}(F,Z)} \tag{10.6}$$

whence

$$\frac{dZ}{dF} = \frac{Z}{1-F}\frac{\mathbf{P}(F,Z)}{\mathbf{Q}(F,Z)} \tag{10.7}$$

where

$$\mathbf{D}(F,Z) = Z - (1-F)^2 \tag{10.8}$$

$$\mathbf{Q}(F,Z) = (j+1)(F - \mathrm{b}]Z - (\mathrm{a} - F)(1-F)F \tag{10.9}$$

$$\mathbf{P}(F,Z) = \mathrm{d}(\mathrm{c} - F\}\mathbf{D}(F,Z) + (\gamma - 1)\mathbf{Q}(F,Z) \tag{10.10}$$

while

$$\mathrm{b} = \frac{\lambda}{(\mathrm{j}+1)\gamma} \tag{10.11}$$

$$\mathrm{c} = \frac{\lambda + 2}{\mathrm{d}} \tag{10.12}$$

$$\mathrm{d} = (\mathrm{j}+1)(\gamma - 1) + 2 \tag{10.13}$$

The structure of the flow field is expressed by the profiles of particle velocity, temperature, pressure and density. For the Eulerian space profiles, the first two are determined directly from the definitions of F and Z, so that

$$\frac{u}{\mathrm{u_n}} = \frac{F}{\mathrm{F_n}}x \tag{10.14}$$

while

$$\frac{T}{\mathrm{T_n}} = \left(\frac{a}{\mathrm{a_n}}\right)^2 = \frac{p}{\mathrm{p_n}}\frac{\rho_\mathrm{n}}{\rho} = \frac{Z}{\mathrm{Z_n}}x^2 \tag{10.15}$$

The evaluation of pressure and density profiles is based on the fact that the field of a self-similar, constant energy blast wave is homogeneously isentropic ("homentropic"), so that

$$\frac{p}{p_n} = \frac{g}{g_n} = (\frac{h}{\mathrm{h_n}})^{\gamma} = (\frac{\rho}{\rho_n})^{\gamma} \tag{10.16}$$

according to which, the second and third terms for $\phi = x$ in Table 9.9 yield the so called adiabatic (in fact isentropic) integral expressed by

$$\frac{p}{p_n}(\frac{\rho}{\rho_n})^{-\gamma} = \frac{g}{g_n}(\frac{h}{\mathrm{h_n}})^{-\gamma} = (\frac{\rho}{\rho_n}\frac{1-F}{1-\mathrm{F_n}}x^{j+1})^{-\nu} \tag{10.17}$$

where

$$\nu = \frac{\lambda}{j+1} \tag{10.18}$$

whence, according to (10.15),

$$\frac{p}{\mathrm{p_n}} = \frac{g}{\mathrm{g_n}} = (\frac{\rho}{\rho_n})\frac{Z}{\mathrm{Z_n}}x^2 \tag{10.19}$$

while, in view of (10.18)

$$\frac{\rho}{\rho_n} = \frac{h}{\mathrm{h_n}} = [\frac{Z}{\mathrm{Z_n}}(\frac{1-F}{1-\mathrm{F_n}})^{\nu}x^{\lambda+2}]^{\frac{1}{\gamma-\nu-1}} \tag{10.20}$$

As evident from the above set, the governing equation is (10.7) – a nonlinear ordinary differential equation for solely the phase coordinates $Z(F)$. Once its integral curve on the phase plane of $Z(F)$ is determined, all the profiles specifying the structure of the blast wave can be evaluated from the algebraic set of equations (10.14), (10.15), (10.19) and (10.20).

10.3 Phase Plane

Properties of integral curves on the phase plane of $Z(F)$ for self-similar blast waves were a subject of extensive studies involving, in particular, their exploration for a number of representative velocity parameters, μ (vid. Sedov 1959b).

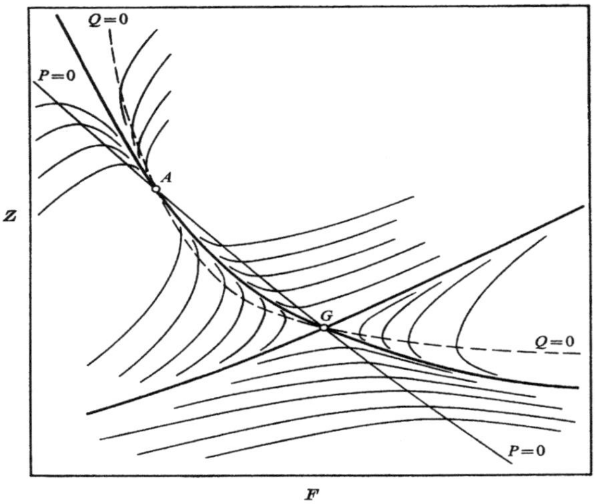

Fig. 10.2. Salient properties of singularities G and A in the phase plane of $Z(F)$

Salient features of the integral curves are revealed by singularities identified by intersections between the loci of numerators and denominators of (10.5), (10.6) and (10.7). As evident from (10.7), besides two fixed singular points at $F = 1$, $Z = 0$ and $Z = \infty$, the coordinates of singularities are identified also by the intersections between the lines of $\mathbf{P} = 0$ and $\mathbf{Q} = 0$. The line of $\mathbf{D} = 0$ is a characteristic of the flow field, since, as evident from the definitions of F, Z and μ, along it $(\frac{\partial x}{\partial t})_\phi = u + a$. It is therefore a locus of singularities.

According to (10.8), the line of $\mathbf{D} = 0$ is a parabola

$$Z = (1 - F)^2 \tag{10.21}$$

while, in view of (10.9) the lines of $\mathbf{Q} = 0$ are specified by

$$Z = \frac{(\mathrm{a} - F)(1 - F)F}{(\mathrm{j} + 1)(F - \mathrm{b})} \tag{10.22}$$

and, by virtue of (10.10), the lines of $\mathbf{P} = 0$ are delineated by

$$Z = \frac{(1 - F)[(\gamma - 1)(\mathrm{a} - F)F + \beta(\mathrm{c} - F)(1 - F)]}{(\mathrm{j} + 1)(\gamma - 1)(F - \mathrm{b}) + \beta(\mathrm{c} - F)} \tag{10.23}$$

Table 10.1. Coordinates of singularities in the phase plane

Singularity	Condition	Co-ordinates F	Co-ordinates Z
A	$P=0, Q=0, D=0$	$\frac{1}{2}\left[\lambda\frac{2-\gamma}{2j\gamma}+1\right] - \left[\frac{1}{4}\left(\lambda\frac{2-\gamma}{2j\gamma}+1\right)^2 - \frac{\lambda}{j\gamma}\right]^{\frac{1}{2}}$	$(1-F_A)^2$
G	$P=0, Q=0, D=0$	$\frac{1}{2}\left[\lambda\frac{2-\gamma}{2j\gamma}+1\right] + \left[\frac{1}{4}\left(\lambda\frac{2-\gamma}{2j\gamma}+1\right)^2 - \frac{\lambda}{j\gamma}\right]^{\frac{1}{2}}$	$(1-F_G)^2$
B	$P=0, Q=0, D\neq 0$	$c=(\lambda+2)/d$	$\frac{(j+1)\gamma(\gamma-1)}{2} \times \frac{(1-F_B)F_B^2}{(j-1)F_B+2}$
D	$Q=\infty, Z=\infty$	$b=\lambda/(j+1)\gamma$	∞
F	$Q=0, Z=0$	$a=\frac{1}{2}(\lambda+2)$	0
O	$F=0, Z=0$	0	0
C	$F=1, Z=0$	1	0
H	$P=0, F=1$	1	–
E	$P=\infty, F=1$	1	∞
I	$F=\infty, Z=\infty$	$\pm\infty$	$\pm\infty$

The coordinates of singularities identified by the intersections between the **Q** = 0 and **P** = 0 lines are provided by Table 10.1. Singularity G on the **D** = 0 line is a saddle point identified by Guderley 1942. Its axis provides the sole trajectory penetrating across the **D** = 0 line between a point at the front of a blast wave at y = 0 and its center at the origin of the phase plane, representing a unique solution for a self-similar implosion in a vacuum. As illustrated in Fig. 10.2, the **Q** = 0 and **P** = 0 lines passing through point G intersect again on the **D** = 0 line at singularity A – a nodal point because G is a saddle point.

If **D** $\neq$ 0, then, as apparent from (10.10), it is possible for **P** = 0, provided that F = c. For that purpose, according to (10.13) and (10.14), the decay coefficient λ = dF – 2 and, upon its elimination from (10.9) taking into account (10.11) and (10.14), the trajectory of singularities corresponding to **P** = **Q** = 0, when **D** ≠ 0, is expressed by

$$\mathbf{B}(F,Z) \equiv [(j-1)F+2]Z - \frac{1}{2}(j+1)\gamma(\gamma-1)(1-F)F^2 = 0 \qquad (10.24)$$

Table 10.2. Loci of the **Q** = 0, **P** = 0, **D** = 0 and **B** = 0 lines on the phase plane

Condition	Equation	$F(Z=\infty)$	$F(Z=0)$
$Q=0$	$Z=\dfrac{(1-F)(a-F)F}{(j+1)(F-b)}$	$b=\dfrac{\lambda}{(j+1)\gamma}$	$0, 1$ or $a=\frac{1}{2}(\lambda+2)$
$P=0$	$Z=(1-F)\times\dfrac{(\gamma-1)(a-F)F+d(c-F)(1-F)}{(j+1)(\gamma-1)(F-b)+d(c-F)}$	$1+\dfrac{\lambda}{2\gamma}$	1 or $\left[\dfrac{8+2j(\gamma-1)+(3-\gamma)\lambda}{4\{j(\gamma-1)+2\}}\right]\pm\left[\left(\dfrac{8+2j(\gamma-1)+(3-\gamma)\lambda}{4\{j(\gamma-1)+2\}}\right)^2-\dfrac{\lambda+2}{j(\gamma-1)+2}\right]^{\frac{1}{2}}$
$D=0$	$Z=(1-F)^2$	$\mp\infty$	1
$B=0$	$Z=\dfrac{(j+1)\gamma(\gamma-1)}{2}\times\dfrac{(1-F)F^2}{(j-1)F+2}$	$\dfrac{\lambda+2}{d}=\dfrac{2}{j-1}$	0 or 1

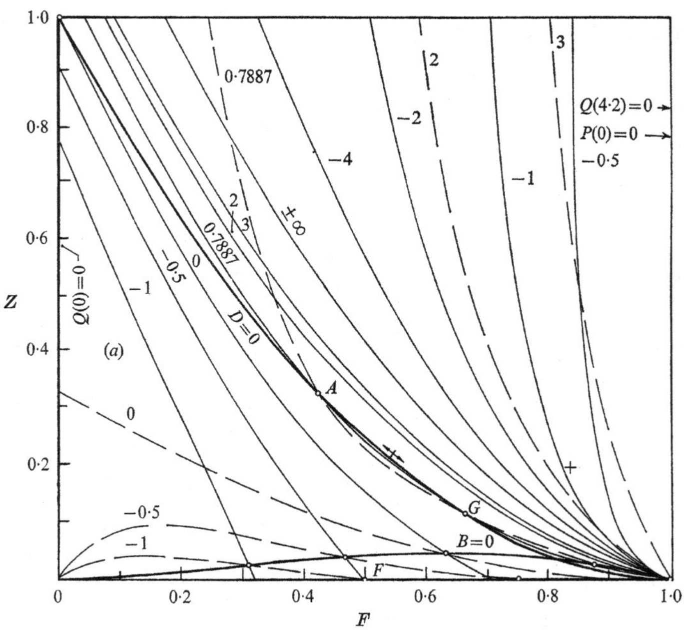

Fig. 10.3. Loci of **Q** = 0 lines (broken) and **P** = 0 lines (continuous) in the phase plane of $Z(F)$ for j = 2 and γ = 1.4.

Singularity D identifies the asymptotes of the P = constant lines at $Z = \infty$. Singularity F identifies their intersections at $Z = 0$.

The **Q** = 0, **P** = 0, **D** = 0 and **B** = 0 lines are specified in Table 10.2 and their loci on the phase plane of $Z(F)$ are displayed in Fig. 10.3 for $\gamma = 1.4$. Each curve is labeled with its decay parameter λ.

Loci of **Q** = 0 are delineated here by broken lines, while those of **P** = 0 are represented by continuous lines, each labeled by the value of the decay parameter, λ. Loci of singularities specified by their intersections, the **D** = 0 parabola and the **B** = 0 curve, are delineated by thick lines. As evident from Fig. 10.3, the singularities A and G are located at $\lambda = 0.7887$ – the classical solution of Guderley 1942.

10.4 Front

Boundary conditions of self-similar blast waves are specified by the coordinates of point corresponding to y = 0 on the Rankine-Hugoniot curve of $P_G = 1$, portrayed by the spike in Fig. 9.2, namely, according to (9.45),

$$F_n = f_n = \frac{2}{\gamma + 1} \tag{10.25}$$

and, according to (9.43),

$$Z_n = \frac{g_n}{h_n} = \frac{p_n}{\rho_n} = \frac{2\gamma(\gamma - 1)}{(\gamma + 1)^2} \tag{10.26}$$

Concomitantly, for $P_G = 1$ at $M_i = \infty$, corresponding to y = 0, according to (8.44'), $\rho_n = \beta_v = \dfrac{\gamma - 1}{\gamma + 1}$, whence, in view of (10.26), $p_n = \dfrac{2\gamma(\gamma - 1)^2}{(\gamma + 1)^3}$. In Fig. 9.2, point n, specified by (10.25) and (10.26), is marked by a cross.

Front trajectories, the loci of states identified by these boundary conditions - are, according to (10.3) and (10.25), power functions with index $\mu = 2/5$ for j = 2. There are two exceptional cases: case 1 corresponding to the decay parameter $\lambda = -2$, and case 2 corresponding to $\lambda = 2/\mu$.

In case 1, according to (10.3), the exponent $\mu = \infty$. However, for a variable μ, instead of (10.3), one has, according to (10.6),

$$\frac{d\ln\mu}{d\ln\xi}=\frac{1}{\mu} \tag{10.27}$$

whence, in terms of the normalized time coordinate, η, introduced by (9.1)

$$\xi=\exp\mu_0(\eta-1) \tag{10.28}$$

- an exponential front trajectory.

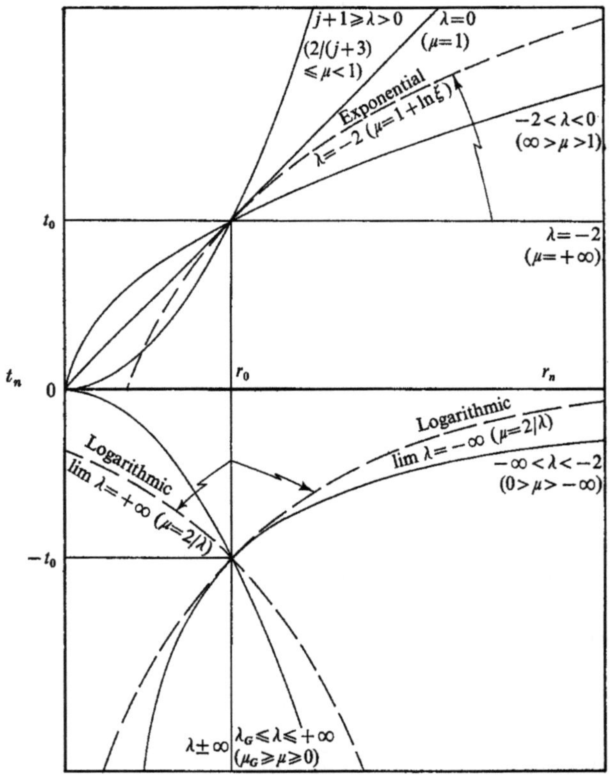

Fig. 10.4. Front trajectories of self-similar blast waves in the time-space plane
Broken lines depict the limiting cases of exponential and logarithmic trajectories.

In case 2 for variable μ, according to (10.6),

$$\frac{d\ln\mu}{d\ln\xi}=-1 \tag{10.29}$$

whence

$$\xi = 1 + \mu_0 \ln \eta \tag{10.30}$$

- a logarithmic front trajectory.

Front trajectories for all these cases are displayed in Fig. 10.4.

10.5 Field

The field of a self-similar wave is expressed by a set of integral curves obtained by solutions of the conservation equations, subject to boundary conditions, displayed in Fig. 10.5 for spherical geometry of j = 2, in a substance whose isentropic index $\gamma = 1.4$.

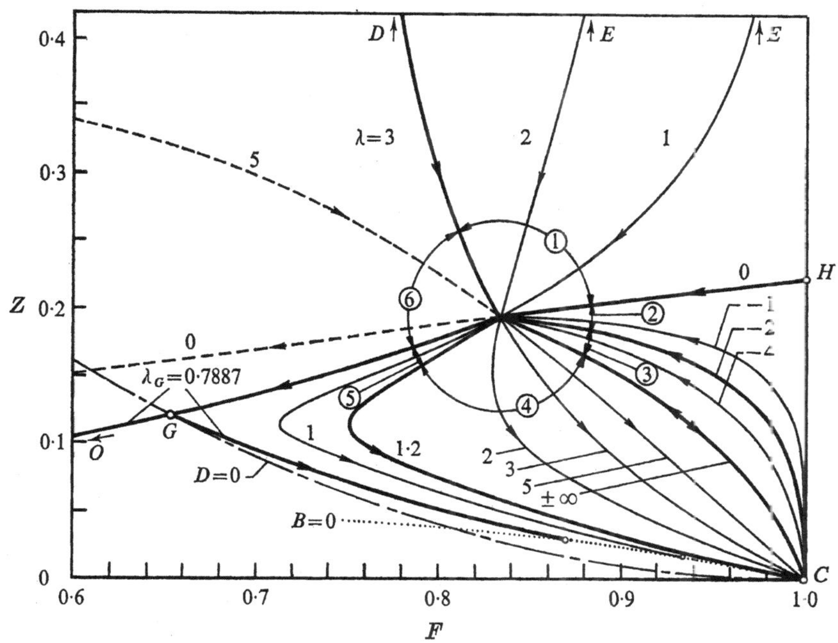

Fig. 10.5. Integral curves in the phase plane for y = 0, j = 2 and $\gamma = 1.4$

All the integral curves culminate at the strong shock boundary, n, whose coordinates, F_n & Z_n,, are specified by (10.25) and (10.26). Curves marked by arrows directed toward it delineate explosions; those for which they are directed away from it denote implosions.

The integral curve for $\lambda = 3$ stems from singularity D at Q= ∞, Z = ∞. The curves in sector (1), corresponding to $3 > \lambda > 0$, stem from singularity E at F = 1, $Z = \infty$ and represent blast waves driven by pistons at zero density. The integral curve for $\lambda = 0$ stems from the saddle point singularity H at P = 0, F = 1 and delineates a wave generated by piston moving at constant speed.

Integral curves in sector (2), corresponding to $0 > \lambda > -2$, and those in sector (3), corresponding to $-2 > \lambda > -\infty$, stem from singularity C at F = 1, $Z = 0$ and represent blast waves driven by pistons at infinite density. The two sectors are separated from each other by the integral curve for $\lambda = -2$, representing the limit of exponential front trajectory. The integral curve for the decay parameter $\lambda = \pm\infty$ delineates the limit of logarithmic front trajectory.

Integral curves in sector (4), corresponding to $+\infty > \lambda > 1.2$ represent implosions into a point of infinite density. The integral curve for $\lambda = a-2 = (j + 1)(y - 1)$, so that, for $j = 2$ and $\gamma = 1.4$, $\lambda = 1.2$, delineates the implosion wave propagating into the singularity $\mathbf{B}(F, Z) = 0$ defined by (10.24), at F = 1, Z = 0.

Sector (5), corresponding to $1.2 > \lambda > \lambda_G = 0.7887$, the axis of the Guderley saddle point singularity G. Since at this point the functions $\mathbf{Q}(F, Z)$ and $\mathbf{D}(F, Z)$ in (10.5), as well as $\mathbf{P}(F, Z)$ and $\mathbf{D}(F, Z)$ in (10.6), change signs, the field co-ordinate of the integral curve passing through this singularity is monotonic and delineates the solution leading to the singular point O at F = 0, Z = 0.

Sector (6), between the lines corresponding to an explosion of $\lambda = j+1$ and the Guderley implosion (Guderley 1942), contains integral curves that have no physical meaning, since, as a consequence of intersecting the $\mathbf{D} = 0$ line, the gasdynamic parameters of the flow field become double-valued functions of the field coordinate.

Profiles of all the gasdynamic parameters evaluated for all the physically meaningful integral curves displayed in Fig. 10.5 - the pressure, the temperature, the density and the particle velocity – by means of, respectively, (10.19), 10.15), (10.20) and (10.14), are presented by Figs. 10.6 – 10.9.

Inserts in these figures depict the solutions for the limit of $\lambda = \pm\infty$, corresponding to the logarithmic front trajectory specified by (10.27), whereas the limit of the exponential front trajectory corresponding to $\lambda = -2$ is not anomalous for the Eulerian space profiles.

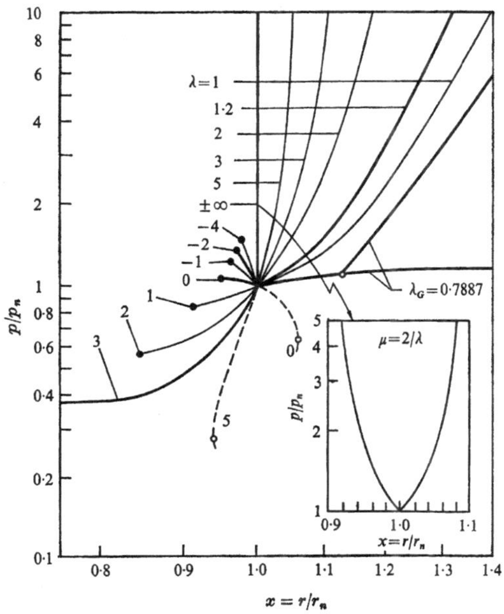

Fig. 10.6. Pressure profiles for y = 0, j = 2 and γ = 1.4

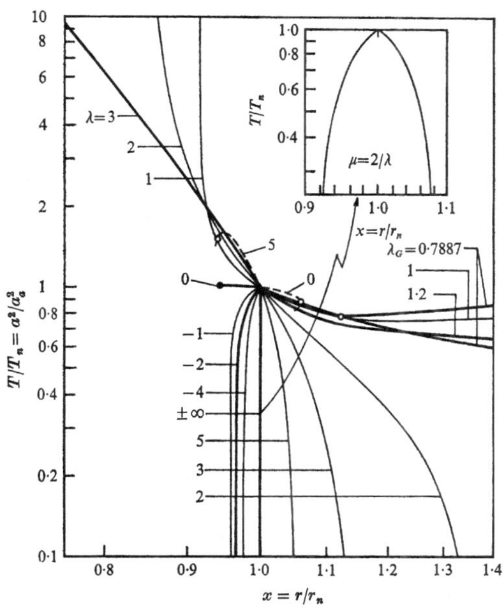

Fig. 10.7. Temperature profiles for y = 0, j = 2 and γ = 1.4

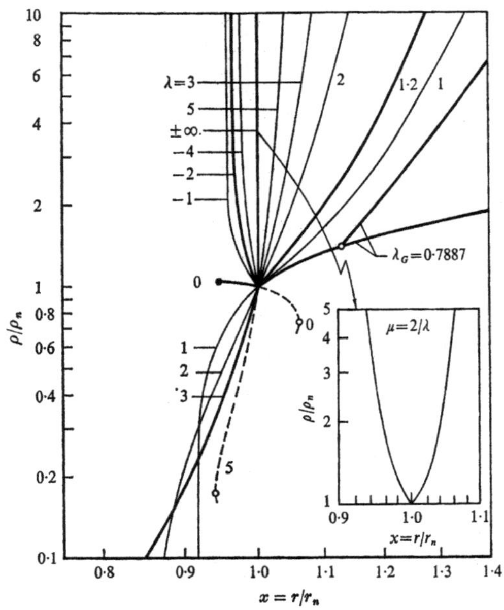

Fig. 10.8. Density profiles for y = 0, j = 2 and γ = 1.4

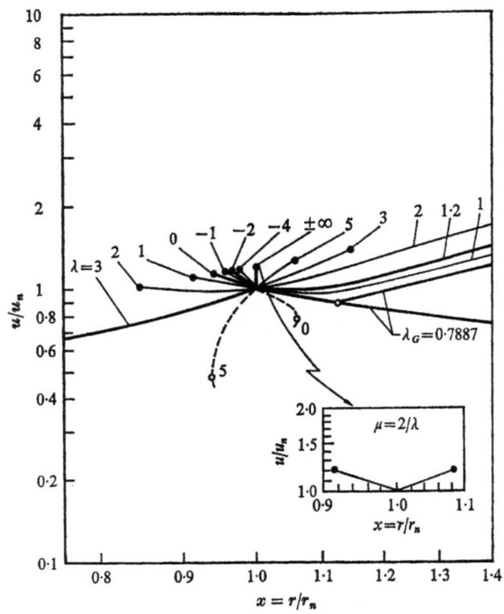

Fig. 10.9. Velocity profiles for y = 0, j = 2 and γ = 1.4

10.6 Analytic Solution

The integral curve for the decay parameter

$$\lambda = j+1 \tag{10.31}$$

portrayed by $\lambda = 3$ for spherical geometry of $j = 2$ in Fig. 10.5, plays the central role. It culminates at singularity **D**(F = b, Z = ∞) and represents the solution for a constant energy blast wave of a point explosion.

With the decay parameter specified by (10.31), the locus of singularity **B** for $\lambda = \mathrm{d}F - 2$ is expressed by

$$Z = \frac{\gamma - 1}{2}\frac{(1-F)F^2}{F - \gamma^{-1}} \tag{10.32}$$

a relationship satisfied by the strong shock boundary condition whose co-ordinates are specified by (10.25) and (10.26). It provides, therefore, a particular *algebraic* solution of the governing equation (10.7) for a constant energy self-similar blast wave whose front propagates at y = 0 with a decay coefficient of $\lambda = j + 1$. A remarkable coincidence!

As a consequence, the structure of this blast wave can be expressed in terms of algebraic equations –a solution published by Sedov 1946, 19559b. Thus, with (10.32), (10.5) yields

$$\frac{\mathrm{dln}\phi}{\mathrm{dlnF}} = \frac{\gamma(\gamma-1)F^2 - 2(\gamma F - 1)(1 - F)}{\alpha(\gamma F - 1)(\mathrm{c} - F)} \tag{10.33}$$

whence, by quadrature,

$$\phi = \left(\frac{\mathrm{F_n}}{F}\right)^{\mathrm{a}}\left(\frac{\gamma F - 1}{\gamma \mathrm{F_n} - 1}\right)^{\beta}\left(\frac{\mathrm{c} - \mathrm{F_n}}{\mathrm{c} - F}\right)^{\chi} \tag{10.34}$$

where

$$\beta = \frac{\gamma - 1}{\mathrm{d}(\gamma \mathrm{c} - 1)} \tag{10.35}$$

and

$$\chi = \frac{2 + (\gamma + 1)(c\gamma - 2)c}{dc(c\gamma - 1)} \tag{10.36}$$

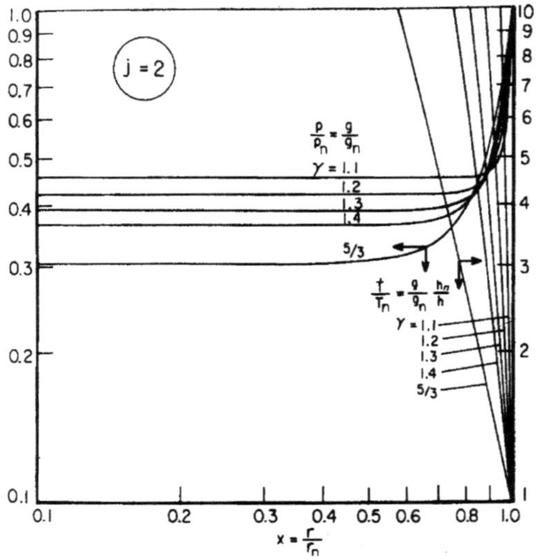

Fig. 10.10. Eulerian space profiles of pressure and temperature in spherical self-similar blast waves for a scope of isentropic indexes

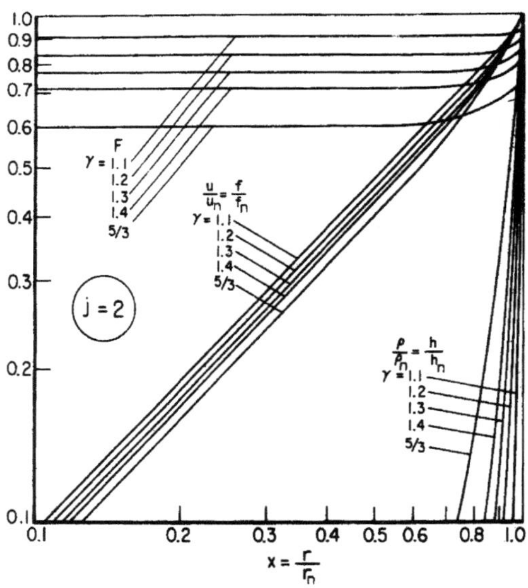

Fig. 10.11. Eulerian space profiles of the normalized coordinate, F, particle velocity and density in spherical self-similar blast waves for a scope of isentropic indexes

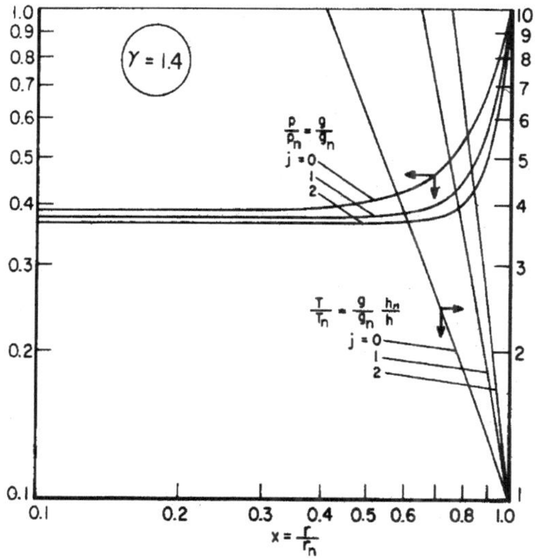

Fig. 10.12. Eulerian space profiles of pressure and temperature in planar, cylindrical and spherical self-similar blast waves for $\gamma = 1.4$

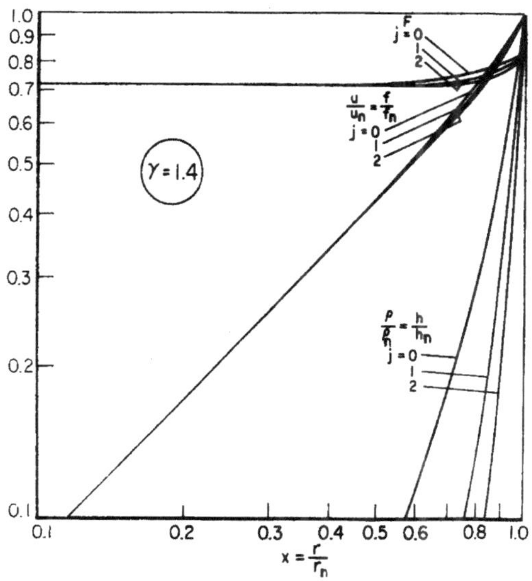

Fig. 10.13. Eulerian space profiles of the normalized coordinate, F, velocity and density in planar, cylindrical and spherical self-similar blast waves for γ=1.4

The algebraic relationship between the field coordinate, x, and the normalized coordinate, F, together with (10.19), 10.15), (10.20) and (10.14) provide a complete description of the blast wave's structure, providing thus convenient means for their parametric investigation. Profiles of all the gasdynamic parameters for spherical self-similar blast waves thus evaluated for a scope of isentropic indexes, γ, are displayed by Figs. 10.10 and 10.11, while those with $\gamma = 1.4$ for planar, cylindrical and spherical geometry are presented by Figs. 10.12 and 10.13.

10.7 Atom Bomb Explosion

The theory of self-similar blast waves was developed as a consequence of interest in the gasdynamic effects of an atom bomb and is therefore of direct relevance to the cinematographic record of the Trinity explosion presented by Fig. 9.1.

In the introduction to Part I of his classical paper, Taylor 1950a, Sir Geoffrey expressed that as follows.

"This paper was written in 1941 and circulated to the Civil Defence Research Committee of the Ministry of Home Security in June of that year. The present writer had been told that it might be possible to produce a bomb in which a large amount of energy would be released by nuclear fission – the name atom bomb had not then been used – and the work here described represents his first attempt to form an idea of what mechanical effects might be expected if such an explosion could occur. In the then common explosive bomb mechanical effects were produced by the sudden generation of a large amount of gas at a high temperature in a confined space*. The practical question which required an answer was: Would similar effects be produced if energy could be released in a highly concentrated form unaccompanied by the generation of gas?"

Thereupon he set up the theory of his self-similar solution with numerical calculations of the density, velocity and pressure profiles normalized respect to their values immediately behind the shock front, for the case of the isentropic index $\gamma = 1.4$, as well as, approximately, $\gamma = 1.666$.

The application of the self-similar blast wave theory to the Trinity atom bomb explosion was presented in Part II of his paper, Taylor 1950b, based on the data presented there by Table 1, whose copy is provided here by Table 10.3. The first section was deduced from the cinematographic records of Mack 1947, presented in this book by Fig. 9.1; the second section, whose record is included in this figure, was obtained by Taylor from the Ministry of Supply, and the remaining sections were reproduced from small images and single photographs in Mack 1947, referred to as MDCC 221, with their glossy prints received appreciatively by Taylor from N.E. Bradburry, Director of Los Alamos Laboratory.

---

* a piston driven blast wave whose theory had been published by Taylor 1946.

Table 10.3. Cinematographic records of Trinity explosion

TABLE 1. RADIUS *R* OF BLAST WAVE AT TIME *t* AFTER THE EXPLOSION

authority	*t* (msec.)	*R* (m.)	$\log_{10} t$	$\log_{10} R$	$\frac{5}{2}\log_{10} R$
strip of small images MDDC 221	0·10	11·1	$\bar{4}$·0	3·045	7·613
	0·24	19·9	$\bar{4}$·380	3·298	8·244
	0·38	25·4	$\bar{4}$·580	3·405	8·512
	0·52	28·8	$\bar{4}$·716	3·458	8·646
	0·66	31·9	$\bar{4}$·820	3·504	8·759
	0·80	34·2	$\bar{4}$·903	3·535	8·836
	0·94	36·3	$\bar{4}$·973	3·560	8·901
strip of declassified photographs lent by Ministry of Supply	1·08	38·9	$\bar{3}$·033	3·590	8·976
	1·22	41·0	$\bar{3}$·086	3·613	9·032
	1·36	42·8	$\bar{3}$·134	3·631	9·079
	1·50	44·4	$\bar{3}$·176	3·647	9·119
	1·65	46·0	$\bar{3}$·217	3·663	9·157
	1·79	46·9	$\bar{3}$·257	3·672	9·179
	1·93	48·7	$\bar{3}$·286	3·688	9·220
strip of small images from MDDC 221	3·26	59·0	$\bar{3}$·513	3·771	9·427
	3·53	61·1	$\bar{3}$·548	3·786	9·466
	3·80	62·9	$\bar{3}$·580	3·798	9·496
	4·07	64·3	$\bar{3}$·610	3·809	9·521
	4·34	65·6	$\bar{3}$·637	3·817	9·543
	4·61	67·3	$\bar{3}$·688	3·828	9·570
large single photographs MDDC 221	15·0	106·5	$\bar{2}$·176	4·027	10·068
	25·0	130·0	$\bar{2}$·398	4·114	10·285
	34·0	145·0	$\bar{2}$·531	4·161	10·403
	53·0	175·0	$\bar{2}$·724	4·243	10·607
	62·0	185·0	$\bar{2}$·792	4·267	10·668

The values of $\frac{5}{2}\log_{10} R$ listed in the last column stem from his self-similarity theory (Taylor 1950a), according to which they should be proportional to time, t. This is, indeed, brought out by the plot of Table 1 in Taylor 1950b, whose copy is provided here by Fig. 10. 14. In accord with it, the velocity modulus of the Trinity blast wave μ =5/2, corresponding to $\lambda = 3$ for j = 2 of a self-similar adiabatic blast wave bounded by a strong front of a shock propagating at y = 0, whose structure is displayed by Figs. 10.10 – 10.13.

On this basis, in (Taylor 1950b) he published the energy of the Trinity blast wave, E, calculated by means of the energy integral (9.48), using variables of the self-similar blast wave structure presented by Figs. 10.10 – 10.13. The calculations were carried this out for, of course, the spherical case of j = 2, taking into account a set of isentropic indices, $\gamma = 1.2$, 1.3, 1.4 and 1.667.

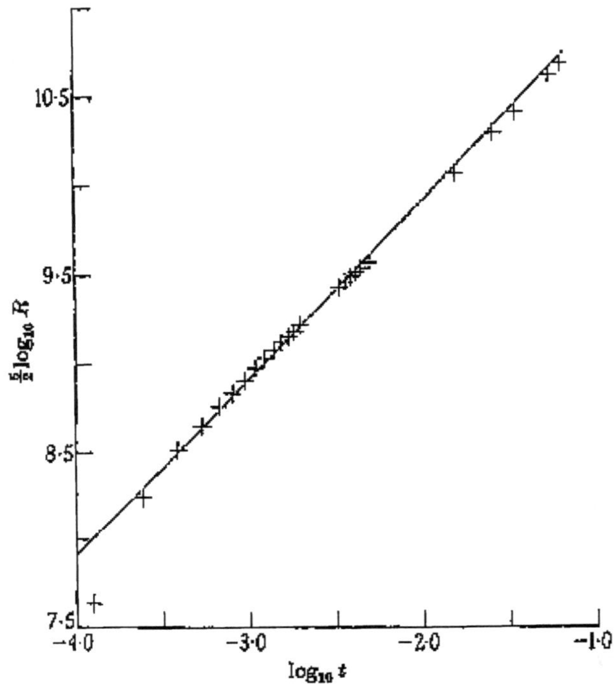

Fig. 10.14. Front trajectory of the Trinity explosion (Taylor 1950)

Thus he found that, if $\gamma = 1.4$, the energy of the Trinity blast wave should be $E = 7.14 \cdot 10^{10}$ MJ, equivalent to, $16.8 \cdot 10^3$ tons of TNT, while, if $\gamma = 1.3$, it would be $E = 9.74 \cdot 10^{10}$ MJ, equivalent to $22.9 \cdot 10^3$ tons of TNT, bracketing thus the value recorded in highly classified military documents of the Trinity explosion.

Sir Geoffrey liked to recount the interrogation, to which he was subjected thereupon by agents of the American intelligence service who wished to know how he gained access to the securely guarded energy of the Trinity atom bomb.

11 Phase Space Method

11.1 Background

As evident from Chapter 10, the structure of a blast wave is associated with the trajectory of its front. Blast wave theory was formulated at the onset of atomic bomb when the problem of a constant energy wave was of prime interest. Since far-field effect of the bomb was then of crucial significance, while the initial discharge of energy was of unprecedented magnitude, of particular interest then were blast waves of point explosions whose initial conditions imply energy deposition at infinite power density.

The first solution of such a blast wave propagating into an atmosphere of finite pressure, was provided by Goldstein and von Neumann (1955) on the basis of its formulation in terms Lagrangian coordinates in the case of a perfect gas with constant specific heats. This was accomplished by the use of a Fourier function to transform the partial differential equations into ordinary differential equations, whereas the position and strength of the shock front at each time step was determined by iteration. Brode (1955 and 1969) and Wilkins (1969), utilized a Lagrangian difference schemes employing artificial viscosity – a method introduced by von Neumann and Richtmyer (1950) whereby the numerical stability of the solution is obtained by spreading the shock front over several mesh steps to ascertain the continuity of flow variables across the front. As pointed out by them, the application of this method is associated with a certain loss of accuracy in "determining the shock position," as well as in "difficulties that arise in multiple front blast waves."

On the other hand, blast wave equations formulated in the Eulerian frame of reference can yield satisfactory data on the evolution of the shock front. Thus, as a prominent example, Korobeinikov and Chushkin (1966) followed by Korobeinikov and Sharovatova (1969) solved the non-self-similar point explosion problem, formulated in Eulerian space coordinates, on the basis of integral relations – a method proposed by Dorodnitsyn (1956a and 1956b) and refined by Belotserkovskii and Chushkin (1965). In spite of its success, the large number of ordinary differential equations that have to be then treated (typically, as much as 27) caused awkward difficulties. It is, in fact, for this reason that this method of approach has not been further developed.

It became therefore of interest to seek approximate solutions for problems where analytical insight into the intrinsic properties of blast waves were of particular interest. The most popular approach to obtain such a solution is based on the method of asymptotic approximations with the dependent variables expressed in terms of the front coordinate, starting with the self-similar solution as the zero order step (Sakurai, 1965 and Korobeinikov et all, 1963). Another way to obtain an approximate solution utilizes the so-called "quasi-similar" method developed by Oshima (1960), where all the terms containing the front coordinate are taken as equal to their values at the front. By virtue of its construction, this method gives exact results at the self-similar limit just behind the front, whereas the accuracy deteriorates rapidly as one proceeds towards the center. In an alternative method, the density is represented by a power law of the field coordinate (Mel'nikova, 1966 and Bach and Lee, 1970).

Approximate analytic methods, applicable to cases when, by virtue of its distance from the center, the front is decoupled from the source of explosion, were developed by Chester (1954), Chisnel (1957), Whitham (1958) and Friedman (1961); these solutions became known as the Whitham rule. The method is based on the postulate that the differential equations for one set of characteristic lines are satisfied by the gasdynamic parameters of the state immediately behind the front. A classical method for, especially, weak non-self-similar regions of point explosion, was formulated by Brinkley and Kirkwood (1947) by seeking a self-consistent set of ordinary differential equations that specify the problem without recourse to partial differential equations expressing the conservation principles.

The Phase Space Method (PSM) is based on the autonomous form of blast wave equations in Eulerian space coordinates presented in Chapter 10. The solution is then obtained by determining a family of integral curves forming a surface in the phase space of blast wave coordinates, rather than directly in the physical space, as is conventionally done by numerical algorithms. It was first presented by Oppenheim, Kuhl and Kamel (Oppenheim et al 1978) and its exposition here has been improved thanks to comments provided recently by Allen Kuhl.

11.2 Formulation

11.2.1 Coordinates

In the phase space, the blast wave problem is formulated in terms of the phase coordinates

$$F \equiv \frac{\tau}{x} f = \frac{t}{\mu r} u \qquad \text{and} \qquad Z \equiv \left(\frac{\tau}{x}\right)^2 \frac{g}{h} = \left(\frac{t}{\mu r}\right)^2 \frac{p}{\rho} \tag{9.29}$$

with respect to $x \equiv \frac{r}{\mathrm{r_n}}$ and $\tau \equiv \frac{t}{\mathrm{r_n}}$ as the independent field variables, with

$$h \equiv \frac{\rho}{\rho_a} \quad g \equiv \frac{p}{\rho_a \mathrm{w}_n^2} \qquad \varepsilon \equiv \frac{e}{\mathrm{w}_n^2} \qquad f \equiv \frac{u}{\mathrm{w}_n} \tag{9.9}$$

are the dependent variables - the gasdynamic parameters of the flow field describing the structure of the flow field.

The exponential trajectory of the front is expressed by $x = t^{\mu}$ or $\xi = \tau^{n}$, where $\xi \equiv \frac{r_n}{\mathrm{r_o}}$, $\eta \equiv \frac{t_n}{\mathrm{t_o}}$, with $\mu \equiv \frac{d \ln r_n}{d \ln t_n} \equiv \frac{d \ln \xi}{d \ln \eta} \equiv \frac{w_n t_n}{r_n} = \frac{\lambda + 2}{2}$, where

$$\lambda \equiv \frac{d \ln y}{d \ln \xi} = -2 \frac{d \ln w_n}{d \ln r_n} = -2 \frac{r_n \dot{r}_n}{\ddot{r}_n^2} \tag{9.4}$$

is the decay parameter.

11.2.1 Constitutive Equations

The constitutive equations for blast waves are cast into an autonomous form of (9.30) that, for Eulerian space profiles, is expressed in terms specified in Table 9.9. On its basis, they are expressed as functions of the phase coordinates as follows

$$\left(\frac{\partial \mathbf{Z}}{\partial F}\right)_y = \frac{Z}{1-F}\frac{\mathbf{P}(F,Z;\Phi^g,\Phi^F,\Phi^Z)}{\mathbf{Q}(F,Z;\Phi^g,\Phi^F)} \tag{11.1}$$

$$\left(\frac{\partial \ln x}{\partial F}\right)_y = -\frac{\mathbf{D}(F,Z)}{\mathbf{Q}(F,Z;\Phi^g,\Phi^F)} \tag{11.2}$$

$$\left(\frac{\partial \ln h}{\partial F}\right)_y = \frac{\mathbf{H}(F,Z;\Phi^h,\Phi^g,\Phi^F)}{\mathbf{Q}(F,Z;\Phi^g,\Phi^F)} \tag{11.3}$$

$$\left(\frac{\partial \ln g}{\partial F}\right)_y = \frac{\mathbf{G}(F,Z;\Phi^g,\Phi^F)}{\mathbf{Q}(F,Z;\Phi^g,\Phi^F)} \tag{11.4}$$

where

$$\mathbf{D}(F,Z) = Z-(1-F)^2 \tag{11.5}$$

$$\mathbf{Q}(F,Z;\Phi^g,\Phi^F) \equiv (\mathrm{j}+1)(F-A)Z - F(1-F)(B-F) \tag{11.6}$$

$$\mathbf{P}(F,Z;\Phi^g,\Phi^F,\Phi^Z) \equiv \mathrm{d}(\gamma-F)\mathbf{D} + (\gamma-1)\mathbf{Q} \tag{11.7}$$

$$\mathbf{H}(F,Z;\Phi^h,\Phi^g,\Phi^F) \equiv \frac{1}{1-F}[\mathbf{Q}-(\mathrm{j}+1)F+\Phi^h)\mathbf{D}] \tag{11.8}$$

$$\mathbf{G}(F,Z;\Phi^g,\Phi^F) \equiv -\gamma[F(B\text{-}F)\text{-}(\mathrm{j}+1)(F\text{-}\Gamma)(1\text{-}F)] \tag{11.9}$$

with j = 0, 1, 2 for, respectively, plane, line and point symmetrical waves, and

$$\mathrm{d} = (\mathrm{j}+1)(\gamma\text{-}1)+2 \tag{10.13}$$

whereas, with $K = F, Z, h, g,$

$$\Phi^k \equiv \frac{\partial \ln K}{\partial \ln \xi} = \lambda \frac{\partial \ln K}{\partial \ln y} \tag{11.10}$$

whereas

$$A \equiv \frac{1}{(j+1)\gamma}(\lambda - \Phi^{g}) \tag{11.11}$$

$$B \equiv \frac{\lambda + 2}{2} - \Phi^{F} \tag{11.12}$$

$$\Gamma \equiv \frac{\lambda + 2 - \Phi^{Z}}{\delta} \tag{11.13}$$

11.3 Procedure

The phase space method is construed by the reduction of the constitutive equations (11.1) – (1.4) into ordinary differential equations by transforming the cross derivatives expressed by (11.10) into algebraic form. This is accomplished by the use of a progress variable, ε, in terms of which

$$\Phi^{k} = \Phi_{i}^{k} + \varepsilon(\Phi_{n}^{k} - \Phi_{i}^{k}) \tag{11.14}$$

where subscript n denotes the outer boundary immediately behind the front at $x = 1$ and i marks the inner boundary at $x = 0$, between which ε varies from 0 to 1.

The expressions for the outer boundary conditions at the front, denoted by subscript n, are provided as functions of y in section 9.9. At the inner boundary in the center of explosion, denoted by subscript i, the Mach number, $M = 0$ and, hence $y = \infty$. In an inviscid substance, the temperature there is infinite and the density is zero, so that $Z = \infty$, while $F = \frac{0}{0}$, forming a saddle point singularity. In order to get an access to it, the constitutive equations, (11.1) - (11.4) and their auxiliary relationships, (11.5) - (11.9), the dependent phase coordinate Z is expressed by its reciprocal $\hat{Z} \cong 1/Z$, in terms of which

$$\left(\frac{\partial \hat{\boldsymbol{Z}}}{\partial F}\right)_y = \frac{\hat{\boldsymbol{Z}}}{1-F}\frac{\hat{\mathbf{P}}(F,\hat{\boldsymbol{Z}})}{\hat{\mathbf{Q}}(F,\hat{\boldsymbol{Z}})} \tag{11.15}$$

$$\left(\frac{\partial \ln x}{\partial F}\right)_y = -\frac{\hat{\mathbf{D}}(F,\hat{\boldsymbol{Z}})}{\hat{\mathbf{Q}}(F,\hat{\boldsymbol{Z}})} \tag{11.16}$$

$$\left(\frac{\partial \ln h}{\partial F}\right)_y = \frac{\hat{\mathbf{H}}(F,\hat{\boldsymbol{Z}})}{\hat{\mathbf{Q}}(F,\hat{\boldsymbol{Z}})} \tag{11.17}$$

$$\left(\frac{\partial \ln g}{\partial F}\right)_y = \frac{\hat{\mathbf{G}}(F,\hat{\boldsymbol{Z}})}{\hat{\mathbf{Q}}(F,\hat{\boldsymbol{Z}})} \tag{11.18}$$

while

$$\hat{\mathbf{D}}(F,\hat{Z}) = \hat{Z} - (1-F)^2 \tag{11.19}$$

$$\hat{\mathbf{Q}}(F,\hat{Z}) \equiv (\mathrm{j}+1)(F-A)\hat{Z} - F(1-F)(B-F^2+\frac{\partial F}{\partial \ln \xi}) \tag{11.20}$$

$$\hat{\mathbf{P}}(F,\hat{Z}) \equiv \delta(\Gamma - F)\hat{\mathbf{D}} + (\gamma-1)\hat{\mathbf{Q}} \tag{11.21}$$

$$\hat{\mathbf{H}}(F,\hat{Z}) \equiv \frac{1}{1-F}\{\hat{\mathbf{Q}} - [(\mathrm{j}+1)F + \Phi^h]\hat{\mathbf{D}}\} \tag{11.22}$$

$$\hat{\mathbf{G}}(F,\hat{Z}) \equiv -\gamma[F(\mathrm{F_b} - F) - (\mathrm{j}+1)(F - \mathrm{F_c})(1-F)] \tag{11.23}$$

The saddle point singularity at i, has to be approached along its axis – a condition that is approximated by an asymptotic solution based on the first term in Taylor's expansion

$$\mathrm{F} = \mathrm{F_i} + \frac{dF_{\mathrm{i}}}{d\hat{Z}}\hat{Z} \tag{11.24}$$

where from (11.15) with (11. 19), (11.20) and (11.21), by l'Hospital's rule,

$$\frac{\partial F_i}{\partial \hat{Z}} = \frac{\mathrm{F_i}(1-\mathrm{F_i})^2(A-\mathrm{F_i})}{\delta(\Gamma-\mathrm{F_i})+(j+1)(1-\mathrm{F_i})} \tag{11.25}$$

Then according to (11.16), (11.17) and (11.18), with subscript o denoting the start of the asymptotic solution in the field,

$$\frac{\mathrm{d}\ln x/\mathrm{x_o}}{\mathrm{d}\hat{Z}} = \frac{\mathrm{c_1}}{\hat{Z}} \tag{11.26}$$

$$\frac{\mathrm{d}\ln \rho/\mathrm{\rho_o}}{\mathrm{d}\hat{Z}} = -\frac{\mathrm{c_2}}{\hat{Z}} \tag{11.27}$$

$$\frac{\mathrm{d}\ln p/\mathrm{p_o}}{\mathrm{d}\hat{Z}} = \mathrm{c_3} \tag{11.28}$$

where

$$\mathrm{c_1} = \frac{1-\mathrm{F_i}}{\delta(\Gamma_i-\mathrm{F_i})} \tag{11.29}$$

$$\mathrm{c_2} = \frac{(\mathrm{j}+1)\mathrm{F_i}+\Phi_i^h}{\delta(\Gamma_i-\mathrm{F_i})\hat{Z}} \tag{11.30}$$

$$\mathrm{c_3} = \frac{\gamma\mathrm{F_i}\mathrm{B_i}(1-\mathrm{F_i})}{\delta(\Gamma_i-\mathrm{F_i})\hat{Z}} \tag{11.31}$$

whence, by quadrature,

$$x = \mathrm{x_o}(\hat{Z}/\hat{Z}_o)^{\mathrm{c_1}} \tag{11.32}$$

$$\rho = \mathrm{\rho_o}(\hat{Z}/\hat{Z}_o)^{\mathrm{c_2}} \tag{11.33}$$

$$p = p_o(\hat{Z}/\hat{Z}_o)^{c_3} \tag{11.34}$$

A numerical solution of this set of equations is obtained for the segment of the integral curve from $\hat{Z} = 0$ to $\hat{Z}_o$, where it is matched with the integral curve evaluated by the integration of (11.1) – (11.4). Each integral curve for a given y depends on four parameters that are unknown a priori:

(1) the decay coefficient, $\lambda(y)$,

(2) the density parameter, $\Phi^h(y)$,

(3) the value of the saddle point singularity, $F_i(y)$

(4) the derivative dF_i/dy at F_i.

These parameters are, in effect, eigenvalues of the problem and their magnitudes are determined by requiring the solution to satisfy, within at least one percent, the mass integral, I_M, specified by (9.47), the energy integral, I_E, expressed by (9.51), its derivative, $dI_E/d\ln\xi$, calculated from (9.56), and the decay coefficient, λ, evaluated by (9.55), by a quadruple iterative procedure.

11.4 Results

Figures 11.1–11.3 present the solution surfaces in the phase space for blast waves generated by point explosions in, respectively, spherical, cylindrical and planar geometry. They are made out of integral curves *Z*(*F*), evaluated for a full decimal set of the front velocity parameter, y, each starting from a point on the Rankine-Hugoniot curve and terminating at the inner singularity, i, of F(y) at $Z = \infty$. On the right, the surface is bounded by the integral curve for $y = 0$ - the self-similar limit of the blast wave- and on the left by the axis of F:= 0 – its acoustic limit.. The shape of the surface is revealed by lines of constant Z. For relatively weak shock fronts of $0.55 < y < 1$, the surface protrudes into the region of negative values of F, corresponding to negative particle velocities that are due to contraction of the flow field into the center upon initial overexpansion produced by the shock front.

Figures 11.4 - 11.6 depict projections of integral curves $Z=Z(F)$ on the plane y = 0, while Figs. 11.7 - 11.9 show their reciprocals, $\hat{Z}(F)$. The slope of the curves is well behaved near the singularity i, as a consequence of the asymptotic approach to it implemented in the phase space method. For comparison, shown by dash curves in Figs. 11.4 and 11.7 are the solutions of Korobeinikov and Sharotova (1969) obtained for the same point explosion by the method of integral relations that does not take into account the effects of the saddle point singularity at the center, which are clearly evident in Fig. 11.7.

The Eulerian space profiles for spherical, cylindrical, and planar explosions are presented by Figs 11.10 – 1.21 in four groups of three for, respectively, pressure, density, temperature and particle velocity. The match points between the integral curves started at the front and the asymptotic solution to the center are denoted by small circles.

The density at the center is always zero, corresponding to an infinite temperature - a characteristic property of inviscid solutions. When transport properties are taken into account, a boundary layer is formed near the center where the temperature is restricted to a finite value. (Kim et al 1975).

Pressures at the center of blast waves for the full scope of the shock strength parameter, y, are displayed in Fig. 10.22 by open circles, triangles and squares in comparison to the results of Korobeinikov and Chushkin (1966) shown by continuous, chain-dash and dash lines, confirming that, in contrast to density, the effect of the saddle point singularity at the center on pressure is quite small – a manifestation of a paradox. On one hand, pressure is the most prominent parameter in fluid dynamics, on the other, it is the least sensitive result of CFD calculations.

Figure 11.23 depicts the decay parameter, λ, of blast waves generated by point explosion plotted as function of y, in comparison to the numerical solutions of Korobeinikov and Chushkin (1966), as well as the classical results of Goldstine and von Neumann (1955), demonstrating an agreement with each other within a maximum deviation of 7% at $y \approx 0.7$.

The display of results obtained by the phase method is culminated by the plots of the energy integral, $I_E(y)$ presented by Fig. 11.24 for spherical, cylindrical and planar geometry. All of them peak at the sonic limit of $y = 1$, where $I_E = [\gamma(\gamma - 1)(j + 1)]^{-1}$. and pass through a minimum at $y \approx 0.3$.

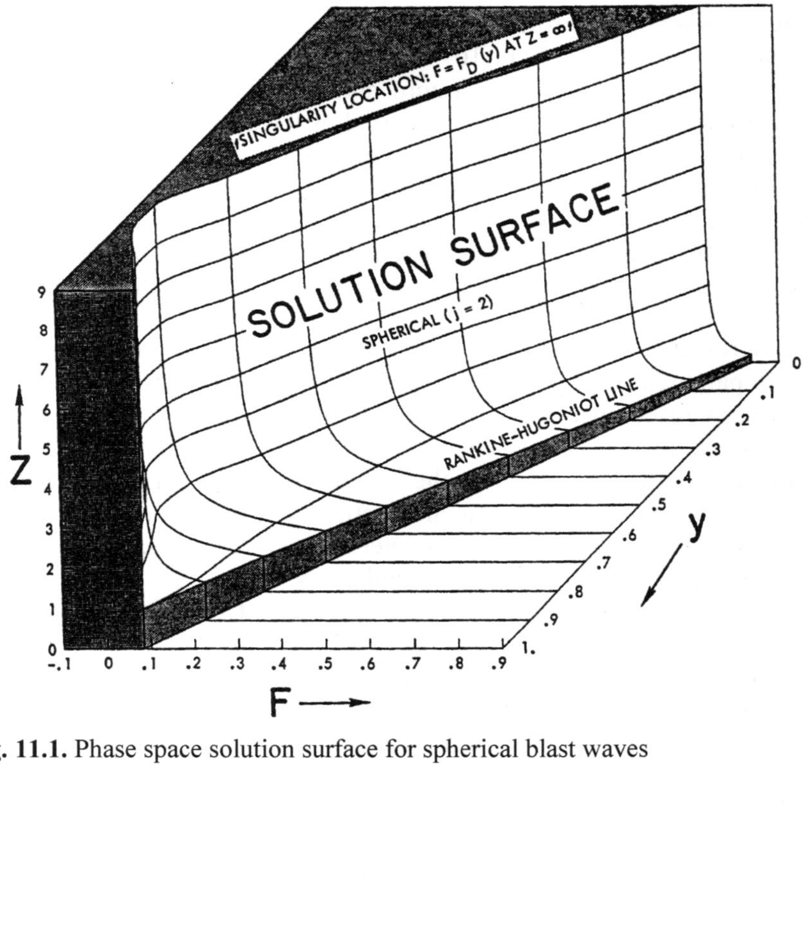

Fig. 11.1. Phase space solution surface for spherical blast waves

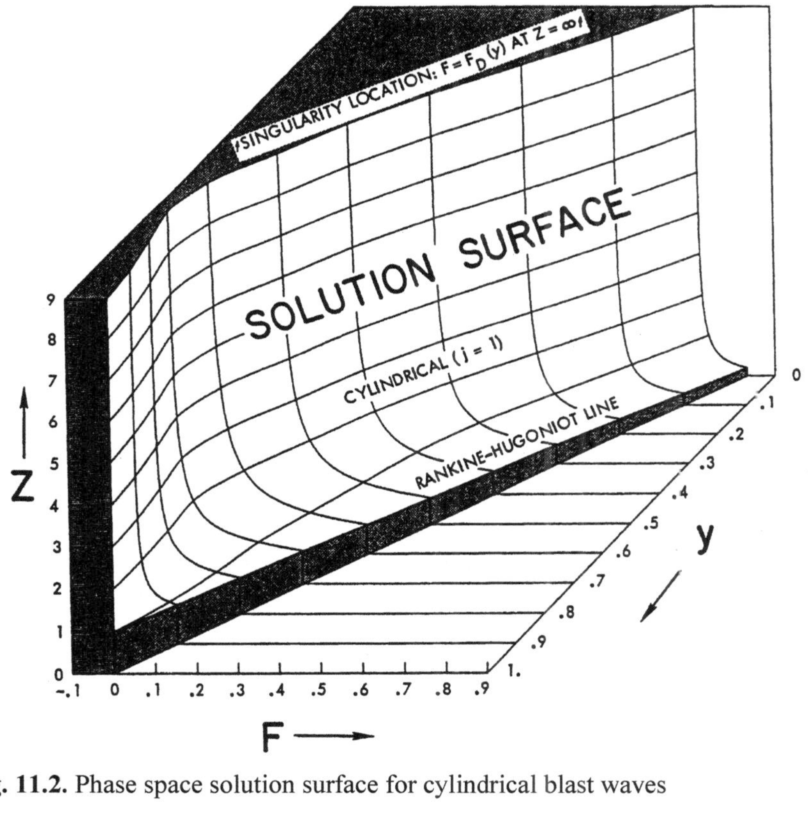

Fig. 11.2. Phase space solution surface for cylindrical blast waves

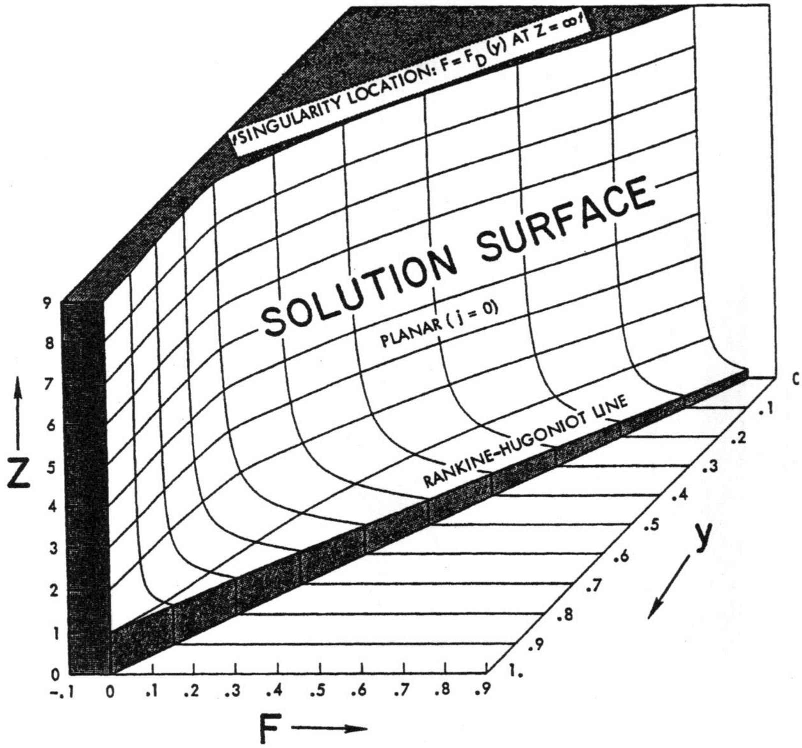

Fig. 11.3. Phase space solution surface for planar blast waves

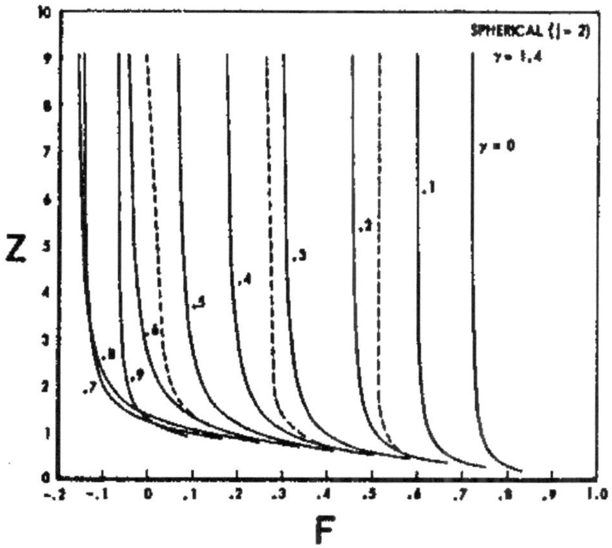

Fig. 11.4. Integral curves of spherical blast waves in the $Z(F)$ plane on y =0

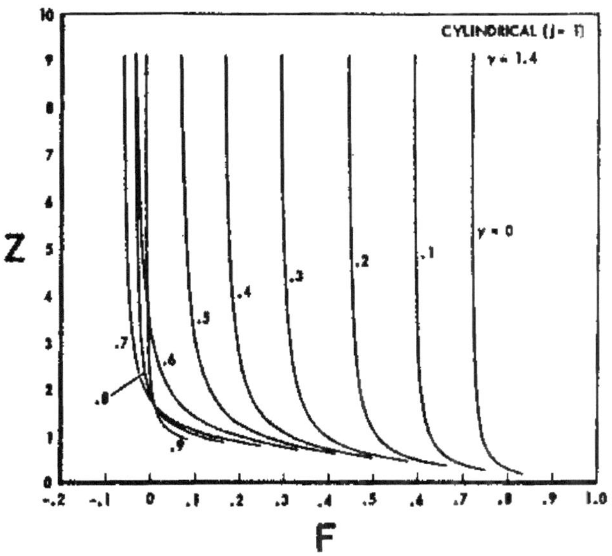

Fig. 11.5. Integral curves of cylindrical blast waves in the $Z(F)$ plane on y =0

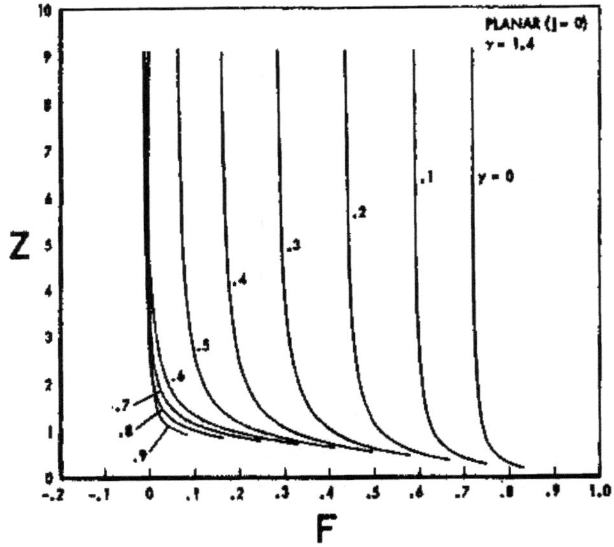

Fig. 11.6. Integral curves of planar blast waves in the $Z(F)$ plane on y =0

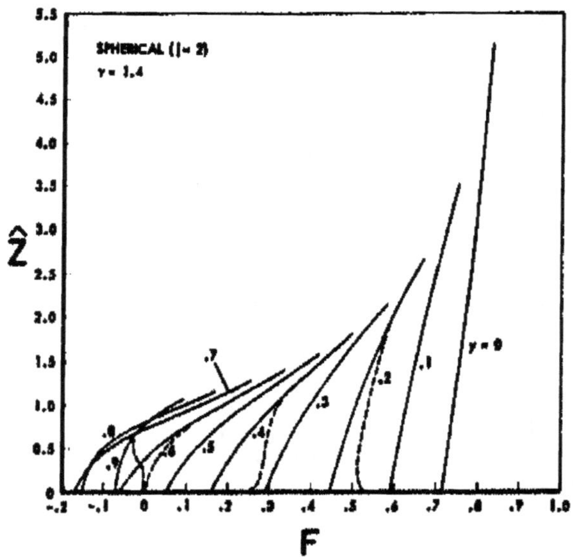

Fig. 11.7. Integral curves of spherical blast waves in the $\hat{Z}(F)$ plane on y =0

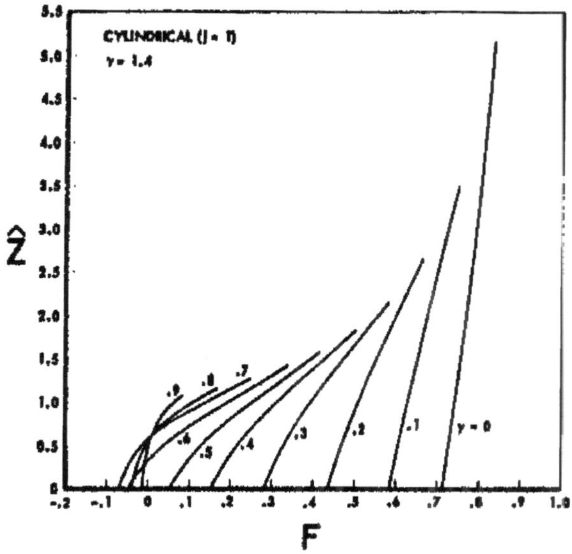

Fig. 11.8. Integral curves of cylindrical blast waves in the $\hat{Z}(F)$ plane y =0

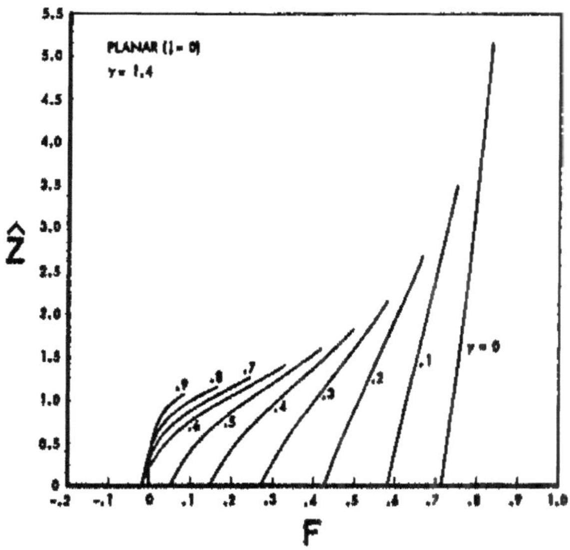

Fig. 11.9. Integral curves of planar blast waves in the $\hat{Z}(F)$ plane on y =0

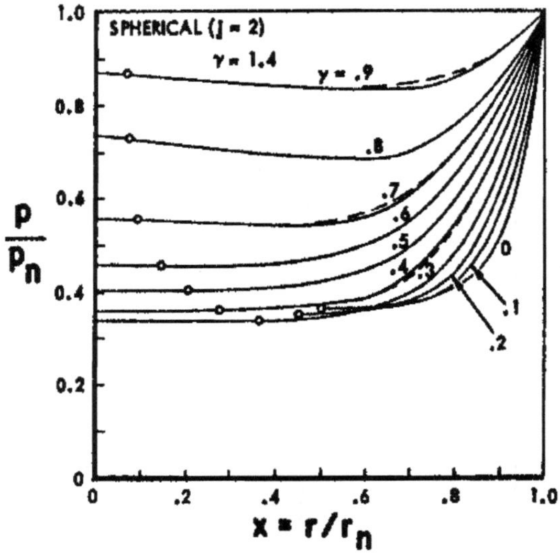

Fig. 11.10. Pressure profiles of spherical blast waves

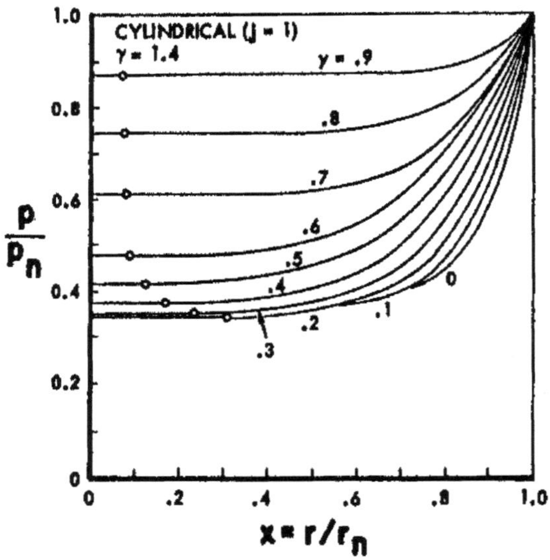

Fig. 11.11. Pressure profiles of cylindrical blast waves

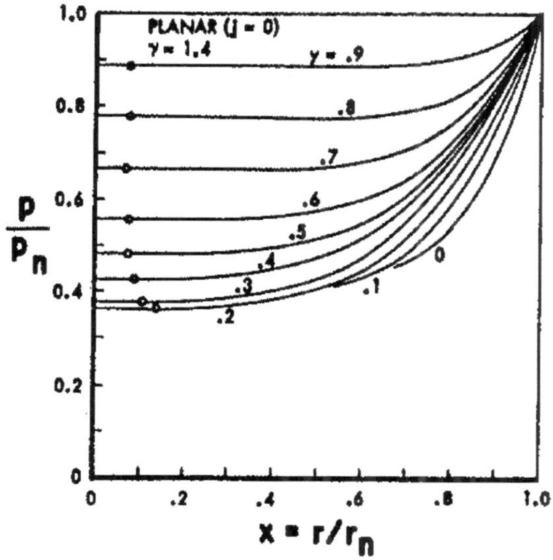

Fig. 11.12. Pressure profiles of planar blast waves

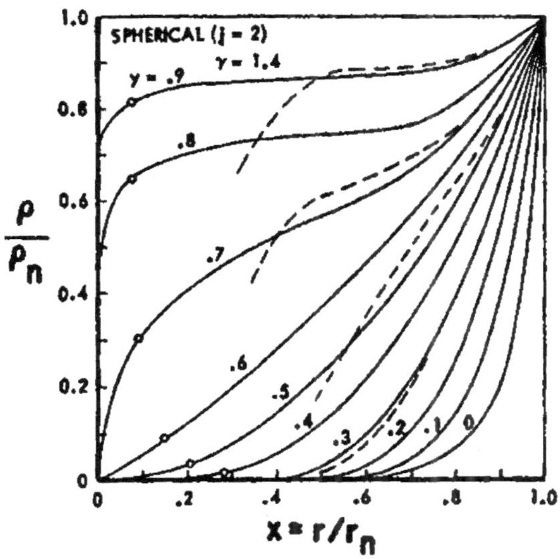

Fig. 11.13. Density profiles of spherical blast waves

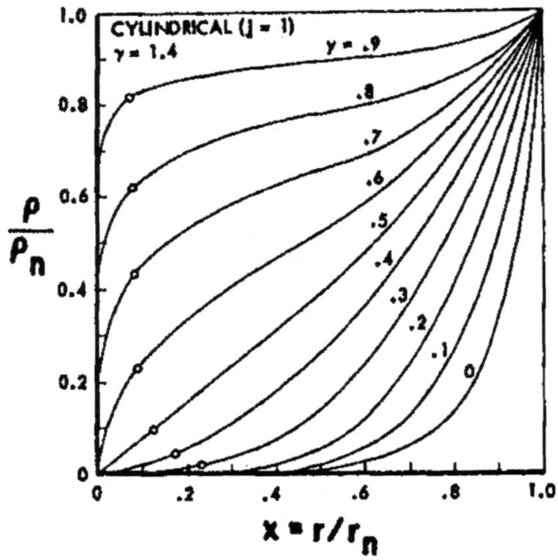

Fig. 11.14. Density profiles of cylindrical blast waves

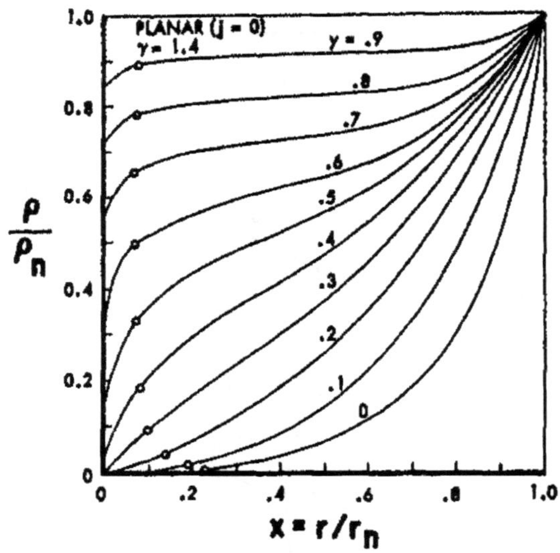

Fig. 11.15. Density profiles of planar blast waves

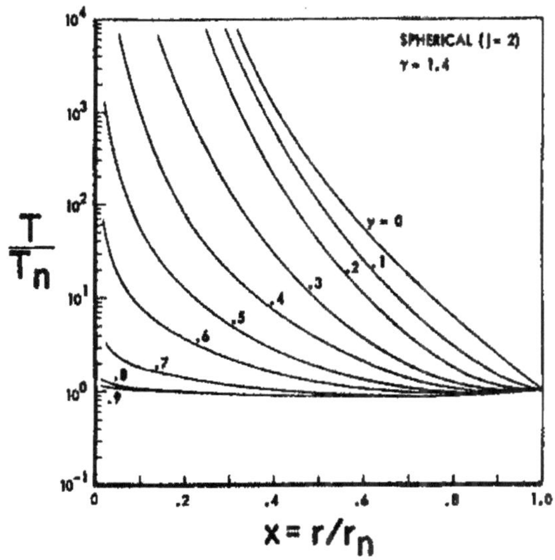

Fig. 11.16. Temperature profiles of spherical blast waves

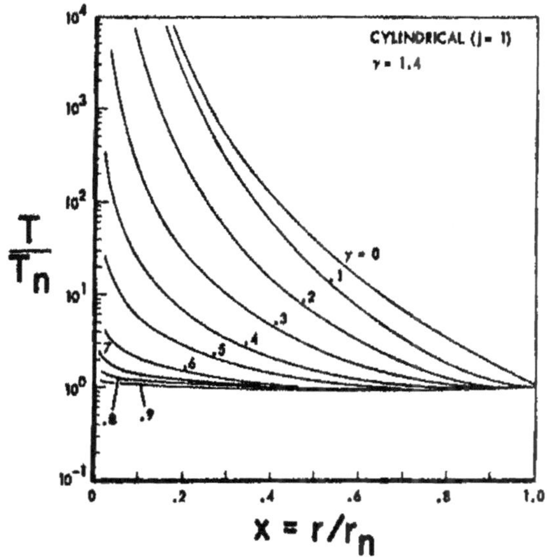

Fig. 11.17. Temperature profiles of cylindrical blast waves

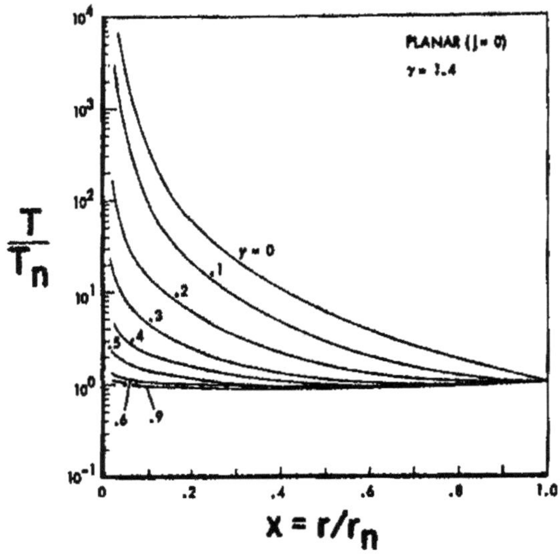

Fig. 11.18. Temperature profiles of planar blast waves

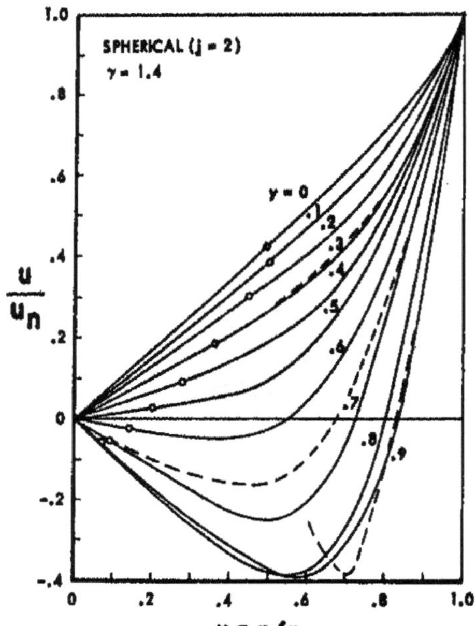

Fig. 11.19. Velocity profiles of spherical blast waves

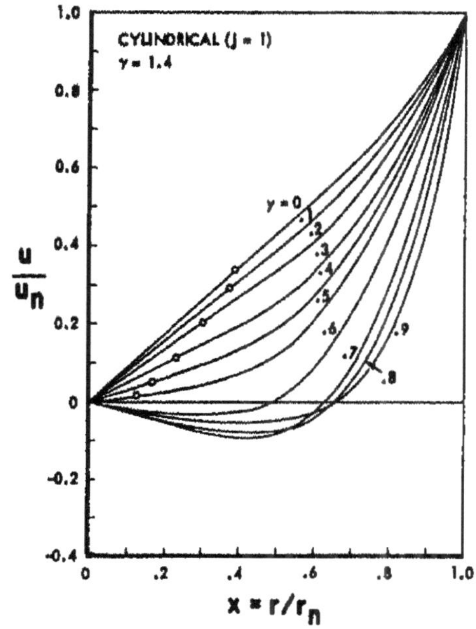

Fig. 11.20. Velocity profiles of cylindrical blast waves

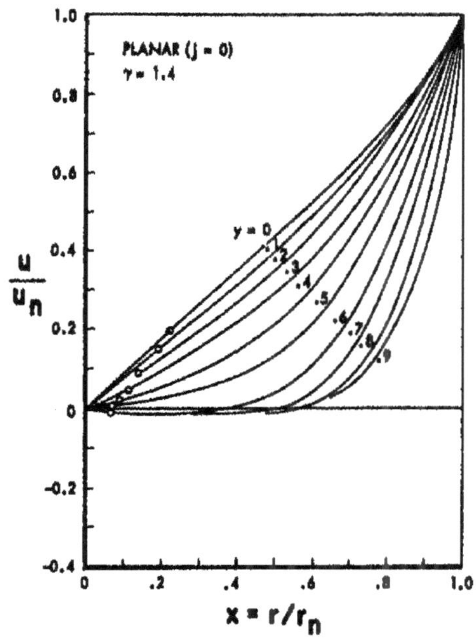

Fig. 11.21. Velocity profiles of planar blast waves

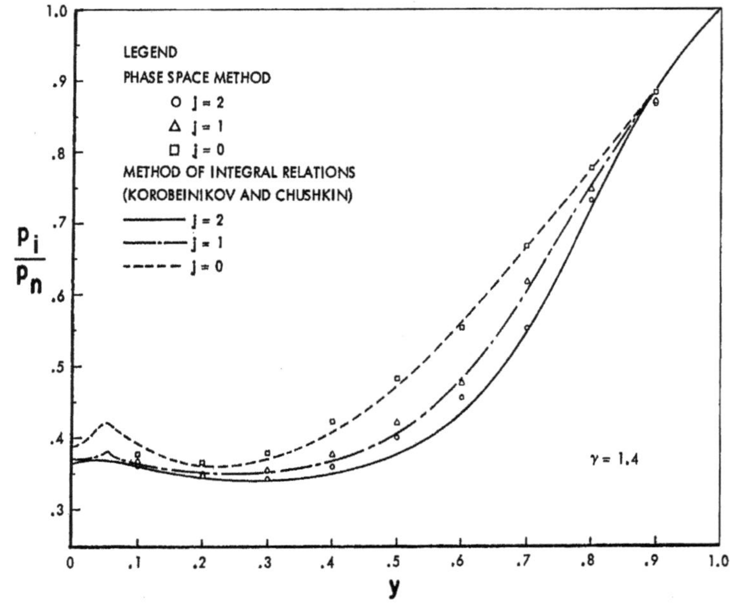

Fig. 11.22. Pressure at the center of blast waves generated by point explosion

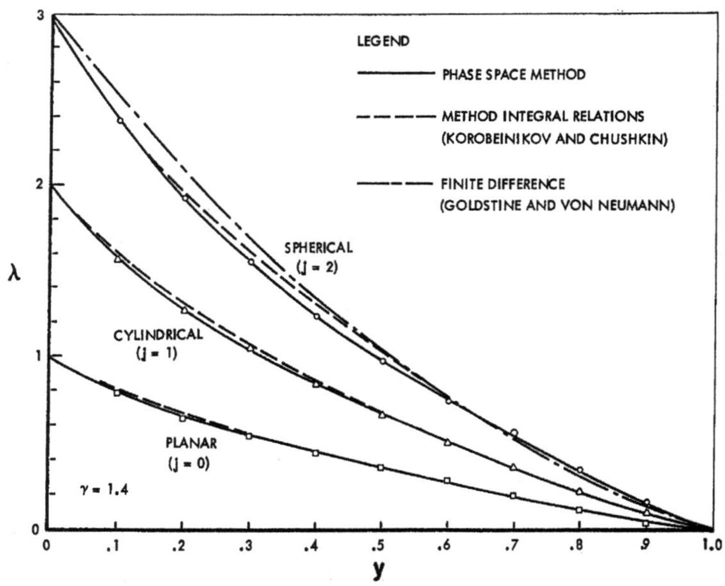

Fig. 11.23. Decay parameter of blast waves generated by point explosion

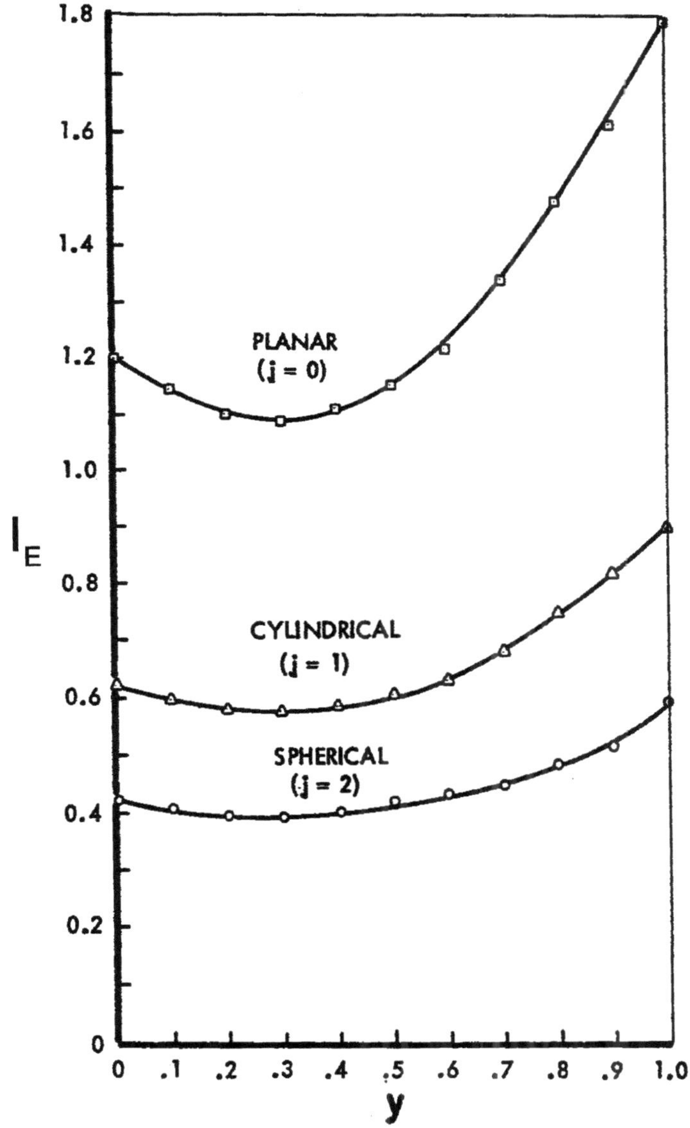

Fig. 11.24. Energy integral of blast waves generated by point explosion

12 Detonations

12.1 Introduction

The dynamic features of combustion are displayed spectacularly in the course of the development of detonation waves, as well as in their structure, – a subjects of research that fascinated combustion scientists for well over a recent century. A list of classical literature, that bears this out is provided here in Section 12.4. Particularly noteworthy among seminal contributions are the publications of Berthelot & Vieille (1822), Mallard & Le Chatelier (1883), Manson (1947), Zeldovich & Kompaneets (1955), Taylor & Tankin (1958).

The development of detonation - a subject that became known as DDT (Deflagration to Detonation Transition) - is recounted here in Section 12.1 and its structure is presented in Section 12.2. In each of them, upon providing a résumé of the background provided by its seminal contributors, the exposition is illustrated by cinematographic schlieren records published by the author with his associates (vid. esp. Oppenheim 1965, 1972, and Oppenheim *et al.* 1968, 1969). The knowledge they disclose is by no means the outcome of just their contributions. It has been attained as a consequence of fundamental studies carried out all over the world in the second half of the last century as one of its most thrilling scientific events. Included among them prominently are the contributions of Soloukhin (1963), Mitrofanov (1962), Shchelkin & Troshin (1963), Voitsekhovsky *et al.* (1963), Bazhenova et al (1968) in Russia, Van Tiggelen (1969) in Belgium; Manson (1947) in France, Edwards et at (1970) in England; Lee (1972, 1977) in Canada; Fujiwara (1970, 1975) in Japan; White (1961) and Strehlow (1968) in United States.

12.2 Development

Formation of detonation by an escalating progress of flame was recorded by the founders of the science of combustion, Mallard & Le Chatelier (1883). Streak photographs of an accelerating flame up to the onset of detonation – a phenomenon they ascribed in a chivalrous manner to the discovery of their competitors, Berthelot & Vieille (1822) – taken by a rotating drum camera that have been featured in their monumental paper as Fig. 1, are displayed here by Fig. 12.1. Onset of detonation was interpreted by them on this basis as an outcome of violent flame vibrations.

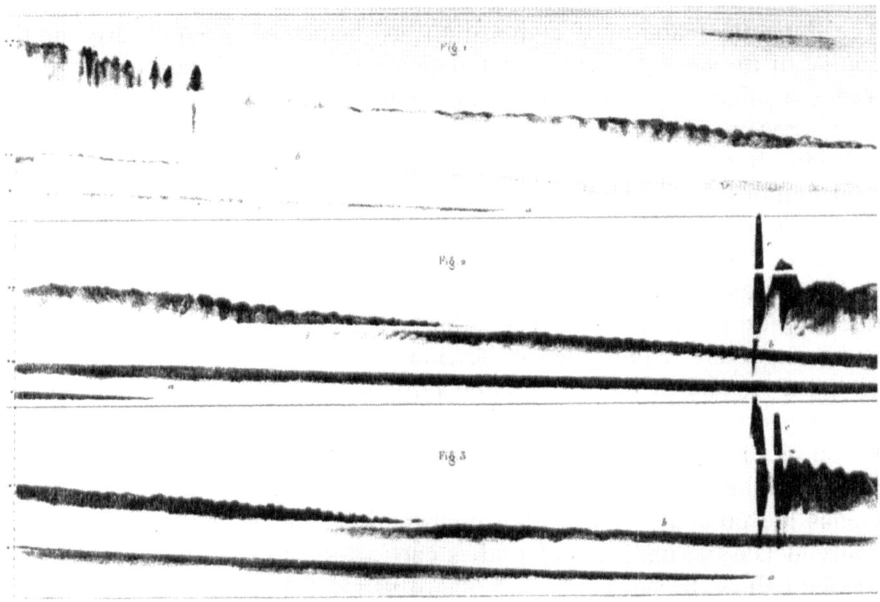

Fig. 12.1. First photographic records of flames accelerating up to the onset of detonation (Mallard et LeChatelie 1881)

A much better insight into the development of detonation was obtained by streak schlieren photographs recorded by rotating drum cameras through slits in the screens of detonation tubes. Particularly noteworthy in this respect are the contributions of Payman & Wheeler (1922), Campbell & Woodhead (1928), and Payman & Titman (1935). An example of this type of records is demonstrated here by Fig. 12.2 – obtained in the author's laboratory.

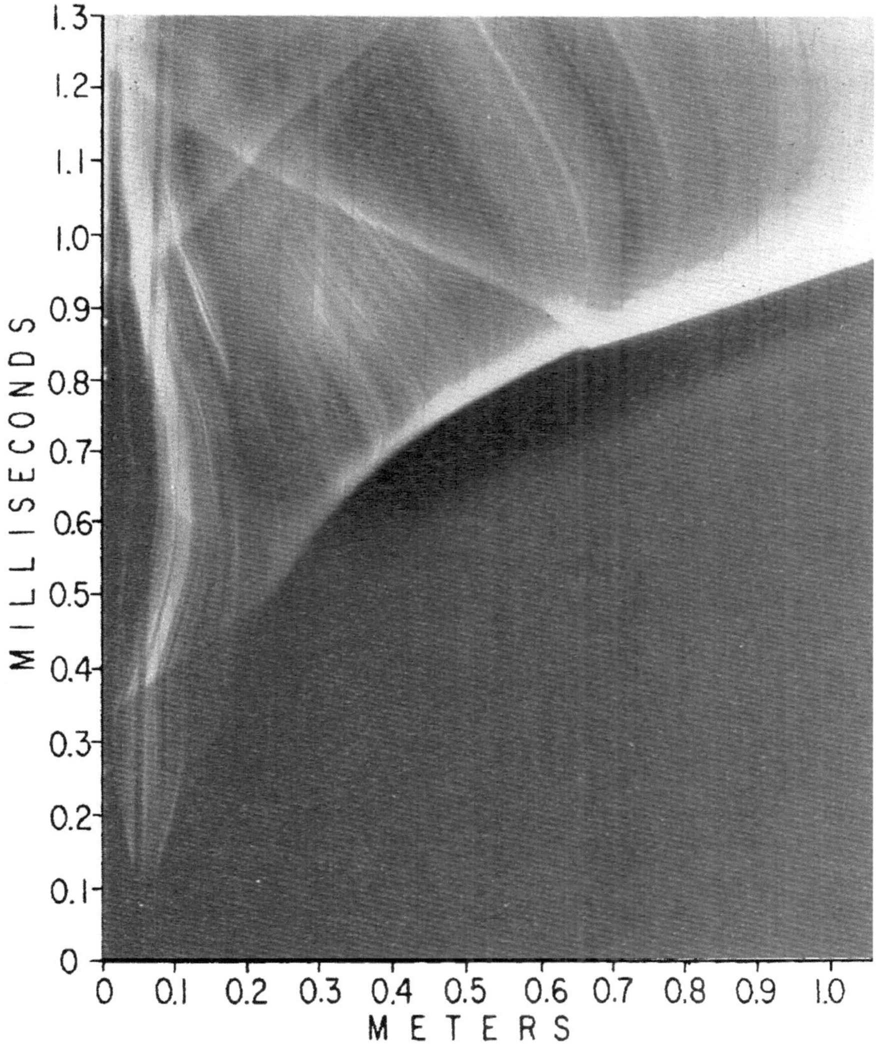

Fig. 12.2. Streak schlieren record of the development of detonation

It is on the basis of such experimental techniques that the domination of wave interactions over the development of a detonation front was realized. The earliest interpretation of these phenomena was published by Bone et al (1935) who ascribed the onset of detonation as a process initiated by ignition of a shock-compressed gas ahead of the flame by radiation from it, as described schematically on their diagram presented here by Fig. 12.3. This conclusion was reached by them in the course of an extensive study on spinning detonation – a concept deduced from the solenoid contours of glass tubes shattered by propagating detonations they contained – hence the cone shape of its front.

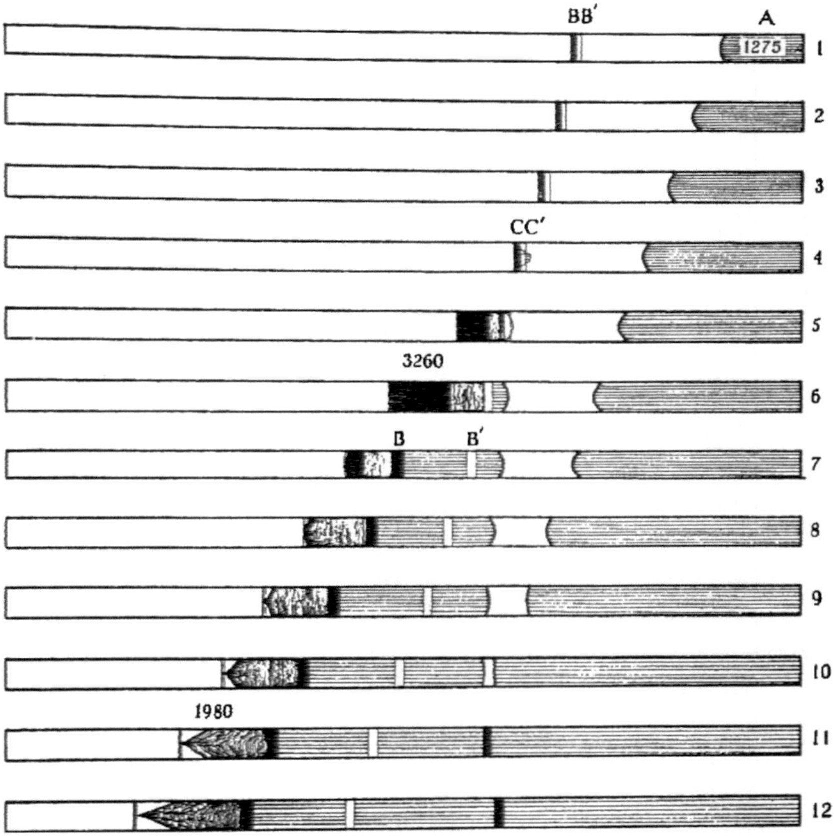

Fig. 12.3. Interpretation of the development of detonation (Bone et al 1935)
Numbers denote front velocities in m/s

A more perceptive insight into the development of detonation was obtained by cinematographic schlieren records. The earliest contribution of this kind was produced by Schmidt et al (1951) in early nineteen forties using three rotating drum cameras operating in conjunction with three pulsating light sources. A record thus obtained is presented by Fig.12.4.

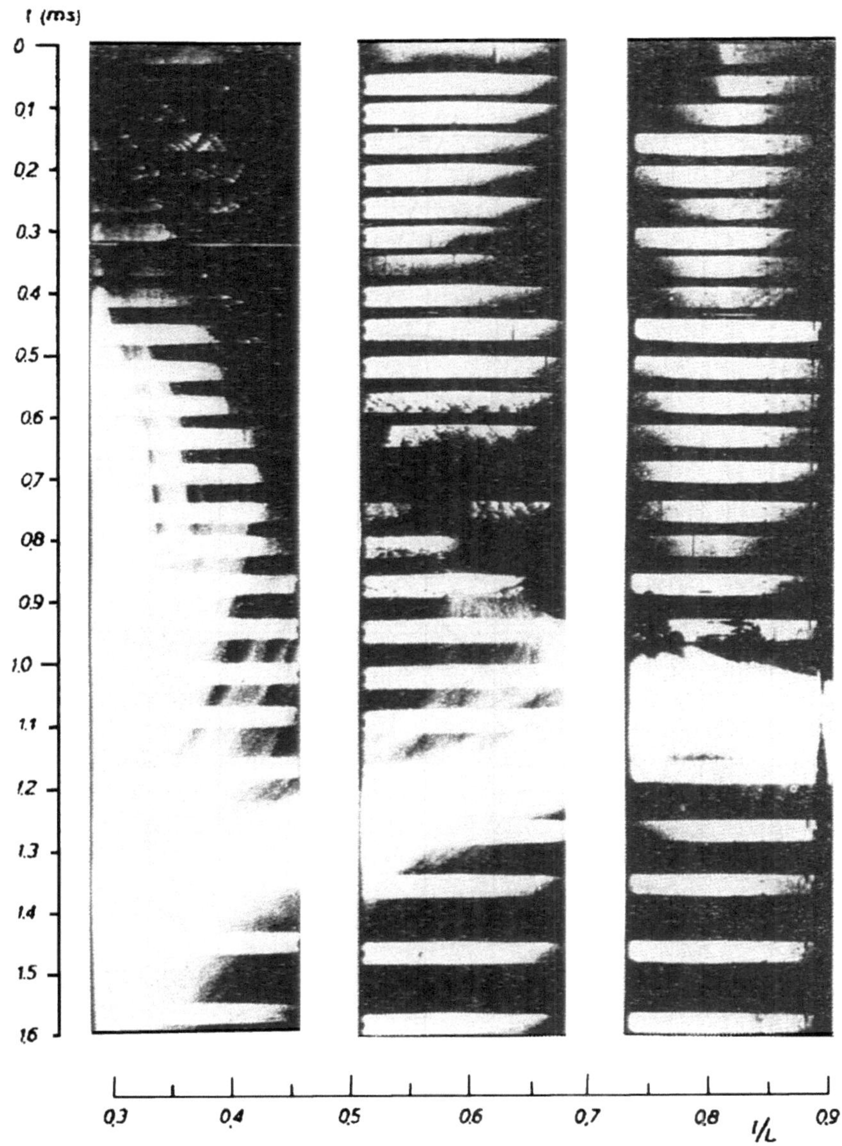

Fig. 12.4. Cinematographic schlieren record of the development of detonation obtained by Schmidt et al (1951)

Its interpretation was obtained by Oppenheim and Stern (1958) with the use of the vector polar method for front intersections, presented in Section 8.11. The time-space wave diagram of their solution is presented by Fig. 12.5, its magnified sector depicting the onset of detonation is provided by Fig. 12.6, and their polar diagram is displayed by Fig. 12.7.

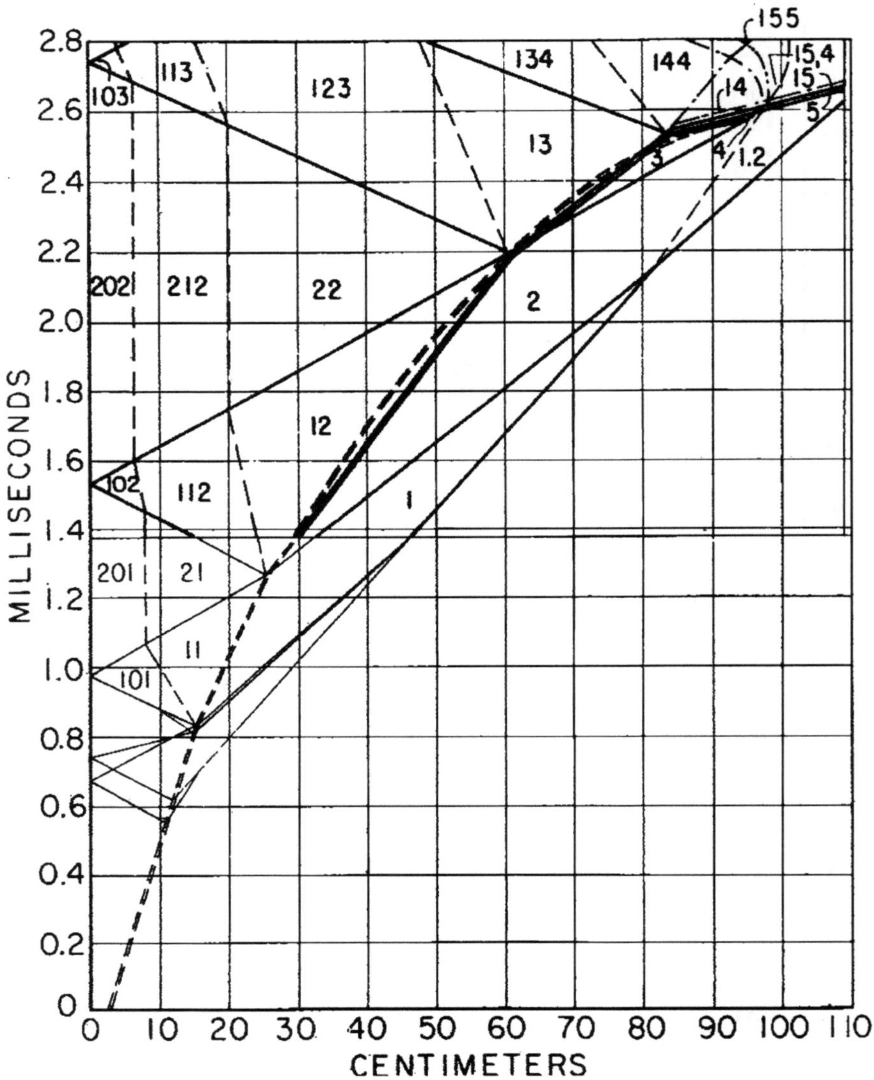

Fig. 12.5. Time-space wave diagram of the development of detonation displayed by Fig. 12.4 (Oppenheim and Stern 1958)

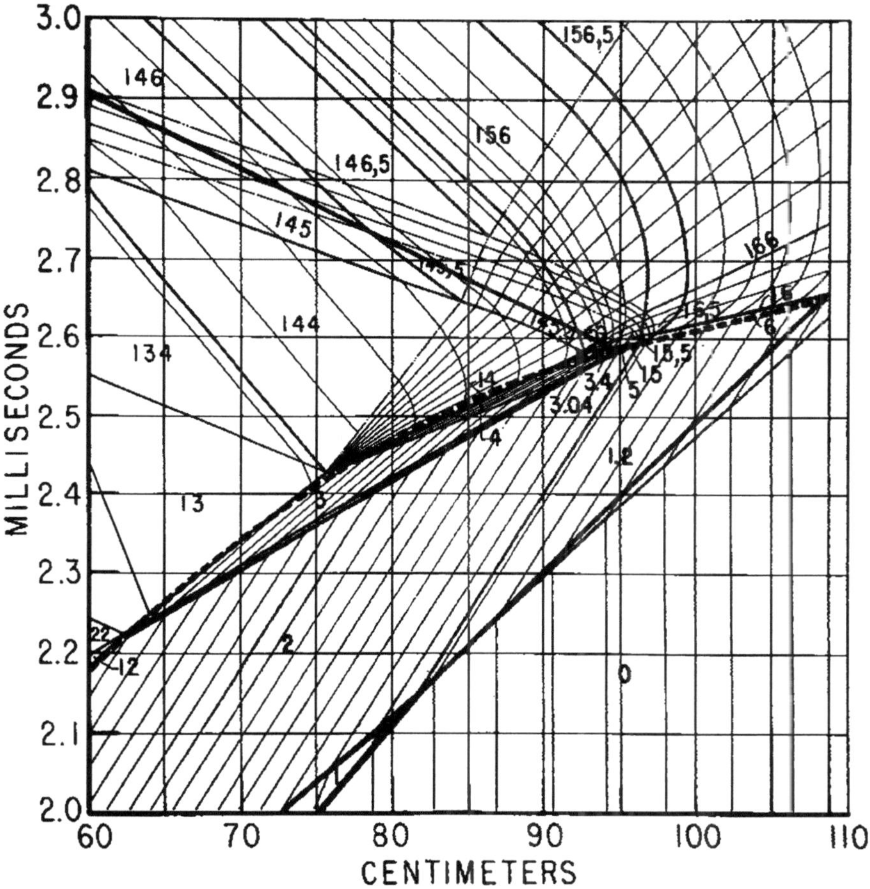

Fig. 12.6. Enlarged sector of Fig. 12.5 demonstrating wave interactions in the vicinity of the onset of detonation (Oppenheim and Stern 1958)

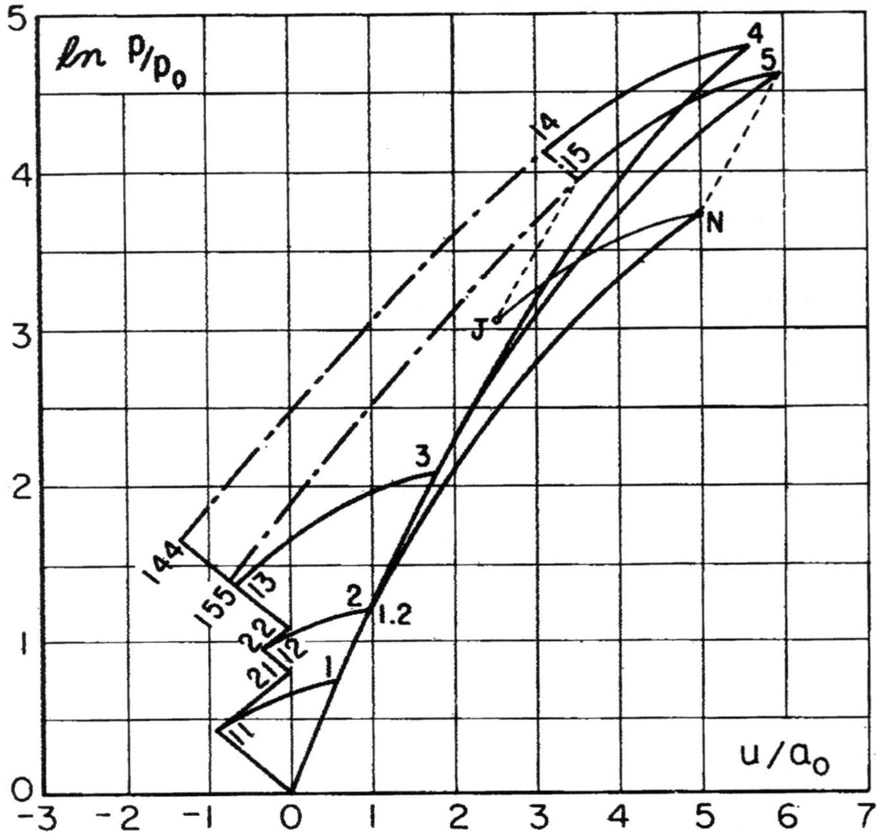

Fig. 12.7. Polar diagram of front intersections presented by Figs. 12.5 and 12.6 (Oppenheim and Stern 1958)

Significant features of an accelerating flame from ignition up to its transition to detonation were revealed by cinematographic schlieren records. The system employed for this purpose utilized an open shutter rotating prism camera with a stroboscopic light source made out of an amplitude modulated, Q-spoiled ruby laser operating at a chronometer controlled frequency of 2×10^5 per second (Oppenheim and Kamel 1972). The detonation tube was of rectangular cross-section, 2.54 cm × 3.81 cm, provided with transparent windows on narrow walls. As the principal test medium, a stoichiometric or equimolar hydrogen-oxygen mixture at room temperature and sub-atmospheric pressures was used. Ignition was performed by an electric spark discharge of a 0.25 μF capacitor, initially at a potential of 6 kV.

The events occurring immediately upon ignition are displayed in Fig. 12.8 portraying a blast wave formed by the spark discharge, followed by a laminar flame front of spheroidal shape. Thereupon, the flame accelerates, due to increase in the volume of the combustion products caused by the growth of its surface area, and becomes wrinkled, bringing about further augmentation of its surface area and, hence, further acceleration illustrated by Fig. 12.9.

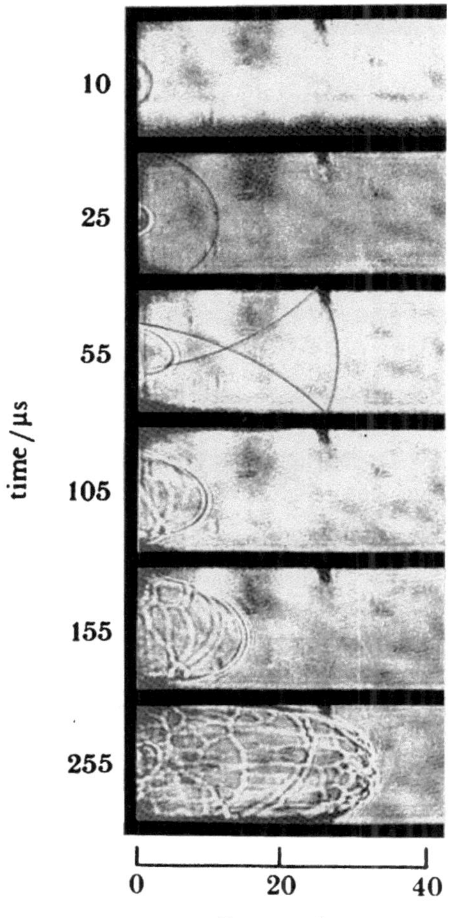

Fig. 12.8. Cinematographic schlieren record of a spark ignited flame in a stoichiometric hydrogen mixture initially at NTP (Oppenheim 1985)

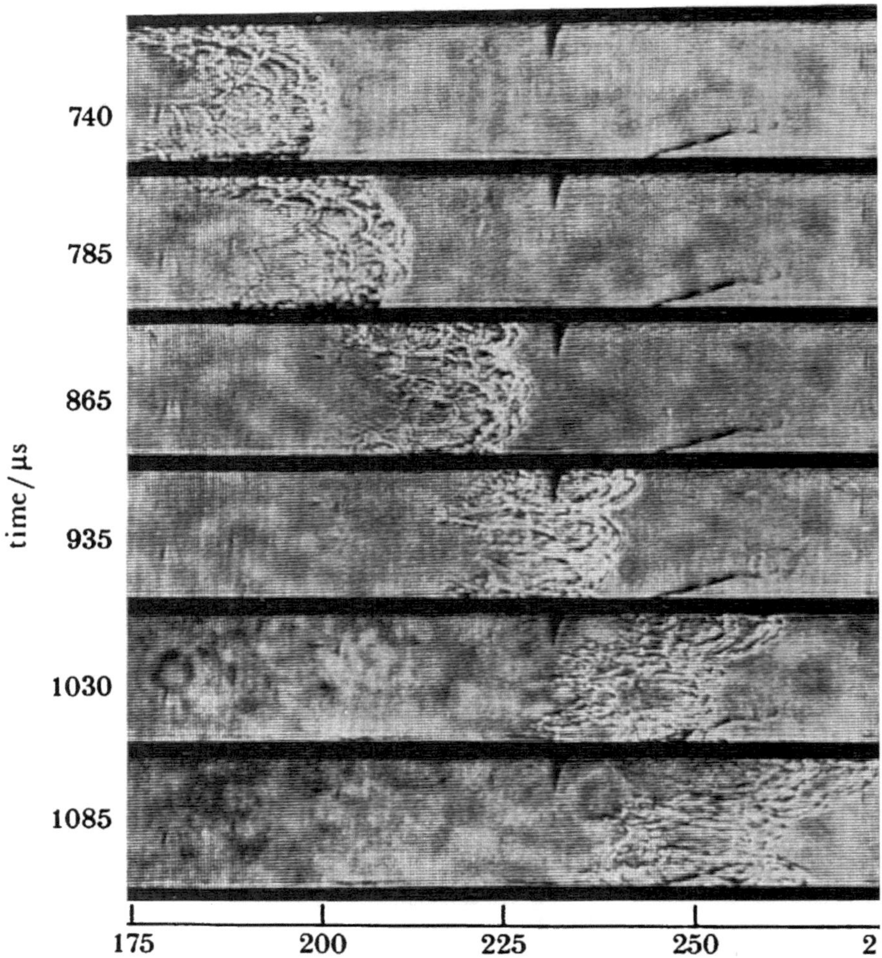

Fig. 12.9. Cinematographic schlieren record of turbulent flame propagation in a stoichiometric hydrogen mixture at NTP (Oppenheim 1985)

The flame acts as an accelerating piston generating a flow field ahead of it forming rolling vortices at the walls, known as the Tollmien-Schlichting waves produced by shear due to the no-slip wall boundary condition.

By entrainment into these waves, the flame front acquires a well known *tulip shape*, depicted by the subsequent sequence of frames in Fig. 12.9. The laminar flame becomes then transformed into turbulent, generating pressure magnifying Mach waves.

A cinematographic schlieren record of the ensuing process is provided by Fig. 12.10 that was obtained with an equimolar hydrogen-oxygen, initially at a pressure of 0.11 bars. As evident there, this pressure wave system collapses into shock fronts that merge, producing transmitted shocks that propagate at a significantly higher Mach number and bring about an appreciable increase in the local gas temperature ahead of the flame (vid. Section 8.11.3). The chemical induction time is thereby reduced leading to the onset of combustion in the shock compressed gas ahead of the flame, where the reacting mixture resided in for the longest time.

As a consequence, a blast wave, referred to as an 'explosion in explosion,' is formed in the kernel of a Tollmien-Schlichting wave at the wall.

This sequence of events recorded between 40 and 65 μsec is shown in a magnified scale by Fig. 12.11. The front of the blast wave appears there first at the top of the gasdynamic interface in frame at 50 μsec. Its interaction with the transmitted shock front is recorded in frame at 55 μsec, and the emergence of a detonation is displayed by frame at 60 μsec. After few oscillations, evident in frames at 60 and 65 μsec, a self-sustained detonation, propagating at a Chapman-Jouguet velocity, is established.

The subsequent events recorded between 80 and 105 μsec are displayed in enlarged scale by Fig. 12.12. The backward propagating front of the blast wave forms the detonation front that collides with the turbulent flame front between 90 and 95 μsec, initiating a second blast wave, evident in the latter frame. Unlike the 'explosions in explosion', the pocket of combustible mixture becomes then consumed on all sides and the second blast wave gets immersed in inert products collapses into a shock front whose sharp record is quite distinct from that of a self-sustained detonation front.

Concomitantly with the cinematographic schlieren records, pressure profiles were measured by transducers of high frequency response at three stations. Their records are presented by inserts in Fig. 12.10, where, on the vertical scale, 1 division = 0.68 bars, the oscilloscope sweep at location 1 leading the cinematographic record by 245 μsec and, at locations 2 and 3, by 175 μsec.

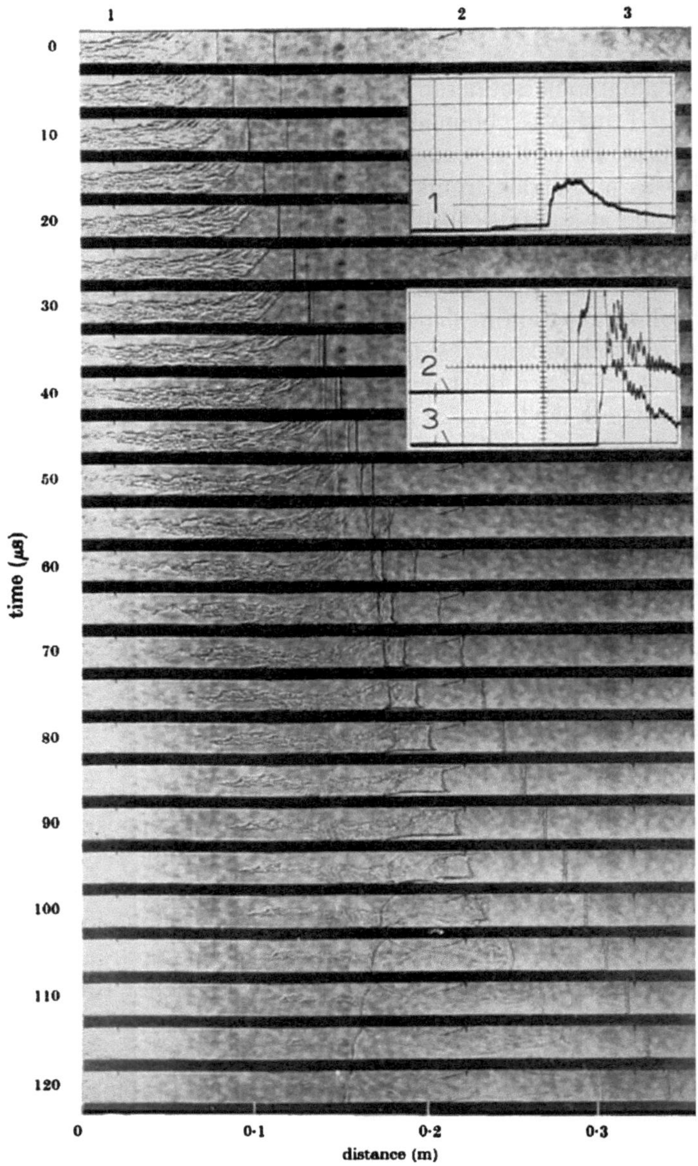

Fig. 12.10. Cinematographic schlieren record of DDT in an equimolar hydrogen-oxygen, initially at a pressure of 0.11 bar, produced by shock merging ahead of flame (Urtiew and Oppenheim 1966)

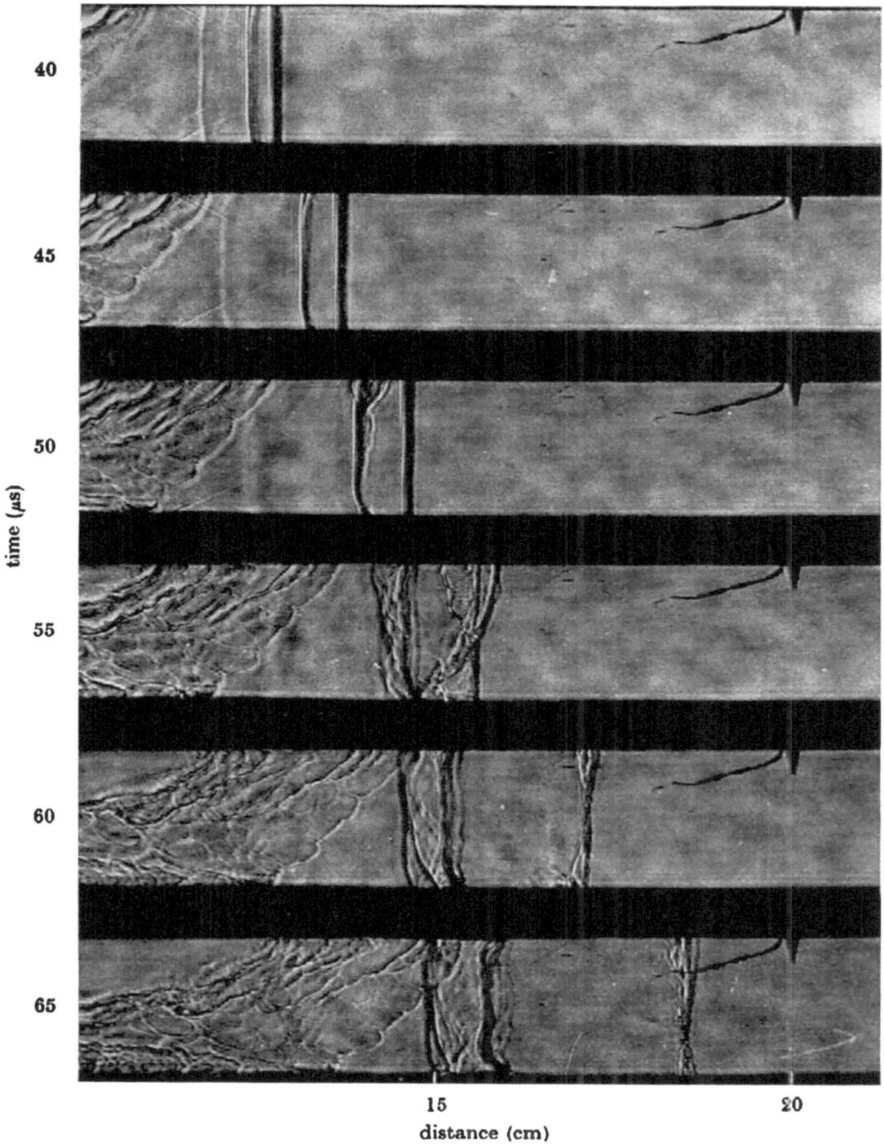

Fig. 12.11. Enlarged frames of Fig. 12.10 from 40 to 65 μsec

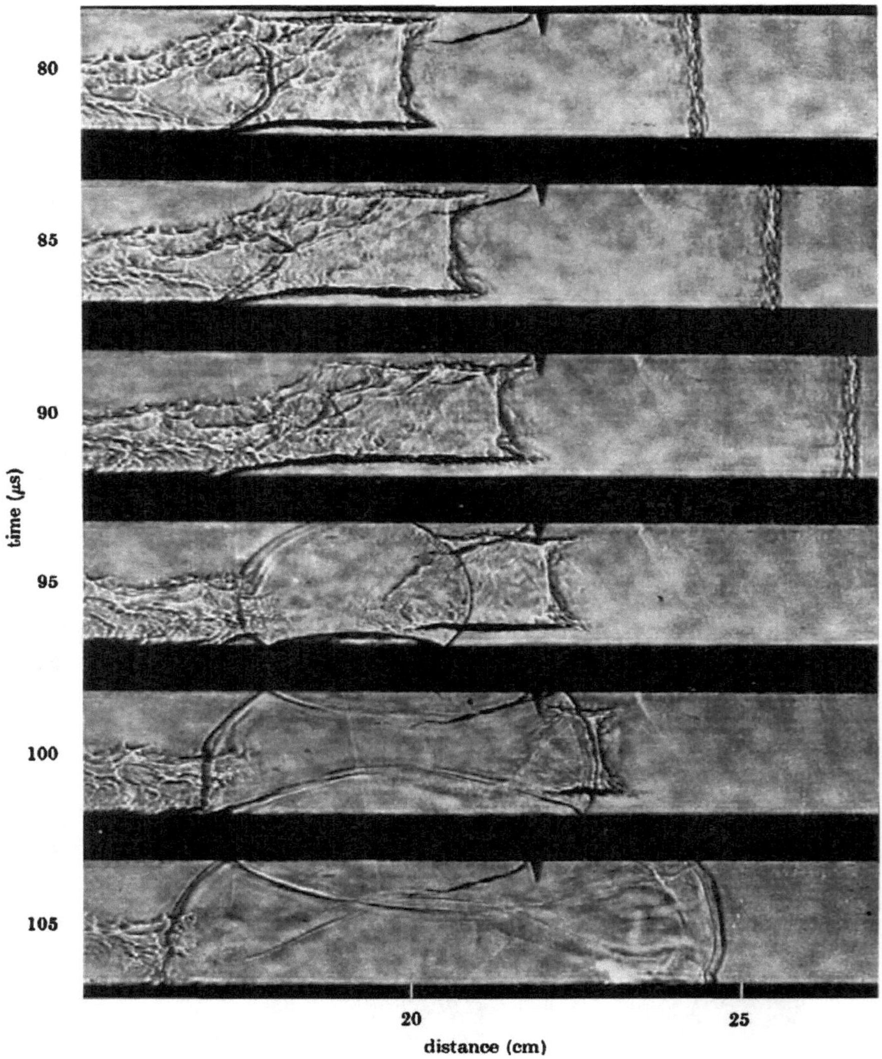

Fig. 12.12. Enlarged frames of Fig. 12.10 from 80 to 105 μsec

At location 1, marked at the top of Fig. 12.10, where the transducer was exposed to the regime created by three shock fronts, the maximum pressure was 2.04 bars. The first two merged, forming a transmitted front propagating at a speed of 720 m/sec (Ma = 1.5). The third propagates at an absolute velocity of 1730 m/sec, corresponding to a relative velocity of 1400 m/s (M = 2.5). The pressure ratio across the transmitted front is 2.5 and that across the third is 7.2 for an overall pressure ratio of 17.3, corresponding to a pressure rise of 1.88 bars. At location 2, the first peak was out of scale, but at location 3, where the transducer was exposed to a fully developed detonation, it was 2.04 bars, corresponding to a pressure ratio of 19.5, while the velocity of the detonation front recorded in Fig. 12.3 was 2380 m/sec. For comparison, the pressure ratio of a Chapman-Jouguet detonation is 17.3 and its velocity is 2240 m/sec.

The mode of DDT presented by the preceding figures is just but one of many in which it can occur (vid. Urtiew and Oppenheim 1966). An example of another kind is provided by Fig. 12.13 portraying its development in a stoichiometric hydrogen-oxygen mixture, initially at a pressure of 0.916 bar, Here the critical blast wave is initiated in the cavity of the tulip-shape turbulent flame, rather than by shock merging ahead of it. There were also two blast weaves recorded in this case, but, unlike the previous case, it was the second blast that overwhelmed the first.

This record was sufficiently complete to trace all the shock interactions, taking place ahead of the accelerating flame right from its inception by the spark discharge. Together with the pressure transducer records taken simultaneously it was then possible to deduce the thermodynamic history of DDT. The results are presented in Fig. 12.14 where, for reference, frame at 715 μsec of Fig. 12.13 is provided on top. Displayed there, in particular, is the path of the critical particle, which was initially at rest and culminated at the center of the 'explosion in explosion' that marks the onset of the blast wave with the detonation front propagating forward and the retonation front propagating backwards.

The evolution of pressure and temperature of the critical particle specified its path on the state diagram of thermal ignition limits (vid. Fig. 3.3) displayed in Fig. 12.15. As apparent there, its terminal point at 720 μsec, when the critical particle is at the center of the DDT's 'explosion in explosion,' is still below the third thermal ignition limit. The gasdynamic compression alone is, therefore, insufficient to produce the blast wave of 'explosion in explosion.' Its formation had to be promoted additionally by radiation from the high temperature of the flame front and active radicals jetted forth from it by the Tollmien-Schlichting waves.

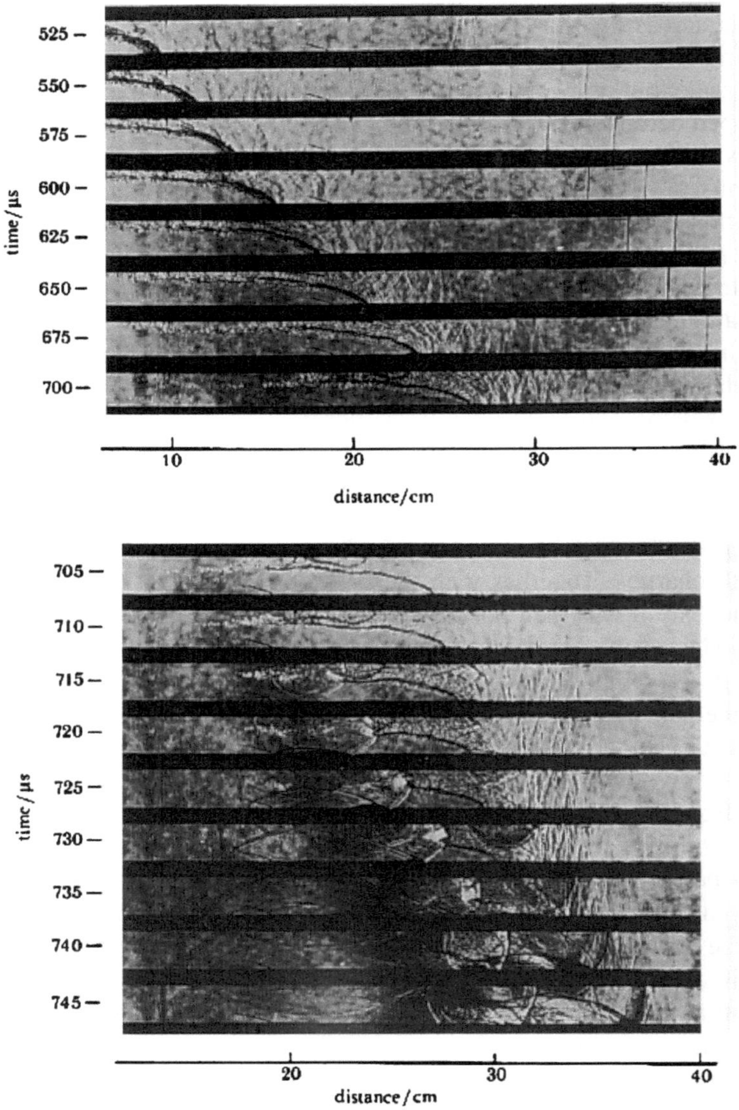

Fig. 12.13. Cinematographic schlieren record of DDT in a stoichiometric hydrogen-oxygen mixture, initially at a pressure of 0.916 bars, triggered in the cavity of a tulip-shape turbulent flame (Meyer *et al.* 1970)

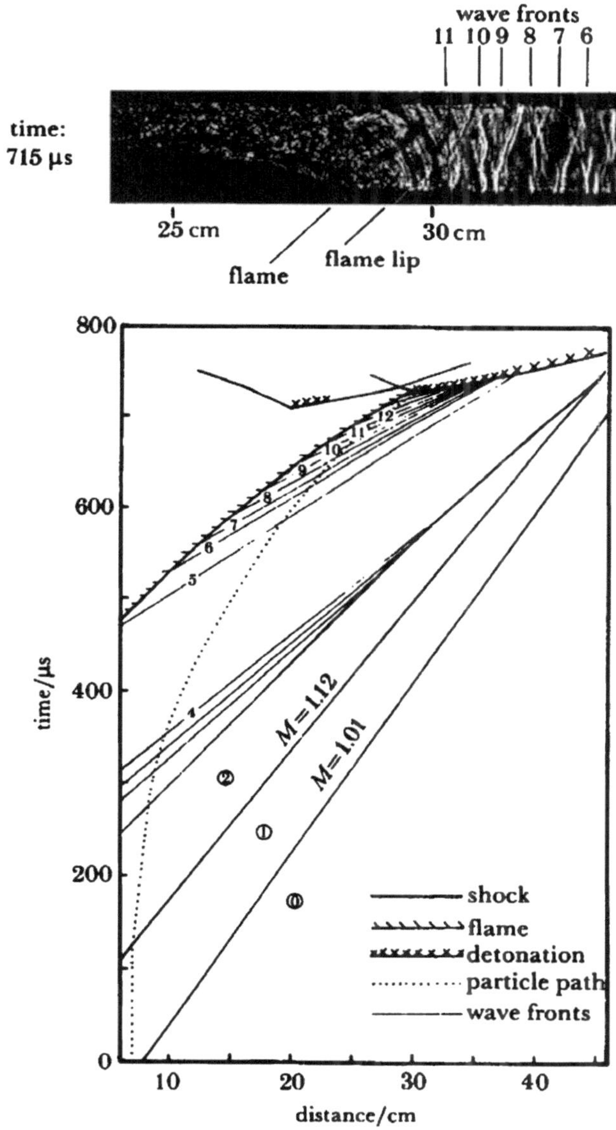

Fig. 12.14. Time-space history of the gasdynamic events culminated by DDT displayed by Figs.12.1-12.3 (Meyer *et at.* 1970)

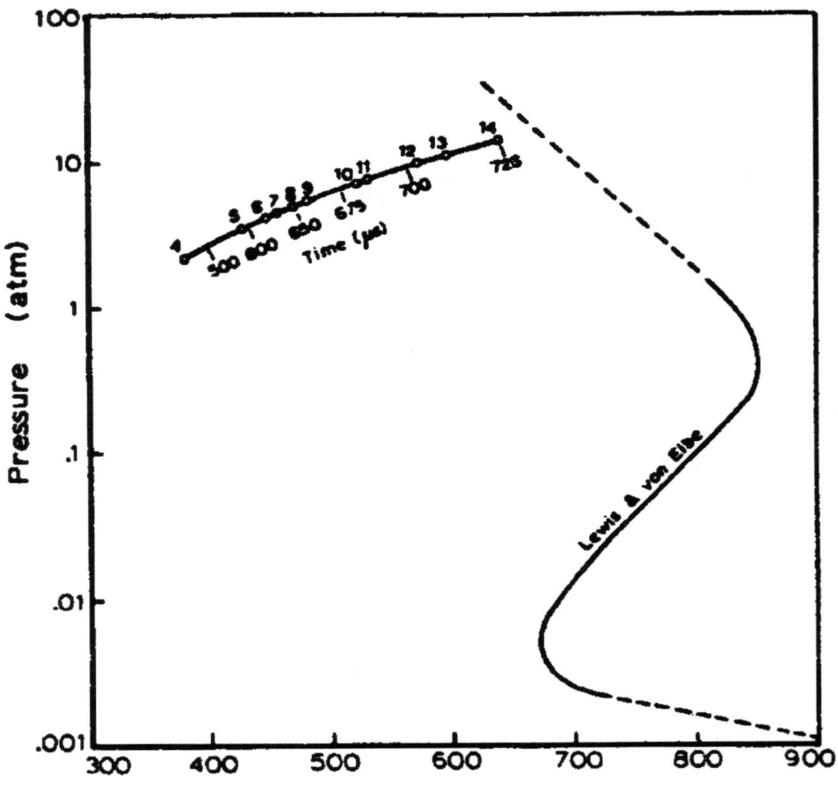

Fig. 12.15. Process of the critical particle on the thermal ignition diagram (Meyer *et at.* 1970)

To provide an analytical interpretation of these records, the gasdynamic flow field produced by an accelerating flame leading to its transformation into detonation was reproduced by numerical analysis carried out by Kurylo et al (1979). The gasdynamic equations presented in Chapter 7 were integrated for this purpose by an explicit, second order accurate, finite difference technique combined with an algorithm for insertion of algebraic expressions for front interaction events. The space-time diagram thus evaluated is presented by Fig. 12.16, while, in order to display the gasdynamic flow field at the onset of detonation, an enlargement of the sector marked by frames in Fig. 12.16 is provided by Fig. 12.17. This solution bears a remarkable similarity to that presented by Fig. 12.5 and 12.6, which was obtained by the cruder vector polar method.

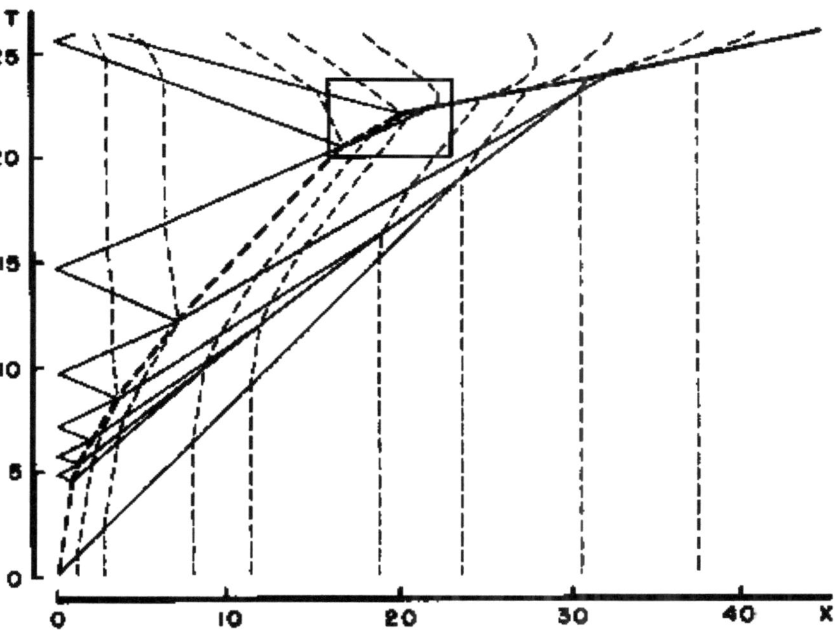

Fig. 12.16. Time-space diagram of flow field produced by an accelerating flame leading to its transformation to detonation (Kurylo et al 1979)

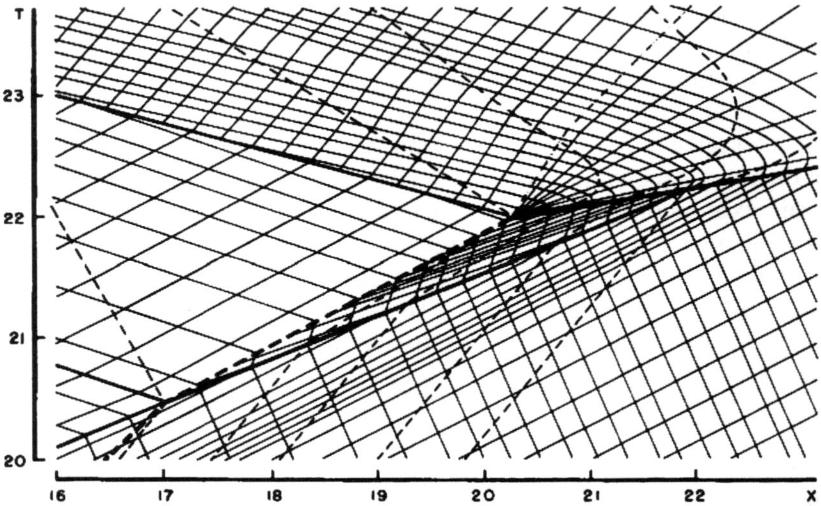

Fig. 12.17. Enlargement of the sector marked by frames in Fig. 12.16 displaying the gasdynamic flow field at the onset of detonation (Kurylo et al 1979)

In the numerical solution presented by Figs. 12.16 and 12.17 the instant of the onset of detonation had to be postulated, just like in Figs. 12.5 and 12 6 obtained by the vector polar method, since a one-dimensional analysis cannot take into account the multidimensional phenomena illustrated by Figs. 12.10-12.13. With the evidence provided by the latter, the physical mechanism of DDT is revealed, while the full history of its development from the start of a flame kernel is accounted for by its one-dimensional interpretation.

The occurrence of such an *explosion in explosion* is of crucial importance to the onset of detonation under any geometrical space constraint. The fundamental reason for its advent is the necessity for attaining a state of critical temperature and pressure beyond the threshold of strong ignition as described in Chapter 4 (vid. Section 4.3.2).

All the instances of the development of detonation described so far pertain to the classical case of media confined in tubes treatable by one-dimensional analysis. There is, nonetheless, a possibility of getting a detonation established in unconfined space, as demonstrated by Figs. 12.18 and 12.19.

Displayed by Fig.12.18 is a cinematographic schlieren record of the onset of detonation by intersection of two strong exothermic centers created consecutively by focused beams of a double, sharp neodymium laser pulse. The medium was an equimolar acetylene/oxygen mixture initially at a pressure of 120 torr. The experimental apparatus was the same as that used by Oppenheim and Karnel 1972. The frequency of the pulsating ruby laser serving as the schlieren light source was 1MHertz, so that the time interval between frames was 1 μsec.

The detonation wave was initiated by Mach intersection created by collision between explosion fronts described in Section 8.12.5. Its structure is clearly evident on the top frames in the second column.

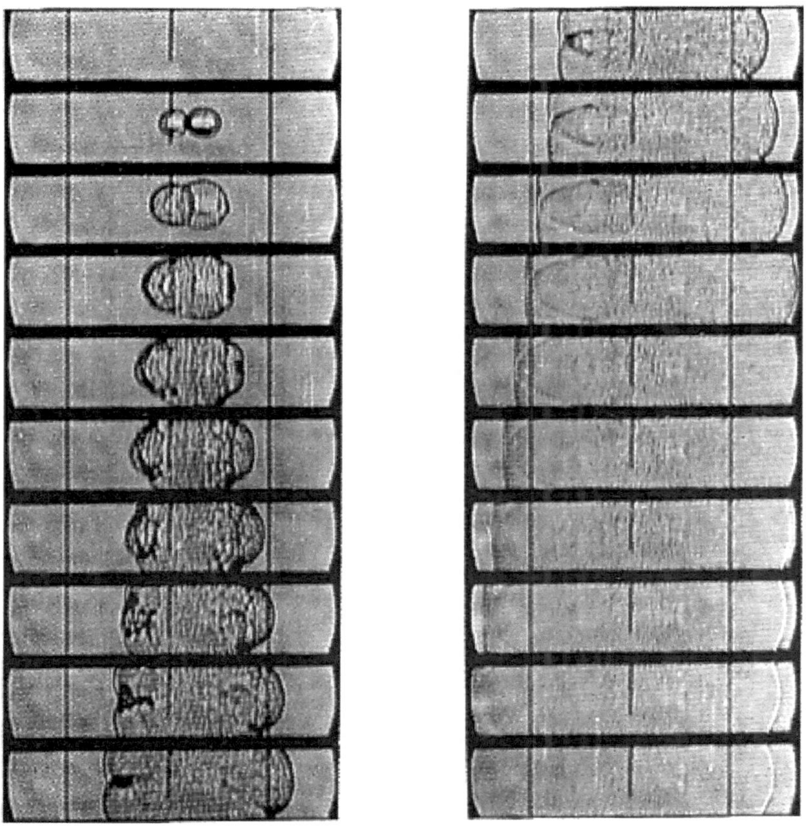

Fig. 12.18. Cinematographic schlieren record of the onset of detonation by intersection of two strong exothermic centers created consecutively by focused beams of a double neodymium laser pulse (Oppenheim and Kamel 1972)
Time interval between frames: 1 μsec

Depicted by Fig. 12.19 is a direct generation of detonation in equimolar acetylene/oxygen mixture initially at a pressure of 100 torr, created by the beam of a sharp neodymium laser pulse focused upon a thin steel wire. The front propagating to the right decays into a shock, while that propagating to the left decomposes into a double-front system described in Section 8.8.

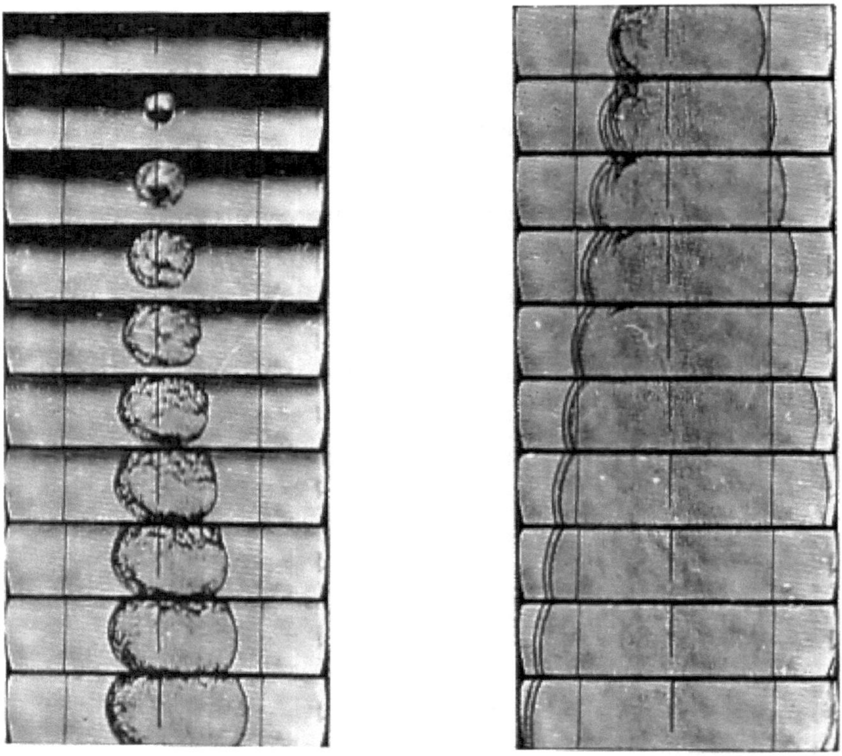

Fig. 12.19. Cinematographic schlieren record of direct generation of detonation by a focused beam of a sharp neodymium laser pulse and its decomposition into a double-front system on one side and decay to a shock front on the other (Oppenheim and Kamel 1972)
Time interval between frames: 1 μsec

12.3 Structure

As a consequence of intensive studies of detonation waves carried out at the outset of the twentieth century (vid. Section 12.4), its structure was modeled as a double front system, a shock followed by deflagration, expressed by the classical NDZ (Neuman-Döring-Zeldovich) model (Von Neumann 1942, Döring and Burkhardt 1944, Zeldovich 1944, cited in Section 12.4). This model could not provide an explanation for the well known, remarkably constant propagation velocity of a self-sustained, Chapman-Jouguet, detonation (Chapman 1899, Jouguet 1906, 1917, cited in Section 12.4).

The inadequacy of the NDZ model was first demonstrated by the Mach-Zehnder interferometer records of White 1961, reproduced by Figs. 12.20 and 12.21. The impression conveyed by them was expressed in the title of his paper: "Turbulent structure of gaseous detonation."

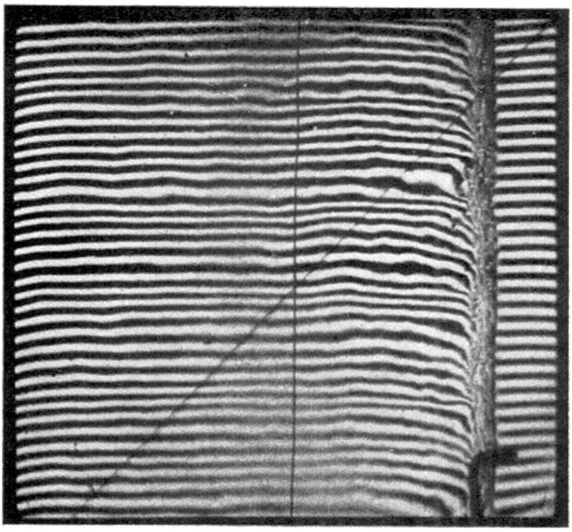

Fig. 12.20. Longitudinal fringe interferometer record of a self-sustained detonation in a H2+O2 mixture at initial pressure of 100.5 torr (White 1961)

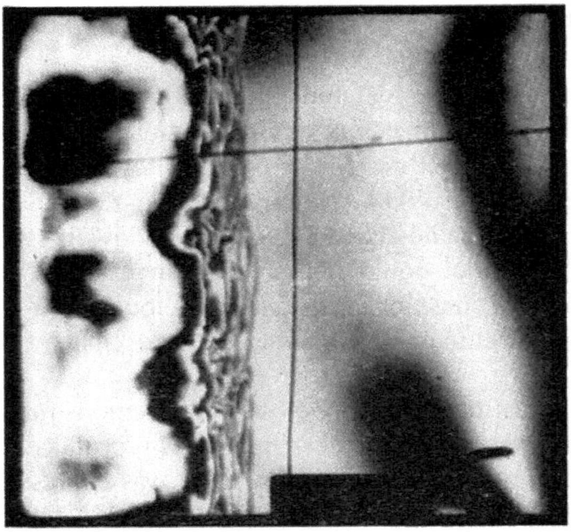

Fig. 12.21. Zero fringe interferometer record of a self-sustained detonation in 2H+O2 mixture at initial pressure of 100.5 mm Hg (White 1961)

Thereupon it was found that what appeared as turbulence was actually due to cell pattern of the footprint created by a self-sustained detonation.

This was made evident by Soloukhin 1963. It has been well known by then that detonation propagating in hydrocarbon/oxidizer mixtures is luminous due to incandescent carbon particles formed at its front, so he carried out detonation experiments with acetylene-oxygen mixtures in a dark room and took open shutter photographs. An example of such a record is given by Fig. 12.22.

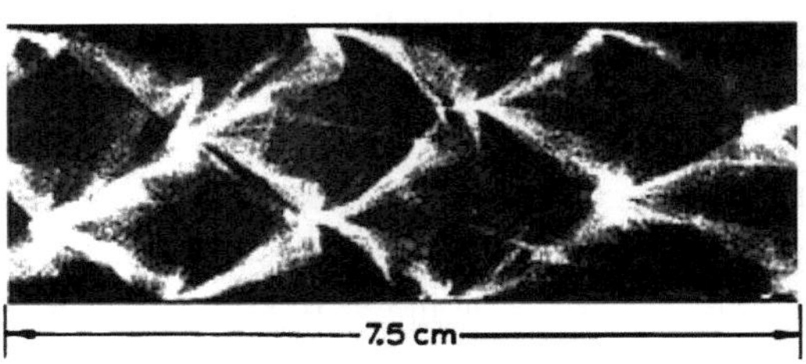

Fig. 12.22. Open shutter photograph of a self-sustained detonation in an equimolar acetylene mixture at 100 torr (Lee et al)

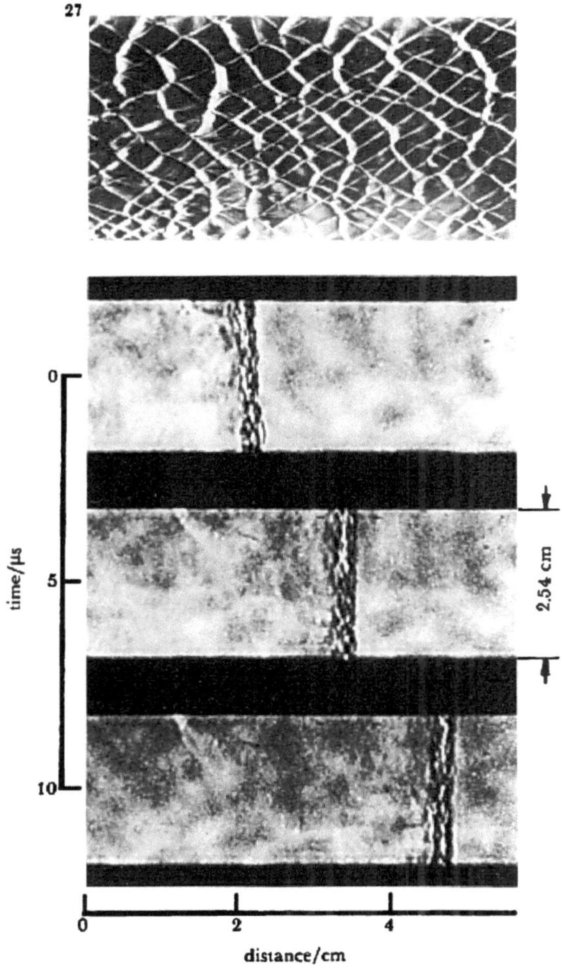

Fig. 12.23. Imprint of a self-sustained detonation front on a soot covered wall and its cinematographic schlieren record (Oppenheim 1985)

The open shutter photographs led Soloukhin 1963 to realize that a self-sustained detonation front ought to have the biblical property of 'writing on the walls.' This fact was recorded by covering the walls of the detonation tube with a thin layer of soot. An example of such a record is provided on the top of Fig. 12.23, while below it is a cinematographic schlieren record taken under the same conditions. The irregular pattern is due to wave fronts that slap across the 1” depth of the rectangular 1”x1.5” detonation tube.

By having one of the glass walls of the test section in a rectangular cross-section tube coated with a soot layer, the detonation front has been caught in the act of ‘writing on the wall.’ A cinematographic schlieren record of this event is displayed by Fig. 12.24.

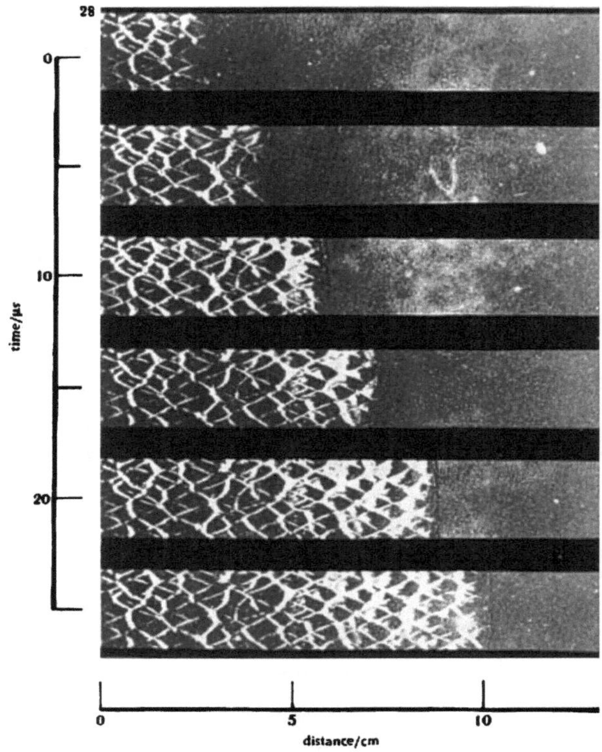

Fig. 12.24. Cinematographic schlieren record of a propagating self-sustained detonation front with its simultaneously recorded imprint on a soot covered wall (Oppenheim 1985)

A more distinct record of ‘writing on the wall’ was obtained with a stoichiometric hydrogen-oxygen mixture initially at a sufficiently low pressure to reduce the power density of deposited exothermic energy to a minimum so that the number of the 'writing heads' is reduced to one, as depicted in Fig. 12.25.

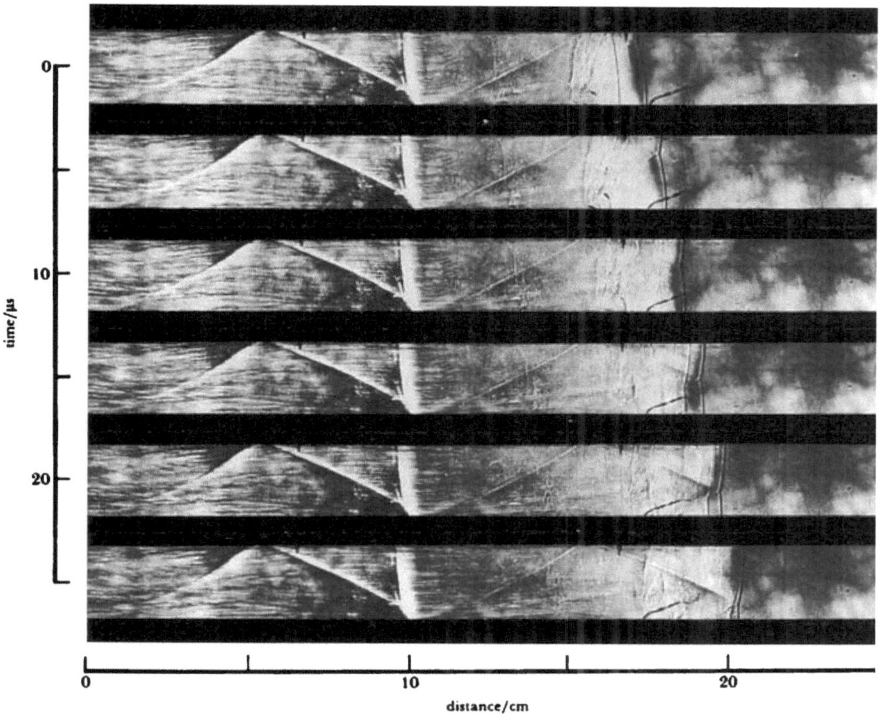

Fig. 12.25. Cinematographic schlieren record of a propagating self-sustained detonation front formed by the triple point of a single Mach intersection with its simultaneously recorded imprint on a soot covered wall (Oppenheim 1985)

A quantitative, experimental insight into gasdynamic mechanism, whereby the cellular structure of a Chapman-Jouguet detonation is created, was provided by Lundstrom & Oppenheim (1969). The experiments were performed with a mixture of $2H_2+O_2+2N_2$, maintained initially at a pressure of 59 mm Hg and room temperature (20°C). A cinematographic schlieren record of a Chapman-Jouguet detonation produced under such circumstances, where, as in Fig. 12.25, there was only one 'writing head' and hence one cell, is presented by Fig. 12.26.

The progress of the shock front across the cell was marked by numbers displayed in white on consecutive frames featured in the analysis.

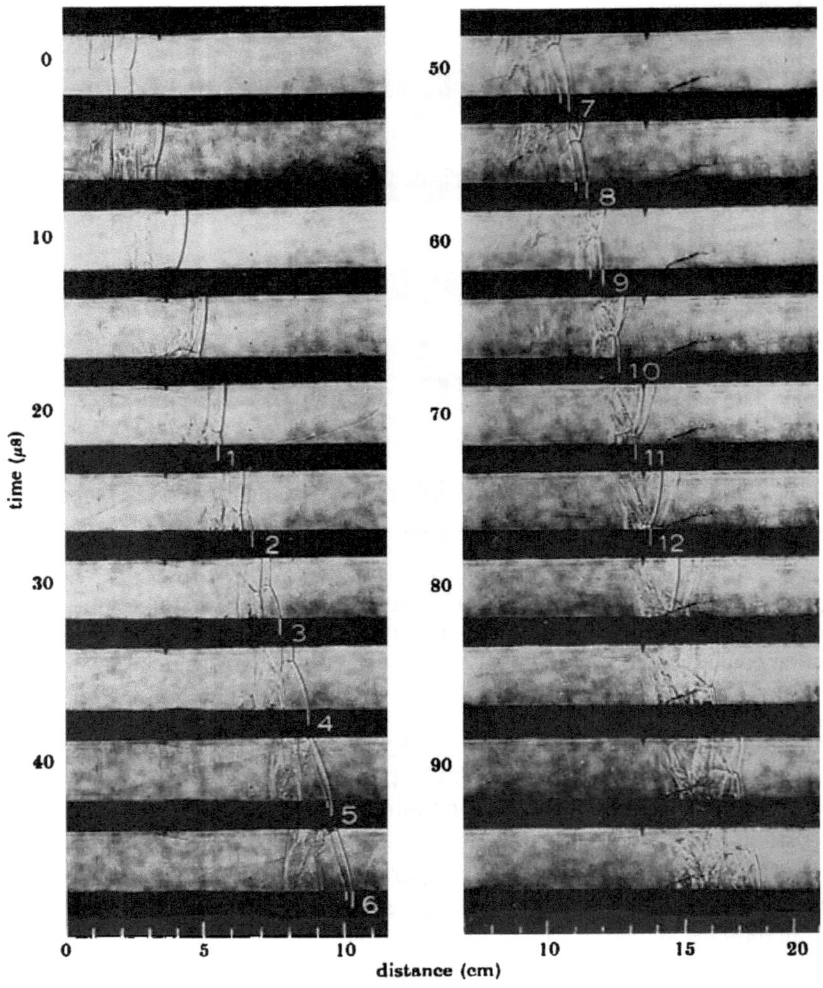

Fig. 12.26. Cinematographic schlieren record of a single Chapman-Jouguet detonation in a mixture of 2H2+02+2N2, initially at a pressure of 59 mm Hg and room temperature with its imprint on a soot-coated wall (Lundstrom and Oppenheim 1968)

A time-space trajectory of the front, $t_i(x_i)$, recorded by Fig. 12.26, is displayed in Fig. 12.27.

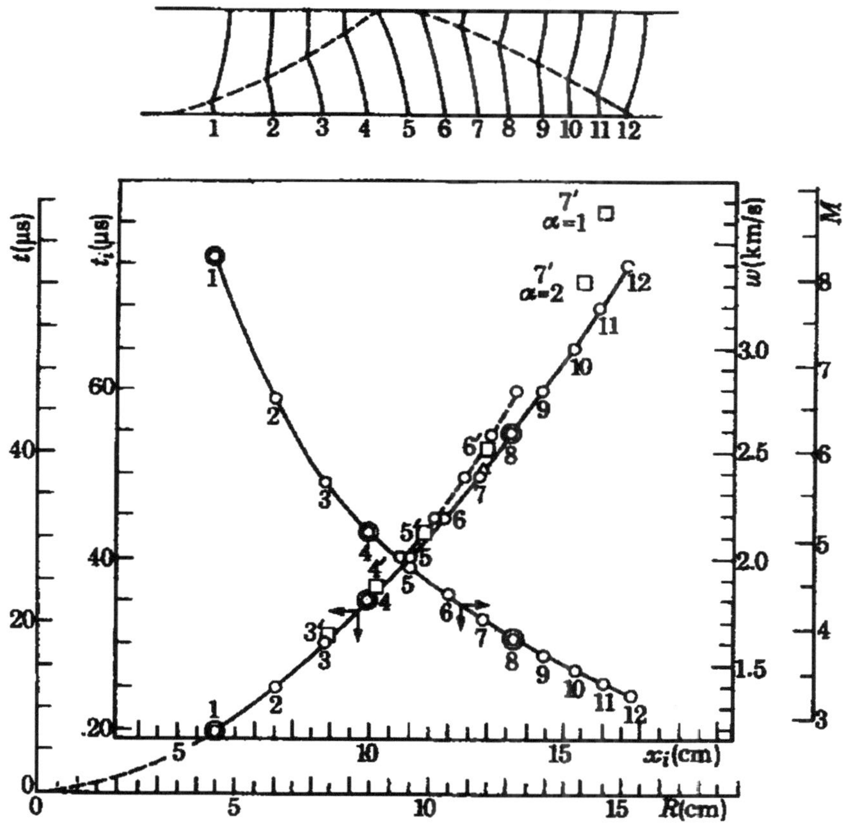

Fig. 12.27. Time-space trajectory of the front of blast wave formed by collision between Mach intersections in the course of its travel across a detonation cell, together with its velocity and Mach number (Lundstrom and Oppenheim 1968)

Its velocity modulus, evaluated thereby, is $\mu = 0.5714$ = constant, corresponding to $\lambda = 1.5$. On this basis, the front trajectory was extrapolated to $t_i = 0$, the origin of the coordinate system where the radius of the front R = 0. As evident from the latter, upon deposition of the exothermic energy at the origin of the blast wave of point explosion, its front decays, as demonstrated by the plot of its velocity, w, and Mach number, M, with reference to the scale on the right side.

Thus, a self-sustained Chapman-Jouguet detonation front has been revealed to possess the following properties:

1. Its elementary component is a Mach intersection

2. Its cellular structure is imprinted on the wall by the chisel-like action of the concentrated vortex formed by shear across the slip interface at the triple point of the intersection, acting as the 'writing head.'

3. It is initiated by point explosions created by collisions between Mach intersections. The dynamic features of this event are presented by Fig. 12.28 – a double version of the phenomenon described in Section 8.12.4.

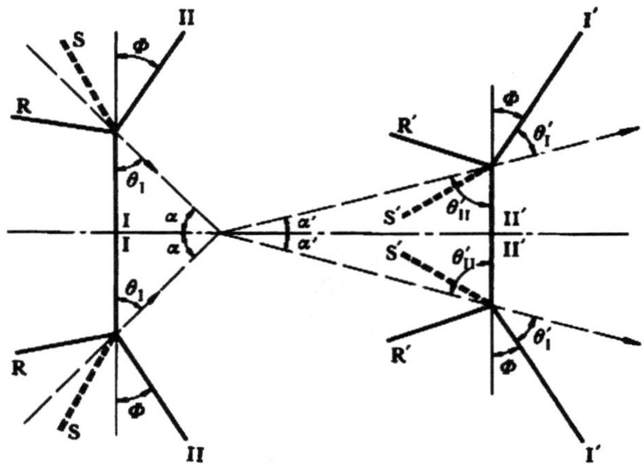

Fig. 12.28. Schematic diagram of a collision between Mach intersections I: *incident shock*; II: *Mach front*; R: *reflected shock*; S: *slip line* (Oppenheim 1985)

A schematic diagram of a self-sustained detonation front, reconstructed on this basis, is presented by Fig. 12.29. As portrayed here, the detonation front is headed by the front of a blast wave, followed by a deflagration. The blast wave is generated by the high power density of exothermic energy deposited in the high temperature and pressure regime created by Mach stems of triple point intersections. Under such circumstances, the exothermic energy of the explosive gas is deposited almost instantaneously. In its progress along the cellular structure, the energy of the blast wave remains invariant and the double-front system of the shock and deflagration is, therefore, progressively decaying. The velocity of the shock is steadily decreasing and so is the temperature and pressure behind its front.

The distance between the deflagration and shock fronts is, concomitantly, increasing, so that, at later stages of its traverse across the detonation cell, the double front system is at the verge of extinction. Propagation of the detonation front depends then crucially on the subsequent collisions between Mach intersections that occur at a frequency commensurate with the eigenvalue of oscillations satisfying the zero normal velocity condition at the walls. It is by the same action that the size of the detonation cells is controlled.

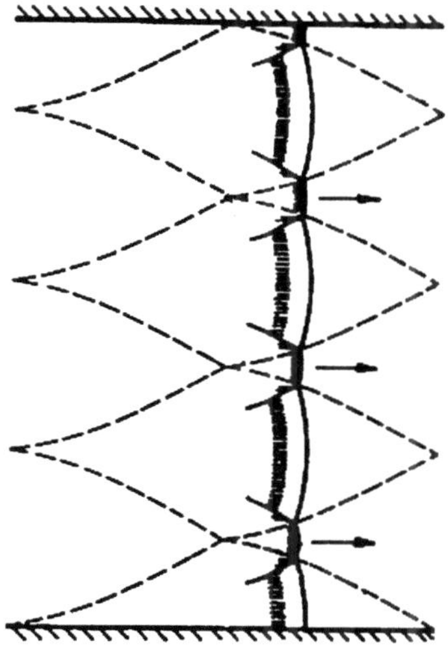

Fig. 12.29. Schematic diagram of a self-sustained detonation front displaying its cellular structures formed by collisions between Mach intersections (Oppenheim 1985)

The maintenance of the remarkably constant propagation velocity of a Chapman-Jouguet detonation front is thus clarified. The velocity is fixed by chocked flow of the Chapman-Jouguet condition, $M_{CJ} = 1$, reached by detonation products behind the front. This constraint is communicated to the front by the expansion wave that is formed whenever it tends to accelerate. The propagation velocity of the self-sustained detonation is therefore established by the mechanism for maintaining the frequency of collisions between Mach intersections at a frequency commensurate with the boundary conditions – a physical modus operandi of a closed-loop control system.

12.4 Detonation Spin

Of particular significance to detonation research is the phenomenon of spin. This mode of propagating detonation occurs close to its extinction limit when the power density of deposited energy is approaching a minimum. It became a subject of intensive studies when, in order to investigate the wave structure, experiments on detonation were carried out at low pressures. It was then found that the edges of broken glass tubes containing the detonation were of spiral shape. The classical study of this phenomenon was carried out Bone et al 1935 (vid. Fig. 12.3).

A soot imprint of a spinning detonation propagating through a mixture of $2C_2H_2+5O_2+3A$, contained initially at a pressure of 9.3 torr in a cylindrical tube 1 inch in diameter, is displayed by Fig. 12.30. The spiral shape of the cellular detonation structure is here evident, but the mechanism by which it is generated remains obscure.

A more revealing record of this phenomenon is provided by the soot imprint in a square tube depicted by Fig. 12.31. The medium in this case was a mixture of $2H_2+O_2+3N_2$ contained initially at a pressure of 21.4 torr in a 3/4x3/4 inch tube.

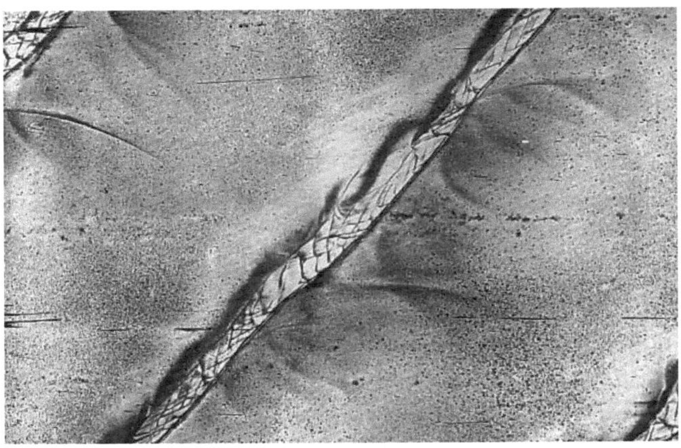

Fig. 12.30. Soot imprint of spinning detonation in a circular tube (Lee et al 1969)

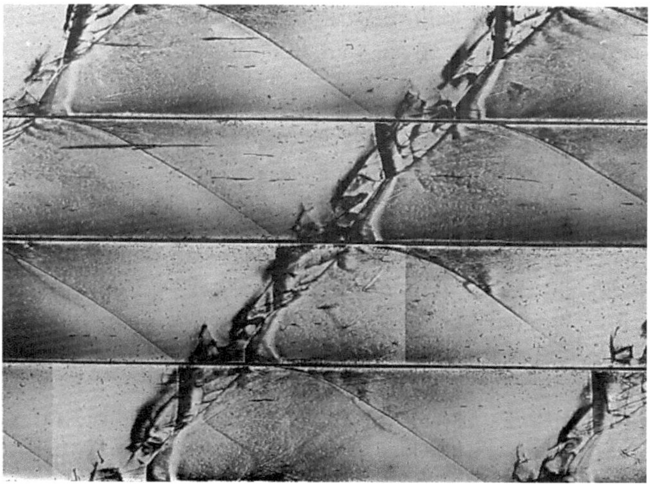

Fig. 12.31. Soot imprint of spinning detonation in a square tube (Lee et al 1969)

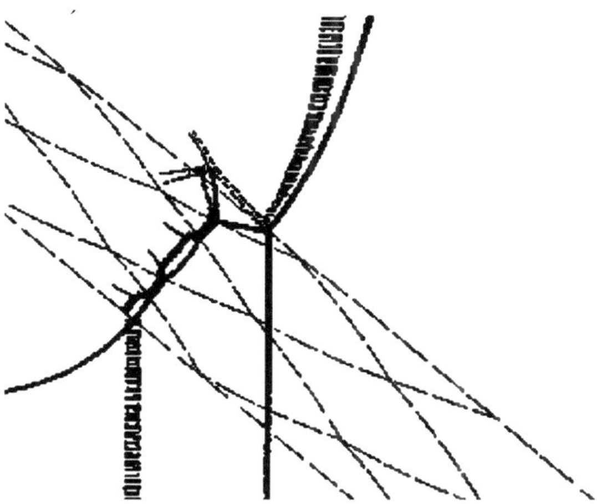

Fig. 12.32. Schematic diagram of the structure of a spinning detonation (Lee et al 1969)

The dynamic features of a spinning detonation deduced from Fig. 12.31 are presented by Fig. 12.32. The state of the medium into which it propagates is evidently inadequate to support a self-sustained detonation. It is brought, however, to a satisfactory level by the incident shock front of the Mach intersection supported by its oblique front.

The most noteworthy feature of spinning detonation is its remarkably constant axial propagation speed that is appreciably below the Chapman-Jouguet velocity.

12.5 Classical Literature

Berthelot, M (1881) Sur la vitesse de propagation des phénomènes explosifs dans les gaz. C. *R. Acad. Sci., Paris* 93: 18-22 out

Mallard E, Le Chatelier H (1881). Sur la vitesse de propagation de l'inflam mation dans les mélanges explosifs. C. *R. Acad. Sci., Paris* 93: 145-148

Berthelot M, Vieille P (1882) Sur la vitesse de propagation des phenomenes explosifs dans les gaz. C. *R. Acad. Sci. Paris* 94: 101-108, 822-823; 95 : 151-157

Berthelot M, Vieille P (1883) L'onde explosive *Ann. Chim. Phys.* Ser. 5 28: 289-332

Mallard E, Le Chatelier H (1883) Recherches expérimentales et théorétiques sur la combustion des mélanges gazeux explosifs. *Ann. Mines* Ser. 4; 8: 274-568

Liveing GD, Dewar J (1884) Spectroscopic studies on gaseous explosions. *Proc.Roy. Soc.* 36: 471-478

Hugoniot H (1887-1889) Propagation du mouvement dans les corps ; *J. Ec. Polyt., Paris,* 57: 3-97; 58: 1-125

Lean B, Dixon HB (1892) Experiments on the transmission of explosions across air gaps. *Mem. Manch. Let. Phil. Soc.* 4: 16-22

Dixon HB (1893) The rate of explosion in gases. *Phil. Trans.Roy.Soc.* A 184: 97-188

Chapman DL (1899) On the Rate of Explosion in Gases *Phil. Mag.* 5th series, 47: 90-104

Vieille P (1899) Sur les discontinuités produites par la détente brusque de gaz comprimes C. *R. Acad. Sci., Paris,* A 129: 1228-1230

Le Chatelier H (1900) Sur le développement et la propagation de I'onde explosive" C. *R. Acad. Sci., Paris,* A 130: 1755-1758

Vieille P (1900) Rôle des discontinuités dans la propagation des phénomènes explosifs. C. *R. Acad. Sci., Paris,* 131 : 413-416

Dixon HB, Bower J; Bradshaw L; Dawson B; Graham E; Jones RH; Strange, EH (1903) On the Movement of the Flame in the Explosion of the Gases. *Phil. Trans. Roy. Soc.* A 200 : 315-352

Jouguet E (1906) Sur la propagation des réactions chimiques dans les gaz. *J. Math. Pures Appl.* 6e Série, Tome 1, 60:347-425; Tome 2, 61: 1-86

Crussard L, Jouguet E (1907) Sur les Ondes de Choc et Combustion. Stabilité de l'Onde Explosive. C. *R. Acad. Sci., Paris,* 144 : 560-563

Crussard L (1907) Ondes de Choc et Onde Explosive. *Bull. Soc. Industr. Min. St. Etienne* 4e Série, 6: 257-364

Dixon HB, Campbel LC, Slater WE (1914) Photographic analysis of explosions in the magnetic field. *Proc. Roy. Soc.* A 90 : 506-511

Jouguet E (1917) Mécanique des explosifs. *Encyclopédie Scientifique,* Doin et Fils., Paris, 516 pp

Becker R (1917) Zur Theorie der Detonation. *Z. Elektrochem.* 23: 40-49; 93-95; 304-308

Becker R (1922) Physikalisches uber feste und gasformige sprengstoffe. *Z. tech. Phys..* 3: 152-159; 249-256.

Becker R (1922) Strosswelle und Detonation. *Z. Phys.* 8: 321-362 (tranl. NACA Technical Memorandum No. 505 and 506

Campbell C (1922) The Propagation of Explosion Waves in Gases Contained in Tubes of Varying Cross-Section. *J. chem. Soc.* 121: 2483-2498

Payman W, Wheeler RV (1922) The Combustion of Complex Gaseous Mixtures. *J. Chem. Soc.* 121, XLVII: 363-379

Laffitte P (1923) Sur la formation de l'onde explosive. C. *R. Accd. Sci., Paris,* 176: 1392-1395

Laffitte P (1923-1924) Sur la propagation de l'onde explosive. C. *R. Acad. Sci., Paris,* 177: 178-180; 179: 1394-1396.

Wendlandt R (1924) Experimental Investigations Concerning the Limits of Detonation in Gaseous Mixtures" Z. *phys. Chem.* 116: 110-227 (transl. (1930) National Advisory Committee for Aeronautics Technical Memorandum No. 553, 25 pp., No. 554, 47 pp.

Dumanois P, Laffitte, P(1926) Influence de la pression sur la formation de l'onde explosive" C. *R. Acad. Sci., Paris* tome 183, 284-285

Campbell C, Woodhead DW (1926) The Ignition of Gases by an Explosion Wave. I. Carbon Monoxide and Hydrogen Mixtures. *J. Chem. Soc.* 125: 3010-3021

Jouguet E (1927) La théorie thermodynamique de la propagation des explosions *Proc. 2nd Int. Congr. Appl. Mech.* pp. 12-22

Egerton A, Gates SF (1927) On Detonation of Gaseous Mixtures of Acetylene and Pentane" *Proc. Roy. Soc.* A 114: 137-151, 152-160; A 116: 516-519.

Bone WA, Frazer RP, Winter DA (1927) The Initial Stages of Gaseous Explosions" *Proc. Roy. Soc.* A 114: 402-419; 420-441

Bone WA, Townsend PTA (1927) Explosions and Gaseous Explosives" *International Critical Tables;* published for the National Research Council by McGraw-Hill, New York, II: 172-195.

Bone WA, Townsend PTA (1927) *Flame and Combustion in Gases;* Longmans, Green, London, 538 pp.

Payman W (1928) The Detonation Wave in Gaseous Mixtures and the Pre-Detonation Period. *Proc. Roy. Soc.* A 120: 90-109

Laffitte P (1929) La physico-chimie du phénomène du 'choc' et des 'antidétonants. *J. Chem. Phys.* 26: 391-423

Lewis B (1930) A Chain ReactionTheory of the Rate of Explosion in Detonating Gas Mixtures. *J. Amer. Chem. Soc.* 52: 3120-3127

Lewis B, Friauf JB (1930) Explosions in Detonating Gas Mixtures. 1. Calculation of Rates of Explosions in Mixtures of Hydrogen and Oxygen and the Influence of Rare Gases. *J. Amer. Chem. Soc.* 52: 3905-3924

Bone WA, Frazer ND (1930) Photographic Investigation of Flame Movements in Gaseous Explosions" IV, V, VI, *Phil. Trans.* Vol. A230: 363-385

Payman W, Woodhead DW (1931) Explosion Waves and Shock Waves. I. The Wave-Speed Camera and Its Application to the Photography of Bullets in Flight" *Proc. Roy. Soc.* A 132: 200-213

Kontrowa TA, Nemann MB (1933) Contribution to the Theory of the Induction-Period" *Phys. Z. Sowjet.* 4: 818-825

Bone WA, Fraser RP, Wheeler WH (1935) A Photographic Investigation of Flame Movements in Gaseous Explosions. VII. The Phenomenon of Spin in Detonation" *Phil. Trans.Roy. Soc.,* A 235: 29-68

Payman W, Titman H (1935) Explosion Waves and Shock Waves. III. The Initiation of Detonation in Mixtures of Ethylene and Oxygen and of Carbon Monoxide and Oxygen" *Proc. Roy. Soc.*A 152: 418-445

Becker R (1936). Uber Detonation" Z. *Elektroch.* 76: 457-461

Rivin M, Sokolik A (1936) Les limites d'explosivité des mélanges gazeux *Acta Phys.-Chem. URSS* 4: 301-306

Sokolik A, Shtsholkin K (1933-1937) Detonation in Gaseous Mixtures" 1. (1933) *Phys.* Z. *Sowjet.* 4: 795-817; 2. (1934) *J. phys. Chem. U.S.S.R.*, 5: 1459-1463, 3. (1937) *Acta Physicochim. U.S.S.R.* 7: 581-596

Langweiler H (1938) Beitrag zur hydrodynamischer detonations theorie. Z. *Techn. Phys.*Ed 21, 9: 271-283

Shtsholkin K (1939) On the Theory of the Development of Detonation in Gases *Dokl. Akad. Nauk SSSR* 23: 636-640

Czerlinsky E (1940) Druck- und Flammengeschwindigkeitsmessungen bei Detonation von Athylather-Luft Gemische. Z. *Techn. Phys* Ed. 21, 4: 77-79

Jost W (1940) *Explosion and Combustion Processes in Gases,* (transl. in 1946 by Croft HO, McGraw-Hill, New York, 621 pp.

Zeldovich YB, Semenov NN (1940) Kinetics of Chemical Reactions in Flames. *Zh. eksp. teoret. fiz.* 10: 1116—1123 (transl. 1946 in National

Advisory Committee for Aeronautics, Technical Memorandum 1084, 15 pp.
Zeldovich YB (1940) On the Theory of the Propagation of Detonation in Gaseous Systems" *Zh. eksp. teoret. fiz.* 10: 542-568 (transl. 1950 in: N C A Technical Memorandum 1261, 50 pp.
Damkohler G, Schmidt A (1941) Gasdynamische beitrage zur auswertung von Flammenversuchen in Rohrstrecken" Z. *Elektrochem* 47: 547-567
Von Neumann J (1942) Progress Report on Theory of Detonation Waves Office of Scientific Research and Development Report 549, 24 pp.
Doring W (1943) The Detonation Process in Gases" *Ann. Phys.* 43: 421-436
Doring W (1949) The Velocity and Structure of Very Strong Shock Waves in Gases *Ann. Phys.*5: 133-150
Doring W, Burkhardt R (1944) Contributions to the Theory of Detonation. VDI Forschungsbericht 1939 [transl. 1949 in Air Material Command, Report F-TS-1277-IA (GDAM A9-T46)] 353 pp.
Manson, N. (1944) Sur le calcul thermodynamique des caracteristiques des ondes explosives dans les mélanges gazeux. C. *R. Acad. Sci., Paris,* 218 : 29-31
Zeldovich YB (1944) Theory of Combustion and Detonation of Gases *Akad. Nauk SSSR,* Institute of Chemical Physics [transl. 1949 in Air Material Command, Wright-Patterson Air Force Base Technical Report FTS-1226-1A (GDAM A9-T-45)], 118 pp.
Schultz-Grunow F (1944) Zur Behandlung nichtstationarer Verdichtungs-stosse und Detonationswellen *Z. Angew. Math. Mech.* 24: 284-288
Payman, W, Shepherd WCF (1946) Explosion Waves and Shock Waves *Proc. Roy. Soc.* A 186: 293-321
Zeldovich YB (1946) Oxidation of Nitrogen in Combustion and Explosions. *Acta Physicochim. SSSR,* 21: 577-625
Zeldovich YB (1946) Theory of Detonation Spin. C. *R. Acad. Sci., U.S.S.R.,* tome 52, pp. 147-150, 1946.
Zeldovich YB (1947) Theory of Detonation Onset in Gases. *Zh. Teknich. Fiz., SSSR,* 17: 3-26
Shchelkin KI (1947) Detonation of Gases in Rough Tubes. *Zh. Teknich. Fiz.SSSR.* 17: 613-618
Manson N (1947) Propagation des détonations et des déflagrations dans les mélanges gazeux" L'Office National d'Etudes et de Recherches Aéronautiques, Paris, Diverse 1,200 pp.(transl. in ASTIA AD No. 132808)
Zeldovich YB, Roslovski A (1947) Conditions for the Formation of Instability in Normal Burning. *Dokl. Akad. Nauk, SSSR* 57 365-368
Courant R, Friedrichs KO (1948) *Supersonic Flows and Shock Waves.* Interscience Publishers, Inc., New York, 464 pp.

Kogarko SM, Zeldovlch YB (1948) Detonation of Gaseous Mixtures. *Dokl. Akad. Nauk, SSSR* 63: 553-556

Ubbelohde AR (1949) Transition from Deflagration to Detonation: The Physico-chemical Aspects of Stable Detonation. Third Symposium (International) on Combustion, Williams and Wilkins, Baltimore, pp. 566-571

Shepherd WCF (1949) The Ignition of Gas Mixtures by Impulsive Pressures. Third Symposium (International) on Combustion, Flame, and Explosion Phenomena, Williams and Wilkins. Baltimore, pp. 301-316

Brinkley SR, Kirkwood JG (1949) On the Condition of Stability of the Plane Detonation Wave. Third Symposium (International) on Combustion, Williams and Wilkins, Baltimore, pp.586-590

Manson N Contribution to the Hydrodynamical Theory of Flame Vibrations. C. *R. Acad.Sci., Paris,* 227: 720-722; Seventh Congress of Applied Mechanics, London, pp. 187-199

Eyring H, Powell RE, Duffey GH, Parlin RB (1949) The Stability of Detonation. *Chem. Revs.* 45: 69-181

Zeldovlch YB, Simonov N (1949) Theory of the Spark Ignition of Explosive Gas Mixtures. *Zh. Fiz. Khimia* 23 : 1361-1374

Manson N (1949) Propagation des déflagrations dans les mélanges gazeux et la naissance des mouvements vibratoires. *La France Energetique* pp. 278-289

Taylor GI (1950) The Dynamics of the Combustion Process Behind Plane and Spherical Detonation Fronts in Explosives. *Proc. Roy. Soc* A 200: 235-247

Berets DJ, Green EF, Kistiakowsky GB (1950) Gaseous Detonations; I. Stationary Waves in Hydrogen-Oxygen Mixtures. II Initiation by Shock Waves. *J. Amer. Chem. Soc.* 72: 1080-1901

Schmidt E, Steinicke H, Neubert U (1951) Flame and Schlieren Photographs of Combustion of Gas-Air Mixtures in Tubes. VDI-Forschungsheft 431, Ausgabe Band 17, Deutscher Ingenieur-Verlag, Dusseldorf, 31 pp.; (1953) Flame and Schlieren Photographs of Combustion Waves in Tubes. In Fourth Symposium (International) on Combustion, Williams and Wilkins, Baltimore, pp.658-666

Lewis B, Von Elbe G (1951) *Combustion, Flames and Explosions of Gases,* Academic Press, New York, 795 pp.

Mooradian AJ, Gordon YE (1951) .Gaseous Detonation. I. Initiation of Detonation. *J. chem. Phys.* 19: 1166-1172

Mooradian AJ, Gordon YE (1951) The Effect of Pressure on the Detonation Velocity in Gases. *Phys. Rev.* 84: 614

Sokolik AS (1951) On the Mechanism of Pre-detonative Acceleration of Flames. *Zh. eksp. teoret. fiz.* 21: 1163-1179

Kistiakowsky GB (1951) Density Measurements in Gaseous Detonation Waves. *J. chem. Phys.* 19: 1611-1612

Kistiakowsky GB (1951) Initiation of Detonation in Gases. *Indust. Engng. Chem.* 43: 2790-2798.

Kistiakowsky GB, Knight HT, Malin ME (1952) Gaseous Detonations. III. Dissociation Energies of Nitrogen and Carbon Monoxide; IV. The Acetylene-Oxygen Mixtures; V. Nonsteady Waves in CO_2-0_2 Mixtures. *J. Chem. Phys.* 20: 876-887, 994-1000

Hirschfelder J0, Curtiss CF, Campbell PE (1953) The Theory of Flames and Detonations. Fourth Symposium (International) on Combustion, Williams and Wilkins, Baltimore, pp. 190-210.

Markstein GH (1953) Instability Phenomena in Combustion Waves. Fourth Symposium (International) on Combustion, Williams and Wilkins, Baltimore, pp. 44-59,

Brinkley SR, Richardson JM (1953) On the Structure of Plane Detonation Waves with Finite Reaction Velocity. Fourth Symposium (International) on Combustion, Williams and Wilkins, Baltimore pp. 450--457.

Manson N, Ferrie F (1953) Contribution to the Study of Spherical Detonation Waves. Fourth Symposium (International) on Combustion, Williams and Wilkins, Baltimore, pp. 486-494

Fay JA (1953) Some Experiments on the Initiation of Detonation in $2H_2$-0_2 Mixtures by Uniform Shock Waves. Fourth Symposium (International) on Combustion, Williams and Wilkins, Baltimore, pp. 501-507

Chu BT (1953) On the Generation of Pressure Waves of a Plane Flame Front. Fourth Symposium (International) on Combustion, Williams and Wilkins, Baltimore, pp. 603-612

Freiwald H, Ude H (1953) Uber die kugelformige ausbreitung der detonation in Acetylen-Luft-Gemischen" Z. *Elektroch.* Ed. 57: 629--632

Shchelkin KI (1953) On the transition of slow burning into detonation. *Zh. eksp.teoret.fiz.* 24: 589--600

Popov VA (1953) Certain laws of the pre-detonation stage of flame development" *Iz. Akad. Nauk, SSSR, Otd. Teknich. Nauk* 10: 1428-1439

Kogarko SM, Novikov AS (1954) Acceleration of flames in the predetonation period" *Zh. eksp. Teort. fiz.* 23: 492-503

Manson N (1954) Formation et célérité des ondes explosives spheriques dans les melanges gazeux. *Rev. Inst. Fran. Pétrole* 9 : 133-143

Guenoche H, Manson N (1954) Etude de l'influence du diamètre des tubes sur la célérité des ondes explosives. *Rev. Inst.franc.petrole* 9 : 214-220

Weir A. Jr, Morrison RB (1954) Equilibrium temperatures and compositions behind a detonation wave. *Industr. Engng. Chem.* 46: 1056-1060

Hirschfelder J0, Curtiss CF, Bird RB (1954) *Molecular theory of gases and liquids;* John Wiley, New York, Chapter II, Sec. 9, pp. 797-813

Steinberg M, Kaskan WE (1955) The Ignition of Combustible Mixtures by Shock Waves. Fifth Symposium (International) on Combustion, Reinhold Publishing Corp., New York, pp. 664—672

Spener G, Wagner HG (1954) Spektralaufnahmen von detonationsflammen. Z. *Phys. Chemie,* Neue Folge, 2: 312-319

Kirkwood SG, Wood WW (1854) Structure of a steady-state plane detonation wave with finite reaction rate. *J. chem. Phys.* 22: 1915-1919

Kistiakowsky GB, Kydd PH (1954) The reaction zone in gaseous detonations. *J. chem. Phys.* 22: 1940-1941

Rudlnger G (1955) *Wave Diagrams for Non-steady Flow in Ducts;* D. van Nostrand Co., New York, 278 pp.

Gilkerson WR, Davidson N (1955) On the Structure of a Detonation Front. *J. chem. Phys.* 23: 687-692.

Zeldovich YB, Kompaneets AS (1955) *Theory of Detonation.* Gosudarstvennoe Isdatel'stvo Tekhnika- Teoreticheskoi Literatury, Moskow, 288 pp.

Stanyukovich KP (1955) T*he Unsteady Motion of Continuous Media.* Gosudarstvennoe Isdatel'stvo Tekhnike Teoreticheskoi Literatury, Moscow, 1955, 304 pp. [transl. (1959) Adashko GJ, ed. Holt M, Pergamon Press, London]

Kistiakowsky GB, Kydd PH (1955) Gaseous Detonation. VI. The Rarefaction Wave" *J. chem. Phys.* 23: 271-274

Kistiakowsky GB, Zinman WG (1955) Gaseous Detonations. VII. A Study of Thermodynamic Equilibrium in Acetylene-Oxygen Waves. *J. chem. Phys.* 23: 1889-1894

Wagner HG (1955) Einige Beobachtungen an Kohlendyoxid-Sauerstoff-Detonationen. *Z. Ektrochem.* 59: 906-909

Troshin YAK, Shchelkin KI (1955) Structure of the Front of Spherical Flames and Instability of Normal Combustion" *Izv. Akad. Nauk, SSSR, Otd. Tekhnich. Nauk,* 9: 160 (transl. Kuvshinoff GW, Applied Physics Lab. Johns Hopkins Univ. TG230T42,10 pp.)

Freiwald H, Ude H (1955) Uber die initierung kugelformiger detonationswellen in gasgemischen" *Z. Elektroch.* 59: 910-913, *C. R. Acad. Sci., Paris, S*eance du 19 September, 736--639.

Shchelkln KI (1955) Phenomena in the Vicinity of Detonation Formation in a Gas. *Zh. eksp. teoret. fiz.* 29:221-226 [transl. (1956) in *Sov. Phys. J. expo theor. Phys.* .2: 296-300]

Evans MW, Given FI, Richeson WE (1955) Effect of Attenuating Materials on Detonation Induction Distance in Gases. *J. Appl. Phys.* 26: 1111-1113.

Troshin YAK (1955) Generalization of Hugoniot Equations for Non-Steady Processes of Flame Propagation in Pipes. *Dokl. Akad. Nauk, SSSR (N.S.)* 103: 465-468

Ubbelohde AR, Copp J (1956) Detonation Processes in Gases, Liquids and Solids. Combustion Processes. In vol. II, *High Speed Aerodynamics and Jet Propulsion,* Princeton Univ. Press, Princeton, N.J., Part 5, 577-612

Troshin YAK (1956) Gas Dynamic Analysis of Non-Stationary Processes of Flame Propagation in Tubes" *!zv. Akad. Nauk, SSSR Otd. Teknieh. Nauk (Bull. Akad. Sci. SSSR, Div. Tech. Sci.)* 1: 80-98

Cook MA, Keyes RT, Filler AS (1956) Mechanism of Detonation.; *Trans. Faraday Soc.* 52: 369-384

Kistiakowsky GB, Mangelsdorf P (1956) Gaseous Detonations. VIII. . Two-Stage Detonations in Acetylene-Oxygen Mixtures. *J. chem. Phys.* 25: 516-519

Kistiakowsky GB, Kydd PH (1956) Gaseous Detonations. IX. A Study of the Reaction Zone by Gas Density Measurements. *J. chem. Phys.* 25: 824-835.

Duff RE, Knight HT (1956) Precision Flash X-ray Determination of Density Ratio in Gaseous Detonation. *J. Chem. Phys.* 25: 1301

Wood WY, Kirkwood JG (1956) On the Existence of Steady-State Detonations Supported by a Single Chemical Reaction. *J. chem. Phys.* 25: 1276-1277

Zeldovich YB, Kogarko SM, Semenov NN (1956) An Experimental Investigation of Spherical Detonation of Gase. *Zh. tekhnich. fiz.* 26: 1744-1768 [translated (1957) in *Sov. Phys.-Tech. Phys.* 8: 1689-1713]

Markstein GH (1957) A Shock Tube Study of Flame-Front Pressure-Wave Interaction. Sixth Symposium (International) on Combustion, Reinhold, New York, pp. 387-398

Sedov LI (1957) Detonation in Media of Variable Density, Sixth Symposium (International) on Combustion, Reinhold, New York, pp. 639-641

Wagner HG (1957) Spectra of the Detonation of Oxygen with H_2, CO and Hydrocarbons. Sixth Symposium (International) on Combustion Reinhold, New York, pp. 366-371

Greifer B, Cooper JC, Gibson FC, Mason CM (1957) Combustion and Detonation in Gases. *J. Appl. Phys.* 28: 289-294

Just T, Wagner HG (1957) Gleichgewichteinstellung in gasdetonationen. *Z. Elektrochem.* 61: 678-685

Just T, Wagner HG (1957) Die reaktionszone in gasdetonationen, *Z. Phys.Chem.* Neue Folge, 13: 241-243

Manson N (1957) La théorie hydrodynamique et le diamètre limite de propagation des ondes explosives. *Z. Electrochemist.* 61: 586-592

Fairbain AR, Gaydon AG (1957) Spectra Produced by Shock Waves, Flames and Detonations. *Proc.Roy. Soc.* A239: 464-475

Peek HM, Thrap RG (1957) Gaseous Detonation in Mixtures of Cyanogen and Oxygen. *J. chem. Phys.* 26: 740-745

Freiwald H, Ude H (1957) Untersuchungen an kugelformige detonationswellen in Gasgemischen. *Z. Elektrochem.* 61: 663-672

White DR (1857) On the Existence of Higher than Normal Detonation Pressures. *J. Fluid Mech.* 2: 513-514..

Penner SS *Chemistry Problems in Jet Propulsion,* Vol. XIV, Pergamon Press, 394 pp.

Sedov LI (1957) *Methods of Similarity and Dimensional Analysis in Mechanics,* 4th Ed., Moscow [transl. (1959) by Cleavor-Hume, Blackwells, Oxford

Millan G (1958) Transition from Deflagration to Detonation, Section 7, Chapter V, Aerothennochemistry; Report of the course conducted by Theodore Von Karman at the University of Paris published by the Air Research Development Command, U.S. Air Force V34-V38

Kantrowitz AR (1958) One-Dimensional Treatment of Nonsteady Gas Dynamics" *Fundamentals of Gas Dynamics,* V. III, *High Speed Aerodynamics and Jet Propulsio,* Princeton University Press pp. 350-416

Taylor GI, Tankin RS (1958) Transformation from Deflagration to Detonation. In *Fundamentals of Gas Dynamics,* V. III, *High Speed Aerodynamics and Jet Propulsion,* Princeton University Press pp. 645-656

Karman Von T (1958) Aerothermodynamic Problems of Combustion. In *Fundamentals of Gas Dynamics,* V. III, *High Speed Aerodynamics and Jet Propulsion,* Princeton University Press, pp. 574-584

Hayes WD (1958) The Basic Theory of Gasdynamic Discontinuities *Fundamentals of Gas Dynamics,* V. III, *High Speed Aerodynamics and Jet Propulsion*, Princeton University Press, pp.417-481

Rudinger G (1958) *Shock Wave and Flame Interactions,* Third AGARD Colloquium on Combustion and Propulsion, Pergamon Press, 153-182

Cook MA (1958) *The Science of High Explosives.* American Chemical Society Monograph Series No. 139, Reinhold, New York, 440 pp.

Manson N (1958) Une nouvelle relation de la théorie hydrodynamique des ondes explosives. *C. R. Acad. Sci., Paris* 246 : 2860-2862

Brossard J, Manson N (1958) Détention comparée des caractéristiques des ondes explosives dans les mélanges gazeux. *C. R. Acad. Sci., Paris,* 247 : 2105-2108

Duff RA (1958) Calculation of Reaction Profiles Behind Steady-State Shock Waves;.1. Application to Detonation Waves. *J. chem. Phys.* 28: 1193-1197

Knight HT, Venable D (1958) Apparatus for Precision Flash Radiography of Shock and Detonation Waves in Gases. Rev. *Sci. Instrum.* 29: 92-98.
Leonas VB (1958) A Study of the Initiation and Propagation of Spherical Detonations. *Zh. Fiz. khim.* 32 [transl.(1959) by Kuvshinoff BW, Applied Physics Laboratory, Johns Hopkins University, Bulletin TG230-T58, 8 pp.
Kogarko SM, Skobelkin VI, Kazakov AN (1958) Interaction of Shock Waves with Flame Fronts. *Dokl. Akad. Nauk, SSSR (phys. Chem. Section)*122: 1046-1048.
Kogarko SM (1958) Investigation of the Pressure at the End of a Tube in Connection with Rapid Nonstationary Combustion. *Zh. tekh. Fiz.* 28: 2041-2045 [transl. (1959) in *Sov. Phys.-Tech. Phys.* 3: 1975-1979]
Martin FJ (1958) Transition from Slow Burning to Detonation in Gaseous Mixtures. *Phys. Fluids* 1: 399-407
Jones H (1958) Accelerated Flames and Detonation in Gases. *Proc.Roy. Soc.* A248: 333-349
Fay JA, Opel G (1958) Two-Dimensional Effects in Gaseous Detonation Waves. (with comments by Duff RE, Knight HT, and Wood WW, Kirkwood JG) *J. Chem. Phys.* 29: 955-958 .
Duff RE, Knight HT, Rink JP (1958) Precision Flash X-Ray Determination of Density Ratio in Gaseous Detonations. *Phys. Fluids* 1: 393-398
Chesick JP, Kistiakowsky GB (1958) Gaseous Detonations X-Study of Reaction Zones. *J. chem. Phys.* 28: 956-961
Cher M, Kistlakowsky GB (1958) Gaseous Detonations XI-Double Waves. *J. Chem. Phys.* 29: 506-511
Zaitsev SG, Soloukhin RI (1958) Combustion in an Adiabatically Heated Gaseous Mixture. *Dokl. Akad. Nauk, SSSR, Phys. Chem.* 122
Hirschfelder JO, Curtiss CF (1958) Theory of Detonation. I. Irreversible Unimolecular Reaction. *J. Chem. Phys.* 28: 1130-1147
Linder B, Curtiss CF, Hirschfelder JO (1958) Theory of Detonation. II. Reversible Unimolecular Reaction. *J. Chem. Phys.* 28: 1147-1151
Kogarko SM (1959) Detonation of Methane-Air Mixtures and the Detonation Limits of Hydrocarbon-Air Mixtures in a Large-Diameter Pipe" *Zh. tekh. Fiz.* 28: 2072-2083 [transl. (1959] in *Sov. Phys. Tech. Phys.* 3: 1904-1914].
Frazer RP (1959) Detonation Velocities in Liquid Fuel Vapors with Air or Oxygen at 100°C and Atmospheric Pressure. Seventh Symposium (International) on Combustion, Butterworths, London, pp. 783-790
Binkley SR, Uwis B (1959) The Transition from Deflagration to Detonation. Seventh Symposium (International) on Combustion, Butterworths, London, pp. 807-811

Cook MA, Pack DH, Gey WA (1959) Deflagration to Detonation Transition. Seventh Symposium (International) on Combustion, Butterworths, London, pp. 820-836

Belles FE (1959) Detonability and Chemical Kinetics: Prediction of Limits of Detonability of Hydrogen. Seventh Symposium (International) on Combustion, Butterworths, London, pp. 745-751

Gordon WE, Mooradian AJ, Harper SA (1959) Limit and Spin Effects in Hydrogen-Oxygen Detonations. Seventh Symposium (International) on Combustion, Butterworths, London, pp. 752-759

Martin FJ, White DR (1959) The Formation and Structure of Gaseous Detonation Waves. Seventh Symposium (International) on Combustion, Butterworths, London

Predvoditelev AS (1959) 1. Concerning Spin Detonation,. 2. On Automodelling Processes in Chemically Active Media, pp. 733-778. 3. Theoretical Examination of Vibratory Movement of the Flame Front in Closed Vessels. Seventh Symposium (International) on Combustion, Butterworths, London, pp. 760—765, 760—765, 779-782

Salamandra GD, Bazhenova, TV, Naboko IM (1959) Formation of Detonation Wave During Combustion of Gas in Combustion Tube. Seventh Symposium (International) on Combustion, Butterworths, London, pp. 851-855

Bazhenova TV, Soloukhin RI (1959) Gas Ignition Behind the Shock Wave. Seventh Symposium (International) on Combustion, Butterworths, London, pp. 866-875

Adams GK, Pack DC (1959) Some Observations on the Problem of Transition between Deflagration and Detonation. Seventh Symposium (International) on Combustion, Butterworths, London, pp. 812-819

Troshin VAK (1959) The Generalized Hugoniot Adiabatic Curve. Seventh Symposium (International) on Combustion, Butterworths, London, pp. 789-798

Popov, VA (1959) On the Pre-Detonation Period of Flame Propagation" Seventh Symposium (International) on Combustion, Butterworths, London, pp. 799-806

Nicholls JA, Dabora EK, Gealer RL (1959) Studies in Connection with Stabilized Gaseous Detonation Waves. Seventh Symposium (International) on Combustion, Butterworths, London, pp. 766-772

Fay JA, Basu S (1959) Ionization in Detonation Waves. Seventh Symposium (International) on Combustion, Butterworths, London, pp. 175-179

Laffitte, P, Bouchet R (1959) Suppression of Explosion Waves in Gaseous Mixtures by Means of Fine Powders. Seventh Symposium (International) on Combustion, Butterworths, London, pp. 504-508

Curtiss CF, Hirschfelder JO, Barnett MP (1959) Theory of Detonations.III. Ignition Temperature Approximation. *J. Chem. Phys.* 30: 470--492

Just T, Wagner HG (1959) Reaktionszone von knall gasdetonationen. II. Zeitschrift fur Physikalische Chemie, Neue Folge, 19: 250-253

Kistlakowsky GB, Tabbutt FD (1959) Gaseous Detonations. XII. Rota tional Temperature of the Hydroxyl Free Radicals. *J. Chem. Phys.* 30: 577-581

Curtiss CF, Hirschfelder JO (1959) Theory of the Structure of Gaseous Detonations. Proc. ARS 14th Annual Meeting, Washington, D.C. pp. 16-20

Brossard J, Manson N (1959) Application de la théorie de la double disco tinuité: caractéristiques des détonations dans les mélanges gazeux. *C. R. Acad.Sci., Paris* 249 : 1033-1035

Brossard J, Manson N (1959) Propriétés des adiabatiques dynamiques dans le cas d'une double discontinuité: choc-onde de combustion. *C. R. Acad. Sci.,Paris,* 249 : 372-374

Penner SS (1959) Selected Analytical Studies on Explosions. In *Explosions, Detonatons, Flammability and Ignition* AGARDograph No. 31, Pergamon Press, pp. 41-72

Markstein GH (1959) Graphical Computation of Shock and Detonation Waves in Real Gases. *J. Amer. Rocket Soc.* 29: 588-590

Fay JA (1959) Two-Dimensional Gaseous Detonations: Velocity Deficit" *Phys.Fluids* 2: 283-289

References

Arpaci VS (1966) Conduction heat transfer (esp. pp. 307-308). Addison-Wesley, Mass.

Ashurst WT (1979) Numerical simulation of turbulent mixing layers via vortex dynamics. Proc. 1st Symp. on Turbulent Shear Flows (ed. Durst et al.),. Springer-Verlag Berlin: pp.402-413

Ashurst WT (1981) Vortex simulation of a model turbulent combustor. Proc. 7th Colloquium on Gas Dynamics of Explosions and Ractive Systems. Prog. Astronaut. Aeronaut. 76: 259-273

Bach GG, Lee JH (1970) An Analytical Solution For Blast Waves. AIAA J., 8, 2, 271-275

Barenblatt, G. I. (1994) Scaling Phenomena in Fluid Mechanics, Cambridge University Press, 50 pp.

Barenblatt, G. I. (1996) Scaling, Self-similarity, and Intermediate Asymptotics, Cambridge University Press, xxii + 386 pp.

Barenblatt, G. I. (2003) Scaling, Cambridge University Press, xiv + 171 pp.

Barenblatt GI, Guirguis RH, Kamel MM, Khul AL, Oppenheim AK, Zeldovich YB (1980) Self-similar explosion waves of variable energy at the front. J. Fluid Mech. 99: 841-858

Batchelor GK (1967) An introduction to fluid mechanics University Press, Cambridge

Baulch DL, Drysdale DD, Horne DG, Lloyd AC (1972) Evaluated kinetic data for high temperature reactions. Butterworths, London

Bazhenova TV, Gvozdeva LG, Nobastov YS, Naboko IM, Nemkov RG, Predvoditeleva OA (1968) Shock waves in real gases. Izdatel'stvo Nauka. Moscow

Becker R (1922) Stosswelle and Detonation Zeitschrift für Physik 8: 321 - 362 (esp. p. 352)

Belotserkoyskii O, Chushkin PI (1965). The numerical solution of problems in gasdynamics, Basic Developments in Fluid Dynamics, (ed. Holt M), Academic Press, New York 1: 1-126

Benson SW (1960) The foundations of chemical kinetics. McGraw-Hill Book Co., New York, xvii +703.

Benson SW (1976) Thermochemical kinetics. John Wiley & Sons, New York, xi + 320 pp.
Berthelot M,. Vieille P (1822) Sur la vitesse de propagation des phenomenes explosifs dans les gaz. C.R. Hebd.Seanc. Acad. Sci. Paris 94: 101-108; 822-823; 95, 151-157
Boddington T, Gray P, Harvey DL (1971) Thermal theory of spontaneous ignition: criticality in bodies of arbitrary shape. Phil. Trans. Roy. Soc. Lond. A270: 467-506
Bone WA, Fraser RP, Wheeler WH (1935) A photographic investigation of flame movements in gaseous explosions Part 7 The phenomenon of spin in detonation, Philosophical Transactions of the Royal Society of London, A 747, 235, 29-66
Bone WA, Fraser RP, Wheeler WH (1935) A Photographic Investigation of Flame Movements in Gaseous Explosions. VII. The Phenomenon of Spin in Detonation Phil., Trans. Roy. Soc., A 235: 29-68
Borisov AA. (1974) On the origin of exothermic centers in gaseous mixtures. Acta Astronautica, Pergamon Press,Oxford, 1: 909-920.
Brinkley SR, Kirkwood JG (1947) Theory of the propagation of shock waves, Physical Review, 71: 606-611
Brode HL (1955) Numerical solutions of spherical blast waves," Journal of Applied Physics 26: 766-775
Brode HL (1969) Gasdynamic motion with radiation: a general numerical method. Astronautica Acta, Pergamon Press, Oxford, 14: 433-444.
Brode HL (1955) Numerical solutions of spherical blast waves, J. Appl. Phys., 26: 766-775.
Bui TD, Oppenheim AK, Pratt DT (1984) Recent advances in methods for numerical solution of O.D.E. initial value problems. J. Computational and Applied Mathematics 11: 283-296
Caratheodory C (1909) Untersuchungen über die Grundlagen der Thermodynamik. Math.Ann. 67: 355-386
Carslaw HS, Jaeger JC (1948) Conduction of Heat in Solids, Oxford Press, Oxford , esp. p.43
Chernyi GG, Korobeinikov VP, Levin VA, Medvedev SA (1970) One-dimensional unsteady motion of combustible gas mixtures associated with detonation waves. Acta Astronautica. Pergamon Press, Oxford 15: 255-266
Chester W (1954) The quasi-cylindrical shock tube, Phil. Mag. 45: 1293-1299
Chisnell RF (1957.) the Motion Of A Shock Wave In A Channel, With Applications To Cylindrical And Spherical Shock Waves, J. Fluid Mech. 2: 286-298

Chorin AJ (1973) Numerical studies of slightly viscous flow. J. Fluid Mech. 57: 785-796

Chorin AJ (1978) Vortex sheet approximation of boundary layers. J. Comp. Phys. 27: 428-442.

Chorin A.J (1980a) Vortex models and boundary layer instability. SIAM J. Scient. Stat. Comput. 1: 1-24

Chorin A.J (1980b) Flame advection and propagation algorithms. J. Comput. Phys. 35: 1-11

Chorin, A.J., Hughes, T.J. R., McCracken, M. F. & Marsden, J. E. 1978 Product formulas and numerical algorithms. Communs pure appl. Math. 31: 205--256

Chorin AJ, Marsden JE (1979) A mathematical introduction to fluid mechanics. Springer-Verlag, Berlin

Clarke JF, Kassoy DR, Riley N (1984) Shocks generated in a continued gas due to rapid heat addition at the. boundary; I. Weak shock waves and II. Strong shock waves. Proc. Roy. Soc. London. A 393: 309-329; 331-351

Cohen LM, Oppenheim AK (1975a) Effects of size and dilution on dynamic properties of exothermic centers Comb. & Flame 25: 207-211.

Cohen LM, Short JM, Oppenheim AK (1975b) A computational technique for the evaluation of dynamic effects of exothermic reactions. Comb. & Flame 24: 319-334

Courant R, Friedrichs KO (1948) Supersonic flow and shock waves. Wiley, New York, XVI + 464

Denisov YuN, Troshin YaK (1959) Pulsiruyushchaia i spinovaia detonatsia gazovikh smeshei v trubakh (Pulsating and spinning detonation of gaseous mixtures in tubes). Dokl. Akad. Nauk SSSR, Moscow 125:110-113

Dixon-Lewis G (1979) Kinetic mechanism, structure and properties of premixed flames in hydrogen-oxygen-nitrogen mixtures. Phil. Trans. Roy. Soc. London. A 292: 45-99.

Dorodnitsyn AA (1956a) On the method of numerical solution of certain non-linear problems in aero-hydrodynamics, Proc. 3rd All-Union Mathematical Congress AN USSR, Moscow, 447-453

Dorodnitsyn AA (1956b) Solution of mathematical and logical problems on high-speed digital computers. Proc. Conf. Develop. Soviet Math. Machines Devices, Part 1, Moscow. Pp. 44-52

Dougherty EP. Rabitz H. (1980) Computational kinetics and sensitivity analysis of hydrogen-oxygen combustion. J. Chem. Phys. 72: 6571-6585.

Edwards DH., Hooper G, Job EM, Parry DJ (1970) The behavior of the frontal and transverse shocks in gaseous detonation waves. Astronautica Acta 15: 323-333.

Ezekoye OA., Greif R (1993) A comparison of one and two dimensional flame quenching: heat transfer results, ASME Heat Transfer Division 250

Frank-Kamenetskii, D. A. 1969 Diffusion and heat transfer in chemical kinetics (transl. N. Thon). Plenum Press, New York

Friedman MP (1961) A simplified analysis of spherical and cylindrical blast waves. J. Fluid Mech. 11: 1-15

Fujiwara T (1970) Plane steady Navier-Stokes detonations of oxy-ozone. J. Phys. Soc. Japan 29: 1350-1364

Fujiwara, T. (1975) Spherical ignition of oxy-hydrogen behind a reflected shock wave. Fifteenth Symposium (International) on Combustion, The Combustion Institute, Pittsburgh pp. 515-1524

Gardiner WC (1972) Rates and mechanisms of chemical reactions. WA Benjamin Inc. Menlo Park, California, x + 284

Gavillet GG, Maxson JA, Oppenheim AK (1993) Thermodynamic and Thermochemical Aspects of Combustion in Premixed Charge Engines Revisited SAE 930432, 20 pp.

Gear CW (1971) Numerical initial value problems for ordinary differential equations, Prentice-Hall, New York, xvii + 253 pp

Ghoniem AF, Gharakhani A (1997) Three-dimensional vortex simulation of time-dependent incompressible internal viscous flows, Journal Computational Physics, 134: 75-95

Ghoniem AF, Chorin AJ, Oppenheim AK (1982) Numerical modelling of turbulent flow in a combustion tunnel. Phil Trans. Roy. Soc. London. A 394: 303-325

Gibbs JW (1875-1878) On the equilibrium of heterogeneous substances. Transactions of the Connecticut Academy, III (1875-76) pp. 108-248; (1877-78) pp 343-524 [(1931) The Collected Works of J.W. Gibbs, Article III, Longmans, Green and Company, New York, 2: 55-353, esp. pp.85-89 and 96-100]

Glass II, Hall JG (1959) Handbook of supersonic aerodynamics- Section 18-Shock Tubes, Bureau of Ordnance, Department of the Navy, Navord Report 1488, 6, XXXVIII + 604 pp

Goldstine H, Neumann von J (1955) Blast wave calculation, Communication on Pure and Applied Mathematics. 8: 327-353; reprinted in (1963) John von Neumann Collected Works (Taub AH, ed.) Vol. VI, Pergamon Press, New York, pp. 386-412

Gordon S, McBride BJ (1994) Computer Program for Calculation of Complex Chemical Equilibrium Compositions and Applications. I. Analysis, NASA Reference Publication 1311, vi + 55 pp

Gray BF, Yang CH (1965) On the unification of the thermal and chain theories of explosion limits. J. Phys. Chem. 69: 2747-2750

Gray P, Lee PR (1967) Thermal explosion theory. Oxidation and Combustion Reviews. Elsevier, Amsterdam. 2, pp. 1-183

Gray P, Scott SK (1990) Chemical oscillations and instabilities (Non-linear chemical kinetics). Clarendon Press, Oxford, xvi+453 pp

Griffiths JF (1990) Thermokinetic interactions in simple gaseous reactions. Ann. Rev. Phys. Chem. 36: 77-104

Grigorian SS (1958) Cauchy's problem and the problem of a piston for one-dimensional non-steady motions of a gas. J. Appl. Math. Mech. 22: 187-197

Gross RA, Oppenheim AK (1959) Recent Advances in Gaseous Detonation, Editorial, ARS Journal, 29: 173-180

Guderley G (1942) Powerful spherical and cylindrical compression shocks in the neighbourhood of a centre of a sphere and of a cylinder axis. Luftfarhriforschung 19: 302-312

Guirguis RH, Oppenheim AK., Karasalo I, Creighton JR. (1981) Thermochemistry of methane ignition. Prog. Astronaut. Aeronaut. AIAA, Washington, 76: 134-153

Hald, OH (1979) Convergence of vortex methods for Euler's equations. II. SIAM J. Numerical. Analysis. 16: 726-755

Heperkan H, Greif R (1981) Heat transfer during the shock-induced ignition of an explosive gas. International Journal of Heat and Mass Transfer, 267-276

Hicks BL (1954) Theory of ignition considered as a thermal reaction. J. chem. Phys. 22, 414-429

Hindmarsh AL (1971) A systematised collection of ODE solvers. ODEPACK Scientific Computing (ed. Stephens RF).IMAC Transactions. on Scientific Computing, North Holland, Amsterdam, pp 55-64

Hirschfelder JO, Curtis CF, Bird RB (1964) Molecular theory of gases and liquids. J. Wiley & Sons, New York, xxvi + 1249 pp

Hofbauer J (1956) The theory of evolution and dynamical systems: mathematical aspects of selection. Cambridge University Press, Cambridge, viii + 341 pp

Hsiao CC, Ghoniem AF, Chorin AJ, Oppenheim AK (1984) Numerical simulation ofa turbulent flame stabilized behind a rearward facing step. In *Proceedings of the Twentieth (International) Symposium on Combustion* Pittsburgh: The Combustion Institute, pp. 495-504

Huang WM, Greif,R, Vosen SR (1987) The effects of pressure and temperature on heat transfer during flame quenching SAE, Paper 872106, 11 pp

Huang WM, Vosen,SR, Greif R (1987) Heat transfer during laminar flame quenching: effect of fuels. Twenty-First International Symposium on Combustion, The Combustion Institute, 1853-1860

Jante A (1960) The Wiebe Combustion Law (Das Wiebe-Brenngesetz, ein Forschritt in der Thermodynamik der Kreisprozesse von Verbrennungsmotoren) Kraftfahrzeugtchnik, v. 9, pp. 340 – 346

Jeffrey A, Taniuti T (1964) Nonlinear Wave Propagation With Applications to Physics and Magnetohydrodynamics, Academic Press, NewYork

Jost, W (1946) Explosion and combustion processes in gases. McGraw-Hill Book Company, New York and London, xv + 621 pp

Jouguet E (1917) Mécanique des Explosifs, Doin et Fils, Paris, XX + 516 (esp. § 193 pp. 278-279)

Kassoy DR, Poland J (1980, 1981) The thermal explosion confined by a constant temperature boundary: I-The induction period solution: J. Appl. Math. 39: 412-430; II-The extremely rapid transient. J. Appl. Math. 41: 231-246

Kee RJ, Miller JA, Jefferson TH (1980) Chemkin: a general-purpose, problem-independent, transportable, Fortran chemical kinetics code package, Sandia Report, SAND80-8003

Kee RJ, Rupley FM, Miller JA (1989) CHEMKIN-II: A Fortran chemical kinetics package for the analysis of gas-phase chemical kinetics. Sandia Report SAND89-8009

Kee R, Rupley FM, Miller JA (1993) The Chemkin thermodynamic data base. Sandia Report SAND87-8215

Keller J0, Vaneveld L, Korschelt D, Hubbard GL, Ghoniem AF, Daily JW,.Oppenheim AK (1982) Mechanism of instabilities in turbulent combustion leading flashback, AIA Aerospace Jl. 20: 254-262

Kestin J (1966-1968) A Course in Thermodynamics. Blaisdell Publishing Co. Waltham, MA, vol.I xix+615 pp. vol.II, xviii+617 pp

Kim KB, Berger SA, Kamel MM, Korobeinikov VP, Oppenheim AK (1975) Boundary Layer Theory for Blast Waves, Journal of Fluid Mechanics, 71: 65-88

Kindelan M, Williams FA (1975) Radiant ignition of a combustible solid with gas phase exothermicity. Acta Astronautica.2: 955-980

Kolmogorov AN, Petrovskii, IG, Piskunov NS (1937) A Study of the Diffusion Equation with Increase in the Amount of Substance and Its Application to a Biological Problem Bull. Moscow Univ., Math. Mech., v.1, no.6, pp.1-26, [(1991) Selected Works of A.N. Kolmogorov, ed: Tikhomirov VM, Kluwer Academic Publishers, Dordrecht, v.1, Mathematics and Mechanics, pp. 242-270]

Kondrat'ev VN (1964) Chemical Kinetics of Gas Reactions (transl. J. M. Crabtree JM, Cation SN, ed. N. B. Slater) Pergamon Press, Oxford. Addison-Wesley Publishing, Reading, xiv + 812 pp

Korobeinikov VP (1969) The Problem of Point Explosion in a Detonating Gas. Astronautica Acta, 14: 411-420

Korobeinikov VP (1985) Problems of Point-Blast Theory, Nauka, Moscow (transl. Adashko G (1991) American Institute of Physics, New York, 382 pp

Korobeinikov VP, Chushkin P1 (1966) Plane, cylindrical and spherical explosions in a gas with counter pressure. (ed. L. 1. Sedov), Izdatel'stvo Nauka, Moscow, pp. 4-33

Korobeinikov VP,.Sharotova KV (1969) Gasdynamic functions of point explosions. Computer Center, USSR Academy of Sciences, Moscow

Korobeinikov VP, Mil'nikova NS, Ryazanov YeV (1961) The theory of point explosion:Fiz.mat.giz. Moscow (transl. by U.S. Department of Commerce, JPRS: 14,334, CSO: 6961-N, Washington, D.C. 1962).

Korobeinikoy VP, Chushkin PI (1963) Calculations for the early stages of points explosions in various gases. Zl.Prikl. Mekhan. i Tekhn: Fiz., 4: 48-57

Korobeinikoy VP, Chushkin PI, Sharoyatoya KV (1963) Tables of gasdynamic functions of the initial stages of point explosions. Probl. Numer. Phys., Computational Center, USSR Academy of Sciences Moscow

Kurylo J, Dwyer HA, Oppenheim AK (1980) Numerical Analysis of Flowfields Generated by Accelerating Flames. AIAA Journal, 18, 3, pp. 302-308, March 1980.

Langweiler H (1938) Beitrag zur Hydrodynamischen Detonationstheorie. Zeitschrift fur Technische Physik, 19:271-283

Laumbach DD, Probstein RF (1969) A Point Explosion in Cold Exponential Atmosphere, Journal of Fluid Mechanics, 35: 53-75

Lee JH (1972) Gasdynamics of detonation. Astronautica Acta 17: 455-466

Lee JH.(1977) Initiation of gaseous detonation. Ann. Rev. Phys. Chem. 28:75-104

Lee JH (1969) Collapsing Shock Waves in a Detonating Gas. Astronautica Acta, 14: 421-425

Lee JH, Soloukhin RI, Oppenheim AK (1969) Current Views on Gaseous Detonation, Astronautica Acta, 14: 565-584

Levin VA, Chernyi GG (1967) Asymptotic laws of behaviour of detonation waves. Prikladnaya Matematika i Mekhanika, 31: 393-405

Lewis B, von Elbe G (1987) Combustion, flames and explosion of gases, (esp. Chapter V, 15, Combustion Waves in Closed Vessels, pp. 381-395), Academic Press, Inc., Orlando, Florida (Revised edition of text published in 1951)

Libouton JC, Dormal M, Van Tiggelen P.J (1981) Re-initiation process at the end of the detonation cell. Gasdynamics of detonations and explosions. Prog. Astronaut. Aeronaut AIAA, Washington, DC, 75: 358-369

Lie S, Engel F (1880) Theorie der Transformationsgruppen. Teubner, Leipzig

Lotka AJ (1924) Elements of Mathematical Biology., Dover Publications reprint xxx + 465 pp

Lu JH, Ezekoye OA, Greif R, Sawyer RF (1991) Unsteady heat transfer during side wall quenching of a laminar flame. Twenty-Third International Symposium on Combustion, The Combustion Institute, pp. 441-446

Lundstrom EA, Oppenheim AK (1969) On the influence of non-steadiness on the thickness of the detonation wave. Proc. Roy. Soc. London, A 310: 463-478

Maas U, Pope SB (1992a) Implementation of simplified chemical kinetics based on intrinsic low-dimensional manifolds. Twenty-Fourth Symposium (International) on Combustion, The Combustion Institute, Pittsburgh, PA, pp. 103-112

Maas U, Pope SB (1992b) Simplifying chemical kinetics: intrinsic low-dimensional manifolds in composition space. Combustion and Flame, 88: 239-264

Mack, J.E. (1947) Semi-popular motion picture record of the Trinity explosion, U.S. Atomic Energy Commission, MDDC221

Mallard E, Le Chatelier H (1883) Recherches experimentales et theoriques sur la combustion des melanges gaseux explosifs. Ann. Mines 8 : 274-568

Manson N (1947) Propagation des detonations et des diflagrations dans les melanges gazeaux. ONERA & l'Institut Francais des Petroles, Paris: (tras!. ASTIA AD. no. 132-808)

McBride BJ, Gordon S (1996) Computer Program for Calculation of Complex Chemical Equilibrium Compositions and Applications, II. Users Manual and Program Description, NASA Reference Publication 1311, vi + 177 pp

McCracken M, Peskin C (1980) Vortex methods for blood flow through heart valves. J. Comput. Phys. 35: 183-205

McDonald H (1979) Combustion modeling in two and three dimensions - some numerical considerations. Prog. Energy Combust. Sci. 5: 97-122

Mellor AM (1979) Turbulent-combustion interaction models for practical high intensity combustors. 17th Symposium (International) on Combustion, The Combustion Institute, Pittsburgh pp. 377-387

Merzhanov AG, Averson AE (1971) The present state of the thermal ignition theory. Combust. Flame, 16: 89-124
Meyer JW, Oppenheim AK (197Ia) On the shock-induced ignition of explosive gases. Thirteenth Symposium (International) on Combustion, The Combustion Institute, Pittsburgh,:pp. 1153-1164
Meyer JW, Oppenheim AK (1971b) Coherence theory of the strong ignition limit. Combust. Flame, 17: 6~68
Meyer JW, Urtiew PA, Oppenheim AK (1970) On the inadequacy of gas dynamic processes for triggering the transition to detonation. Combust. Flame, 14: 13-20
Melnikova NS (1966) On a point explosion in a medium with variable initial density, Non-steady motion of compressible media associated with blast waves, (ed. L. I. Sedov), Proceedings of the V. A. Steklov lnst. Math., lzd. Nauka, Moscow, pp. 66-85
Mitrofanov VV (1962) Struktura detonatsonnoi volny v ploskom kanale (Structure of detonation wave in a flat channel) Zh. Prikl. Mekhan i Techn. Fiz. 4, 100-105
Neumann von J (1941) The point source solution. NDRC, Div. B. Report AM-9; revised in (1944) Shock Hydrodynamics and Blast Waves (Bethe HA, ed.); reprinted in (1963) John von Neumann Collected Works (Taub AH, ed) Vol. VI, Pergamon Press, New York, pp. 219—237
Neumann von J, Richtmyer D (1950) A method for the numerical calculations of hydrodynamic shocks, J. Appl. Phys., 21: 332
Oppenheim AK (1952) A Contribution to the Theory of the Development and Stability of Detonation in Gases, Journal of Applied Mechanics, 19: 63-71
Oppenheim AK (1953) Water Channel Analogue to High Velocity Combustion," Journal of Applied Mechanics, 20: 115-121.
Oppenheim AK, Stern RA (1959) On the Development of Gaseous Detonation - Analysis of Wave Phenomena. Seventh Symposium (International) on Combustion, Butterworths Scientific Publications, London, pp. 837-850 [French transl. (1959) Notes Techniques du Ministere de l'Air, No. 85, pp. 103-132]
Oppenheim AK (1961) Development and structure of plane detonation waves. Fourth AGARD Combustion and Propulsion Colloquium, Milan, Italy, 1960,:Pergamon Press London. pp. 186-258
Oppenheim AK (1964) On the Dynamics of the Development of Detonation in a Gaseous Medium, Archiwum Mechaniki Stosowanej (Archives de Mechanique), 16: 403-424.
Oppenheim AK (1965) Novel insight into the structure and development of detonation. Acta astronaut. 11: 391-400

Oppenheim AK (1970) (2nd edition 1972) Introduction to gasdynamics of explosions. Courses and Lectures no. 48, The International Centre for Mechanical Science, Udine. Springer-Verlag, Wien-New York, VI + 220 pp

Oppenheim AK (1982) Dynamic effects of combustion. Proceedings of the Ninth U.S. National Congress of Applied Mechanics, The American Society of Mechanical Engineers, New York, pp. 29-40

Oppenheim AK (1985) Dynamic features of combustion. Phil. Trans. Roy. Soc. London A 315: 471-508

Oppenheim AK (2004) Combustion in Piston Engines. Springer-Verlag, XI + 160 pp

Oppenheim AK, Barton JE, Kuhl AL, Johnson WP (1997) Refinement of Heat Release Analysis. SAE 970538, 23 pp

Oppenheim AK, Cohen LM., Short JM, Cheng RK, Hom K. (1975a) Dynamics of the exothermic process in combustion. Fifteenth Symposium (International) on Combustion, The Combustion Institute, Pittsburgh, pp. 1503-1513

Oppenheim AK, Cohen LM, Short JM, Cheng RK. Hom K (1975b) Shock tube studies of exothermic processes in combustion. Modern Developments in Shock Tube Research, Proceedings of the Tenth International Shock Tube Symposium. Kyoto, pp. 557-574

Oppenheim AK, Ghoniem AF (1983) Aerodynamic features of turbulent flames. AIAA Aerospace Paper no.83-0470, 10 pp

Oppenheim AK, Kamel MM (1972) Laser cinematography of explosions. Courses and Lectures no. 100. The International Centre for Mechanical Sciences, Udine, Springer-Verlag, New York, 226 pp

Oppenheim AK, Kuhl AL (1998) Life of Fuel in Engine Cylinder. SAE 980780, Modeling of SI and Diesel Engines. SAE SP-1330: 75-84; SAE Transactions, Journal of Engines 103, pp. 1080-1089

Oppenheim AK, Kuhl AL (2000a) Energy Loss from Closed Combustion Systems. Proceedings of the Combustion Institute, Pittsburg, 28:1257-1263

Oppenheim AK, Kuhl AL (2000b) Dynamic Features of Closed Combustion Systems. Progress in Energy and Combustion Science 26: 533-564

Oppenheim, A.K., Kuhl, A.L. and Kamel, M.M. 1978 On the Method of Phase Space for Blast Waves, Archives of Mechanics (Archiwum Mechaniki Stosowanej) Warsaw, 30: 553-571

Oppenheim AK, Kuhl AL, Lundstrom EA, Kamel MM (1972) A parametric study' of self-similar blast waves. J. Fluid Mech., 52: 657-682

Oppenheim AK, Lundstrom EA, Kuhl AL, Kamel MM (1971) A systematic exposition of the conservation equations for blast waves. J.Appl. Mech., 783-794

Oppenheim AK, Maxson JA (1994) A thermochemical phase space for combustion in engines. Twenty-Fifth Symposium (International) on Combustion, The Combustion Institute, Pittsburgh, Pennsylvania, pp. 157-165

Oppenheim AK, Smolen JJ, Kwak D, Urtiew PA (1970) On the dynamics of shock intersections 5th Symposium (International) on Detonation, Pasadena, California, 17pp

Oppenheim AK, Smolen JJ, Zajac LJ (1968) Vector Polar Method for the Analysis of Wave Intersections. Combustion and_ Flame,_12: 63-76

Oppenheim AK, Soloukhin RI (1968) Gas Phase Detonations. Recent Developments, Combustion and Flame, 12: 81-101

Oppenheim AK and Soloukhin RI (1973) Experiments in Gasdynamics of Explosions, Annual Review of Fluid Mechanics, 5: 31-58, 1973

Oppenheim AK, Stern RA (1958) On the Development of Gaseous Detonation - Analysis of Wave Phenomena," Seventh Symposium (International) on Combustion, London and Oxford, August 28-September 38; (1959) Butterworths Scientific Publications, London, pp. 837-850 (French translation: (1959) Notes Techniques du Ministere de l'Air, No. 85, pp. 103-132)

Oppenheim AK, Urtiew PA, Laderman AJ (1964) Vector Polar Method for the Evaluation of Wave Interaction Processes, Archives of Machines (Archiwum Budowy Maszyn) Warsaw, XI, 3: 441-495

Oran ES, Bonis JP, Young T, Flanigan M, Barks T, Picone M (1981) Numerical simulation of detonations in hydrogen-air and methane-air mixtures. 18th Symposium (International) on Combustion.: The Combustion Institute, Pittsburgh, pp.1641-1649

Oshima K (1960) Blast waves produced by exploding wires, Aero. Research lnst., Univ. Tokyo, Report No 358 (reprint:. Exploding Wires, (Chace W, Moore H, eds) Plenum Press, New York, 2: 159-174, 1962)

Oswatitsch K (1956) Gas Dynamics, Academic Press Inc., New York, XV + 610 pp

Peters N (2000) Turbulent combustion, Cambridge University Press, xvi + 304 pp

Poincaré H (1892) Thermodynamique, Gothiers-Villars, Paris, xix + 432 pp [1908 edition, xix + 458 pp]

Poland J, Hindash IO. Kassoy DR (1982) Ignition processes in confined thermal explosions. *CombustionScience and Technology* 27, 21~227.

Rashevsky N (1948) Mathematical Biophysics The University of Chicago Press, xxiii+669 pp

Reynolds WC (1996) STANJAN interactive computer programs for chemical equilibrium analysis, Department of Mechanical Engineering, Stanford University, Stanford, California, 48 pp

Roshko, A. (1976) Structure of turbulent shear flows: a new look. AIAA Journal 114: 1349-1357.

Rotman, D. A., Pindera, M. Z., and Oppenheim, A. K (1989) Fluid Mechanical Properties of Flames Propagating in Closed Channels. Dynamics of Reactive Systems Part I: Flames, Progress in Astronautics and Aeronautics, American Institute of Aeronautics and Astronautics, New York, 113: 251-265.

Rudinger G (1955) Wave Diagrams for Non-steady flow in Ducts,_D. Van Nostrand Company, Inc., New York, XI + 278 pp

Sakurai A (1965) Blast wave theory. Basic developments in fluid dynamics (ed. Holt M), Academic Press, New York, 1: 309-375

Samarski AA (1962) An efficient method for multi-dimensional problems in an arbitrary domain. Zh. Vych. Matmat. Fiz. 2: 787-811 [trans!. (1964) U.S.S.R. Comp. Math. & Math. Phys. 63,894-896]

Saytzev SG, Soloukhin RI (1962) Study of combustion of an adiabatically-heated mixture. 18th Symposium (International) on Combustion, The Williams and Wilkins Co. Baltimore, pp. 344-347

Sedov LI (1946) Rasprostraneniya sil'nykh vzryvnykh voln (Propagation of intense blast waves) Prikladnaya Matematika i Mekhanika, 10: 241-250

Sedov LI (1959a) Propagation of strong explosion waves. Prikl. Mat. Mekh. 10: .241-250

Sedov LI (1959b) One dimensional unsteady motion of a gas. Similarity and dimensional methods in mechanics, 4th edn. Moscow, Gostekhizdat [trans!. by Friedman M, ed. Holt M (1967) IV: 146-295]

Sell GR (1937) Dynamics of evolutionary equations. Springer, New York, xiii + 670 p.

Semenoff NN (1934) Chain Reactions Goskhimtekhizdat, Leningrad, [transl.: Chemical Kinetics and Chain Reactions (1935) Oxford University Press]

Semenov NN (1958-59) Some Problems in Chemical Kinetics and Reactivity" Princeton University Press, Princeton, 1: xii+239 pp; 2: v+331 pp.

Semenov NN. (1943) On types of kinetic curves in chain reactions. I. Laws of the autocatalytic type. Dokl. Akad.Nauk SSR XLII, 8: 342-348

Semenov, N. N. (1944a) On types of kinetic curves in chain reactions. II. Consideration of the interaction of active particles. *Dokl. Akad. Nauk SSSR* XLIV, 2, 62-66.

Semenov, N. N. (1944b) On types of kinetic curves in chain reactions. Allowance for chain rupture on walls of reaction vessel in the case of oxidation of hydrogen. Dokl. Akad. Nauk SSSR XLIV, 6: 241-245.

Semenov, N. N. (1958,1959) Some problems in chemical kinetics and reactivity (trans!. by Boudart M) Princeton University Press. 1: xii+239; 2: ix+331

Shchelkin KI, Troshin, YaK (1963).Gazodynamika Gorenia (Gasdynamics of Combustion), Izdatelstvo Akademii Nauk SSSR, Moscow, 255pp [transl. (1965) Mono Book Corp., Baltimore, pp VI + 222

Sokolik AS (1960) Self-ignition flame and detonation in gases. Moscow: Izdatel'stvo Akademii Nauk SSSR. [trans!. by Kaner N (1963). Jerusalem: Israel Program for Scientific Translations. Available from Washington: U.S. Department of Commerce, Office of Technical Services, OTS 63-11179; NASA TT F-125)

Soloukhin R1 1963 Detonatsionnye volny v gazakh Uspekhi Fiz. Nauk 80, 4, 525-551 (transl: Detonation waves in gases 1964. Soviet Physics 6, 4, 523-541)

Soloukhin R1 (1963b) Udarnye volny i detonatsia v gazakh .(Shock waves and detonations in gases) Moscow: Gosudarstvennoye Izdatel'stvo Fizyko-matematycheskoi Literatury. [trans!. by Kuvshinoff BW (1966) Baltimore: Mono Book Corp.]

Strehlow RA (1963) Detonation initiation AIAA Journal 2, 4, 465-480

Strehlow RA (1968) Gas phase detonations: recent developments. Combustion and Flame 12: 81-101

Spalding B (1957) I. Predicting the laminar flame speed in gases with temperature-explicit reaction rates. II. One-dimensional laminar flame. Theory for temperature-explicit reaction rates" Combustion and Flame, I: 287-295; II: 296-307

Stanyukovich KP (1955) Unsteady Motion of continuous Media, Gostekhizdat, Moscow, [transl. ed. Holt M (1960) Pergamon Press, New York]

Steinfeld JI, Francisco JS, Hase WL (1989) Chemical Kinetics and Dynamic. Prentice Hall, New Jersey, x + 518

Stull DR, Prophet H (1971) JANAF Thermochemical tables. National Bureau of Standards (now National Institute of Standards and Technology), US Department of Commerce) Report NSRDS-NBS 37, pp 1141

Taki S, Fujiwara T (1981) Numerical simulation of triple shock behavior of gaseous detonation. In Eighteenth Symposium (International) on Combustion, The Combustion Institute, Pittsburgh: pp 1671-1681

Taylor GI (1946) The air wave surrounding an expanding sphere. Proc. Roy. Soc. London, A, 186: 273-292

Taylor, G.I. (Sir Goeffrey) (1950a). The formation of a blast wave by a very intense explosion. I. Theoretical Discussion, first published in British Report RC-210 (1941); revised version in Proceedings of the Royal Society, London, Series A, Vol. 201, pp.159-174.

Taylor, G.I. (Sir Goeffrey) (1950b). The formation of a blast wave by a very intense explosion. II The explosion of 1945, Proceedings of the Royal Society, London, Series A, Vol. 201, pp.175-186.

Taylor G1, Tankin RS (1958) Gas dynamical aspects of detonation. In Gasdynamics of Combustion and Detonation, ed. Emmons HW, Princeton University Press.

Urtiew PA, Oppenheim AK (1965) Gasdynamic effects of shock-flame interactions in an explosive gas, AIAA Journal, 3: 876-883

Urtiew PA, Oppenheim AK (1966) Experimental observations of the transition to detonation in an explosive gas. *Proc. R. Soc. Lond.* A 295, 13-28.

Van Tiggelen PJ (1969) On the minimal initial size of an explosive reaction center. Combust. Sci. Technol, 1:225-232

Vaneveld L, Hom K, Oppenheim AK (1984) Secondary effects in combustion instabilities leading to flashback. AIAAerospace Journal 122: 81-82.

Van't Hoff JH (1896) Studies in chemical dynamics. Williams & Norgate, London.

Vermeer DJ, Meyer JW, Oppenheim AK (1972) Auto-ignition of hydrocarbons behind reflected shock waves. Combust. Flame 18: 327-336.

Vibe II (1956) Semi-empirical expression for combustion rate in engines (Полуэмпирическое Уравнение Скорости Сгорания В Двигателях) Proceedings of Conference on Piston Engines, USSR Academy of Sciences, Moscow, pp. 185-191

Vibe II (1970) Progress of combustion and cycle process in combustion engines (Новое о равочем цикле двигателей: Скорость сгорания и рабочий цикл двигателя) (transl. Heinrich J Brennverlauf und Kreisprozess von Verbrennungsmotoren VEB Verlag Technik,Berlin, 286 pp.

Voitsekhovsky BV, Mitrofanov VV, Topchian ME (1963) Struktura fronta detonatsii v gazakh (Structure of the detonation front in gases.) Novosibirsk: Izd. Sib. Otd. ANSSSR [trans!. by B. W. Kuvshinov L (1965) Mono Book Corp. Baltimore)

Volterra V (1937) Principes de biologie mathematique (Principles of biological mathematics) Acta Biotheoretica 3

Vosen SR, Greif R, Westbrook CK (1985) Unsteady heat transfer during laminar flame quenching. Twentieth International Symposium on Combustion, The Combustion Institute, 75-83.

Voyevodsky VV, Soloukhin R1 (1965) On the mechanism and explosion limits of hydrogen-oxygen chain self-ignition in shock waves. In Tenth Symposium (International) on Combustion, The Combustion Institute, Pittsburgh, pp. 279-283.:.

Warnatz, J. 1984 Survey of rate coefficients in the *CHO* system. In Chemistry of combustion (ed. Gardiner WC Jr, Springer-Verlag, New York Chapter 5: 197-360

Warnatz J, Maas U, Dibble RW (1996) Combustion. Physical and chemical fundamentals, modeling and simulation, experiments, pollutant formation. Springer, Berlin, x + 265 pp.

Westbrook C K, Pitz WJ (1984) A comprehensive chemical kinetic reaction mechanism for oxidation and pyrolysis of propane and propene. Combustion Science and Technology, 37: 117-152

White DR (1961) Turbulent structure of gaseous detonation. Phys. Fluids 4: 465-480.

Whitham GB (1958) On the propagation of shock waves through regions of non-uniform area or flow, J. Fluid. Mech., 4: 337-360

Wilkins ML (1969) Calculations of elastic-plastic flow, University of California Radiation Laboratory, Report No 7322'

Williams FA. (1974) A review of some theoretical considerations of turbulent flame structure. Specialists meeting on Analytical and Numerical Methods for Investigation of Flow Fields with Chemical Reaction, Especially Related to Combustion. AGARD PEP 43rd Meetings, Liege, Belgium, vo!. II, pp. 1-125

Williams FA (1985) Combustion Theory. The Benjamin/Cummings Publishing Company, Menlo Park, California (2nd edition) xxiii + 680 pp

Williams FA, Libby PA, editors (1980) Turbulent reacting flows. Topics in Applied Physics 4. Springer-Verlag, Berlin, Heidelberg, New York

Yang CH, Gray BF (1967). The determination of explosion limits from a unified thermal chain theory. Eleventh Symposium (International) on Combustion, The Combustion Institute, Pittsburgh, pp 1099-1106

Zajac LJ, Oppenheim, AK (1971) Dynamics of an explosive reaction center. AIAAerospace Journal, 19: 545-553

Zel'dovich YaB (1941) On the Theory of Thermal Intensity: Exothermic Reactions in a Jet. Zhurnal Techniskoi Fiziki 11: 493-500

Zel'dovich YaB, Barenblatt GI., Librovich VB, Makhviladze GM (1980) Matematicheskaya teoriya goreniya i vzryva. Nauka, USSR Academy of Sciences, Moscow, 478 pp [trasnsl. McNeil DH (1985) The mathematical theory of combustion and explosions. (vid. esp. Chapter 6, Combustion in Closed Vessels. pp. 470-487) Consultants Bureau, New York and London, pp xxi + 597]

Zel'dovich YaB, Frank-Kamenetskii DA (1938) Teoriya teplovogo rasprotraneniya plameni (A theory of thermal flame propagation) Zhurnal Fizicheskoikhimii, 12: 100-105

Zel'dovich YaB, Kompaneets AS (1955) Teoriya detonatsii. Gostekhizdat. Moscow: [trans!. (1960) Theory of detonation. Academic Press, New York]

Zel'dovich, Ya. B. and Rayzer, YuP (1963). Fizika udarnykh voln i vysokotemperaturnykh hydrodinamicheskikh yavlenyi (Physics of shock waves and high-temperature hydrodynamic phenomena), {1963} ;Gos. Izd. Fiz. Mat. Literatury, Moscow, 686 pp [transl. edited by Hayes WD, Probstein RF.), (1966-67) Academic Press, New York., I & II, XXIII + XXIV, 916 pp

Zel'dovich YaB, Barenblatt G1, Librovich VB, Machviladze GM. (1980) Mathematical theory of combustion and explosion Izdatel'stvo Nauka, Moscow, 478 pp

Nomenclature

Symbols

A	area
C_K	$(\partial e_K/\partial w_K)_p$ gradient of a vector in the state phase diagram
c_{Kp}	$(\partial h_K/\partial T)_p$ specific heat at constant volume
c_{Kv}	$(\partial e_K/\partial T)_v$ specific heart at constant pressure
D	mass diffusivity
e_K	internal energy
ev	constant internal energy and specific volume
G	Green's function
g	$\dfrac{p}{\rho_a w_n^2}$ normalized pressure
F	$\dfrac{\tau}{x} f = \dfrac{t}{\mu r} u$ velocity coordinate of phase space for blast waves
f	$\dfrac{u}{w_n}$ normalized velocity
h	$\dfrac{\rho}{\rho_a}$ normalized density
h_K	enthalpy
hp	constant enthalpy and pressure
M	mass
M_K	molecular mass
m_k	$1 - n_k^{-1}$
n	polytropic index
p	pressure
P	p/p_i normalized pressure

q_{Ro}	$u_{Ro} - u_{Po}$ - reference exothermic energy
q_R	exothermic energy
q_W	energy expended by heat transfer to the walls
R	reactants, universal gas constant
R_K	$\mathbf{R}/M_K$
R	radius
t	time
T_K	temperature
U	velocity normal to exothermic front
u	internal energy in thermodynamic tables
v	specific volume
V	v_K/v_i
w	pv dynamic potential
w_w	energy expended by work on the surroundings
W_K	w_K/w_{Si}
x	progress parameter, $\frac{r}{r_n}$
Y_K	mass fraction
Y_R	mass fraction of reactants
Z	$(\frac{\tau}{x})^2 \frac{g}{h} = (\frac{t}{\mu r})^2 \frac{p}{\rho}$ velocity of sound coordinate
z_K	$e_{K,,} w_K$, generalized state coordinate
α	thermal diffusivity; coefficient of the life function
χ	power index of life function
Δ	dilatation
δ	Dirac delta function; index in life function,
ε	$\frac{e}{w_n^2}$ normalized internal energy
Φ	scalar potential of (irrotational) velocity
γ	isentropic index
κ	bulk viscosity
λ	air-equivalence ratio
λ	$\frac{d \ln y}{d \ln \xi} = -2 \frac{d \ln w_n}{d \ln r_n} = -2 \frac{r_n \ddot{r}_n}{\dot{r}_n^2}$ decay parameter of blast waves
μ	$\frac{d \ln r_n}{d \ln t_n} \equiv \frac{d \ln \xi}{d \ln \eta} \equiv \frac{w_n t_n}{r_n} = \frac{\lambda + 2}{2}$ velocity modulus of blast waves,
μ	shear viscosity

ν	v_S/v_c normalized volume, stoichiometric coefficient, kinematic viscosity
π	pv^n polytropic function
θ	crank angle
ρ	density
σ	air/fuel mass ratio
τ	$\frac{t-t_i}{t_f-t_i}$, $\frac{\Theta-\Theta_i}{\Theta_f-\Theta_i}$ progress parameter of time, $\frac{t}{r_n}$
ξ	$\frac{r_n}{r_o}$ normalized front radius of a blast wave
ζ	exponent of the life function

Vectors

B	vector potential of (rotational) velocity
$\mathbf{n}_F$	unit vector normal to front
s	unit vector unidirectional to front
u	velocity vector
$\mathbf{u}_\Delta$	dilatational velocity component
$\mathbf{u}_\omega$	vortical velocity component
U	component of velocity vector normal to exothermic front
$\mathbf{W}_F$	exothermic front velocity
x	space co-ordinate
ω	vorticity vector

Subscripts

A	air
a	atmosphere of surroundings
c	compression
E	effective part of generated products, or of consumed fuel
f	final state
F	front, fuel
i	initial state
I	ineffective part of generated products or of consumed fuel
n	front
p	piston, effective part of consumed fuel

P	products
R	reactants
st	stoichiometric
S	system

Designations

A	air
B	inert component
C	charge
c	compression
e	expansion
E	effective
F	fuel
f	final
i	initial
I	ineffective
K	A, F, R, B, C, P
R	reactants
P	products
s	surroundings
S	system
t	terminal

Subject Index

MIX
Papier aus verantwortungsvollen Quellen
Paper from responsible sources
FSC® C105338

If you have any concerns about our products,
you can contact us on
ProductSafety@springernature.com

In case Publisher is established outside the EU,
the EU authorized representative is:
Springer Nature Customer Service Center GmbH
Europaplatz 3, 69115 Heidelberg, Germany

Printed by Libri Plureos GmbH
in Hamburg, Germany